Applying Systemic-Structural Activity Theory to Design of Human–Computer Interaction Systems

Ergonomics Design and Management: Theory and Applications

Series Editor
Waldemar Karwowski
Industrial Engineering and Management Systems
University of Central Florida (UCF) – Orlando, Florida

Published Titles

Application of Systemic-Structural Activity Theory to Design and Training
Gregory Z. Bedny

Applying Systemic-Structural Activity Theory to Design of Human–Computer Interaction Systems
Gregory Z. Bedny, Waldemar Karwowski, and Inna Bedny

Ergonomics: Foundational Principles, Applications, and Technologies
Pamela McCauley Bush

Aircraft Interior Comfort and Design
Peter Vink and Klaus Brauer

Ergonomics and Psychology: Developments in Theory and Practice
Olexiy Ya Chebykin, Gregory Z. Bedny, and Waldemar Karwowski

Ergonomics in Developing Regions: Needs and Applications
Patricia A. Scott

Handbook of Human Factors in Consumer Product Design, 2 vol. set
Waldemar Karwowski, Marcelo M. Soares, and Neville A. Stanton

Volume I: Methods and Techniques

Volume II: Uses and Applications

Human–Computer Interaction and Operators' Performance: Optimizing Work Design with Activity Theory
Gregory Z. Bedny and Waldemar Karwowski

Human Factors of a Global Society: A System of Systems Perspective
Tadeusz Marek, Waldemar Karwowski, Marek Frankowicz, Jussi I. Kantola, and Pavel Zgaga

Knowledge Service Engineering Handbook
Jussi Kantola and Waldemar Karwowski

Trust Management in Virtual Organizations: A Human Factors Perspective
Wiesław M. Grudzewski, Irena K. Hejduk, Anna Sankowska, and Monika Wańtuchowicz

Manual Lifting: A Guide to the Study of Simple and Complex Lifting Tasks
Daniela Colombiani, Enrico Ochipinti, Enrique Alvarez-Casado, and Thomas R. Waters

Neuroadaptive Systems: Theory and Applications
Magdalena Fafrowicz, Tadeusz Marek, Waldemar Karwowski, and Dylan Schmorrow

Safety Management in a Competitive Business Environment
Juraj Sinay

Self-Regulation in Activity Theory: Applied Work Design for Human–Computer Systems
Gregory Bedny, Waldemar Karwowski, and Inna Bedny

Forthcoming Titles

Organizational Resource Management: Theories, Methodologies, and Applications
Jussi Kantola

Applying Systemic-Structural Activity Theory to Design of Human–Computer Interaction Systems

Gregory Z. Bedny
Waldemar Karwowski
Inna Bedny

CRC Press is an imprint of the
Taylor & Francis Group, an **informa** business

CRC Press
Taylor & Francis Group
6000 Broken Sound Parkway NW, Suite 300
Boca Raton, FL 33487-2742

Printed on acid-free paper
Version Date: 20140611

International Standard Book Number-13: 978-1-4822-5804-2 (Hardback)

Visit the Taylor & Francis Web site at
http://www.taylorandfrancis.com

and the CRC Press Web site at
http://www.crcpress.com

Contents

Section II Design

Section III Quantitative Assessment of Computer-Based Task

Preface

Today, human–computer interaction (HCI) is not limited to trained software users. People of all ages use all different kinds of gadgets such as mobile phone, tablets, and laptops. The levels of computer proficiency of computer interface users vary in a wide range. How do we make HCI user friendly? How do we shorten the training process for new kinds of software and for constantly changing interfaces?

Regardless if the interface is used for communication, entertainment, or production operations, human activity should be broken down to individual tasks, performance of which can be efficiently designed. Such efficient design should be performed first at the analytical level. The sooner the improvements are made, the cheaper is their implementation. The flexibility of activity during task performance is significantly greater for computer-based tasks. In the software design one should strive to reduce unnecessary explorative activity and, associated with it, abandoned (unnecessary) actions to shorten the performance and acquisition times of a task. Psychologists in the fields of *situated concept of action* and *situated cognition* recognize that human cognition and behavior in general are not only flexible but also have social and situated features. In the introduction, we will not go into a detailed discussion of these concepts but only state that SSAT clearly demonstrates that situated aspects of human activity and dependence on cognition of outside-of-the-head world cannot be understood without considering the analysis of activity self-regulation and the concepts of cognitive and behavioral actions as basic elements of activity. In SSAT, thanks to the concept of self-regulation, activity and its basic components, cognitive and behavioral actions, are described in the context in which they occur. SSAT views activity and its cognitive components as a self-regulated system rather than as a linear sequence of information stages as it is described in cognitive psychology or as an aggregation of responses to multiple stimuli as described in behaviorism. SSAT views activity as a goal-directed rather than a homeostatic self-regulative system. Activity during task performance is adapting to a situation. A subject utilizes various strategies to achieve his or her task's goal. The process of self-regulation is described as various stages of processing information that involves different psychological mechanisms. Each stage is called a function block, because it performs a particular function in activity regulation during task performance. The interaction between the different function blocks is a critical factor in developing strategies of task performance.

HCI is an interdisciplinary field that gained recognition as one of the critically important fields in ergonomics. It is an area of study that draws on ideas and theoretical concepts from computer science, psychology, industrial design, and other fields. For ergonomics, psychological aspects of HCI analysis are specifically relevant. Presently, psychological aspects of HCI studies are based on the application of the information processing branch of cognitive psychology. However, there is no single, unified approach in cognitive psychology for resolving basic issues of software design. A general opinion is that scientists and practitioners should pick the methods that they see fit in each specific situation. However, all existing methods of analysis should be clearly defined for design. Such design processes have various stages of description and analysis of collected data. Usually a design process starts with qualitative analysis that is fairly flexible and gives specialists an opportunity to select the most adequate method. In the subsequent stages, the design process is transferred into more formalized and standardized methods of analysis. Based on such methods, it becomes possible to create models of not yet existing objects for the purpose of materializing these models into a ready product. Formalized methods usually are combined with qualitative methods, which help to optimize design solutions. All these stages of design are well known by designers in the field of engineering. However, this ideology of design is largely ignored in mainstream human–computer interaction studies that are based on cognitive psychology. Currently, there are not only no standardized principles of design but there is also no unified and standardized terminology in this field that could be utilized for the description of human activity during task performance. Software designers often make decisions based on intuition and do not use methods derived from human information processing approaches. Thus, there is a well-known gap between research results and practical design. In the literature, one can find some radical proposals to go beyond applying the human information processing approach in HCI studies. We do not support this point of view. In systemic-structural activity theory (SSAT), cognitive approach is considered as one qualitative stage of activity analysis. This stage is used for the analysis of separate cognitive processes that are involved in the performance of specific tasks. However, human information processing methods should not be used in isolation for HCI studies.

Creation of design models is central for any design method. However, cognitive psychology does not have method of creation of such models. We can find various models of human-information processing systems in cognitive psychology. However, they are not design models of activity during task performance. Such models are suggested only in SSAT. Computer-based tasks are very flexible, but cognitive psychology does not offer methods for analyzing flexible human activity. This is reflected in the analysis of two approaches to design solutions. The first, known as the instruction-based approach, is based on the idea that there is only one best method of task performance. This approach ignores flexibility of human performance. The other approach, known as the constraint-based approach, gives importance to defining constraints, where a user has to independently decide how to perform a task within the existing constraints. This approach is based on the idea that our activity is very flexible and there is no one right way of task performance. The last approach in fact entirely rejects the idea of design because there is no opportunity to create models of activity and compare them with configured interfaces or other types of equipment. The first approach is inadequate because contemporary work activity is very flexible, which is specifically true for HCI.

SSAT eliminates this contradiction. The concepts of activity self-regulation described in this book are very powerful tools for the analysis of various strategies of task performance at the stage of qualitative analysis. The book contains various examples of activity analysis from the standpoint of the theory of activity self-regulation. It is also shown that existing models of self-regulation outside of SSAT are not applicable in solving problems of ergonomic design and work analysis. The method of morphological analysis of activity, including its algorithmic description, developed in SSAT allows describing the structure of variable activity during the study of various computerized tasks in a formalized manner. Design involves selecting appropriate units of activity analysis and standardized methods for their description. In this regard, the book presents the justified method of classification and standardized description of cognitive and motor actions. The unique method of eye movement analysis during task performance is developed. This method highlights the principles of cognitive actions extraction and description of their structural organization, which is especially important for analyzing computer-based tasks where visual information is particularly important.

A system known as MTM-1 offers descriptions of the motor components of activity. This system has been developed for the analysis of production operations with predetermined sequence of motor components and has not been adapted for the analysis of flexible motor activity, which is usually combined with cognitive components. In order to adapt this system for contemporary task analysis, it has been analyzed from the standpoint of activity self-regulation. It is shown that without analyzing the expected activity strategies, it is impossible to correctly select standardized motions in flexible motor activity. The other drawback of MTM-1 is the fact that it considers external motor behavior as a system of motions. However, the external behavior is a hierarchically organized system that includes various units of physical activity. Motions are components of motor actions that are in turn the basic elements of external behavior. Each motor action includes several motions that are integrated by conscious goal of this action. Thus, MTM-1 ignores the concept of motor action. SSAT demonstrates how the concept of motor action can be used along with MTM-1. As a result, the fundamentally new and more efficient method of using MTM-1 in contemporary task analysis is presented. Formalized methods of activity analysis help create unique principles of quantitative assessment of computerized tasks.

The book suggests quantitative methods for assessing the psychological complexity of computer-based tasks. Evaluation of the complexity is multidimensional. On the basis of such an evaluation, it is possible to optimize task performance according to complexity criteria. The book also provides a method for assessing the reliability of task performance. The final chapter of the book presents methods of qualitative and quantitative analysis of exploratory activity during interaction with the computer. Exploratory activity is the most flexible activity and it requires special methods of analysis.

Authors

Dr. Gregory Z. Bedny worked as a professor in several Ukrainian universities and taught at Essex County College in New Jersey after his arrival in the United States. He has now retired and is a research associate at Ergologic, Inc. He earned his doctorate degree (PhD) in industrial organizational psychology from Moscow Educational University and his postdoctorate degree (ScD) in experimental psychology from the National Pedagogical Academy of Science of the Soviet Union. Dr. Bedny is a board-certified professional ergonomist (BCPE). He is also an honorary academician of the International Academy of Human Problems in Aviation and Astronautics in Russia and an honorary doctor of science at the University of South Ukrainian. He has been awarded an honorary medal by the Ukrainian Academy of Pedagogical Sciences for his achievement in psychology and his collaborative works with Ukrainian psychologists.

Dr. Bedny is the founder of the systemic–structural activity theory (SSAT). SSAT is a high-level generality theory or framework that is the basis for unified and standardized methods of studying human work. He authored a number of original scholarly books and multiple articles in this field. He has applied his theoretical study in the field of human–computer interaction, manufacturing, merchant marines, robots systems, work motivation, training, fatigue reduction, etc.

Waldemar Karwowski, PhD, DSc, PE, is professor and chairman of the Department of Industrial Engineering and Management Systems at the University of Central Florida, Orlando, Florida. He is also executive director, Institute for Advanced Systems Engineering, University of Central Florida, Orlando, Florida. He holds an MS (1978) in production engineering and management from the Technical University of Wroclaw, Poland, and a PhD (1982) in industrial engineering from Texas Tech University, United States.

Dr. Karwowski was awarded DSc (dr habil.) in management science by the State Institute for Organization and Management in Industry, Poland (2004). He also received honorary doctorates from three European universities. He is past president of the Human Factors and Ergonomics Society (2007) and the International Ergonomics Association (2000–2003). Dr. Karwowski served on the Committee on Human Systems Integration, National Research Council, the National Academies, United States (2007–2011). He is a coeditor of the *Human Factors and Ergonomics in Manufacturing & Service Industries,* and editor-in-chief of *Theoretical Issues in Ergonomics Science.* He is an author or editor of over 400 scientific publications in the areas of human systems integration, cognitive engineering, activity theory, systems engineering, HCI, fuzzy logic and neuro-fuzzy modeling, applications of nonlinear dynamics to human performance, and neuroergonomics.

Dr. Inna S. Bedny is a computer professional with a PhD in experimental psychology. Her research involves the application of SSAT to HCI. She is also the author or coauthor of over 10 scientific publications in the field of HCI.

Section I

Concept of Self-Regulation in Psychology and Ergonomics

1

Concept of Self-Regulation Outside of Activity Theory

1.1 Concept of Self-Regulation versus Input/Output Task Analysis

Currently, there are no effective methods of task analysis that can be efficiently used in the study of variable human activity. This is reflected in the two polarized approaches to task analysis: instruction-based approach and constraint-based approach. The instruction-based approach strictly determines all the required procedures of task performance. It is considered by some scientists as not being very efficient in contemporary task analysis. Vicente (1999) suggests resolving the apparent conflict between these approaches by introducing the concept of *constraints,* which specifies *what should not be done by a performer.* According to the second approach, performers independently decide how to perform the task within the existing constraints.

However, any design implies some constraints even when there is one best way of performing the task, such as time constraints, safety constraints, etc.

Moreover, a performer can utilize multiple methods of task performance inside existing constraints. Some of them can be efficient, and others inefficient. Constraints can be introduced using different parameters. Constraints exist in traditional engineering design of equipment. Even in the presence of the same constraints, there is still a possibility of utilizing different versions of equipment design solutions. Some of them are more efficient than others. Similarly, when performing a task with the same constraints, a subject can use both efficient and inefficient methods of task performance. The formalized stage of ergonomic design involves the creation of models of human activity during task performance. Such models should describe in a standardize manner the structure of activity during task performance, and at the next stage, this structure should be compared with a configuration of the designed equipment. Based on such comparison the optimal equipment design solution can be found. In cognitive psychology, there are no principles of creation of analytical models of human activity. Mentalistic models in cognitive psychology cannot be considered as design models. There are no units of analysis or language of description of human activity in cognitive psychology. In human–computer interaction (HCI) tasks, human activity is extremely flexible, and there are no clearly defined constraints that exist in traditional kids of work. Thus, the statement "a worker decides how to achieve the goal" (Vicente, 1999), within existing constraints, assumes that there is no design in ergonomics.

In cognitive psychology, there are no principles of analysis or description of flexible human behavior. However, such principles are critically important for ergonomic design. As we will be seeing in the course of the book, control theory models of self-regulation of human behavior are too mechanistic. They cannot describe flexible strategies of human performance. Self-regulation cannot be presented as a homeostatic process. In systemic-structural activity theory (SSAT), self-regulation is a goal-directed process. Thanks to self-regulation of activity, humans create a conscious goal and develop strategies to achieve this goal. Therefore, the description of these strategies during task performance is an important stage of task analysis, specifically for computer-based tasks.

The concept of self-regulation becomes critically important in the design of HCI systems. So, we start our discussion considering this concept as it is presented outside of activity theory (AT). This will allow us to compare various approaches to self-regulation from an AT, and specifically SSAT, perspective.

All of our behavior or activity, except for automated involuntary reactions, is organized based on principles of self-regulation, according to which we can arbitrarily create goals and choose appropriate strategies to achieve them. We always act differently in different circumstances due to self-regulation. Self-regulation is one of the central concepts in this book. This concept is widely used in contemporary psychology, but it often has a totally different meaning or is even used incorrectly. For example, self-regulation has been made synonymous with such notions as willpower, ego strength, and volition or as a motivational mechanism (Kanfer, 1996; Kuhl, 1992). Bandura has identified self-regulation as a self-reflective process that involves a comparison of current performance with subjective criteria of success from previous behavior. Subjects' self-efficacy levels affect the choice of a subjective criterion of success (Bandura, 1978, 1997). The basic elements of self-regulation systems are a goal as a standard. This standard is a mechanism that provides a comparison and an error correction routine. The main idea is based on the fact that a person controls her/his behavior by eliminating deviations from a goal that is acting as a standard. This system actually adjusts a person's behavior using a trial-and-error process. Thus, one approach reduces the process of self-regulation to studying isolated psychological mechanisms of behavior. However, the complex nature of self-regulation cannot be reduced to examining its individual mechanisms such as goal, motivation, self-efficacy, volition, etc. Bandura (1977, 1978) described self-regulation as the following sequence of the steps: observing oneself → judging oneself → rewarding oneself → regulating oneself. The self-regulation process cannot be presented as a linear sequence of such steps.

Self-regulation is a process that characterizes not only living beings but also nonliving systems. From this was concluded that there is another approach to the study of self-regulation in psychology. This approach derives from control theory. Presently, there are different viewpoints on self-regulation in psychology that derive from control theory. In most of the presented theories of self-regulation that are based on control theory, human behavior is considered as a discrepancy reduction process, the main mechanism of which is negative feedback. Self-regulation is reduced to a homeostatic process. However, human activity is not just an adaptive process but a goal-directed system. Adaptive and goal-directed self-regulative processes are not the same thing.

Knowledge of the principles of self-regulation of human activity is crucial for conducting adequate task analysis and ergonomic design. We will raise various aspects of self-regulation in our further discussion in this book. In this chapter, we dwell on the general aspects of this area of study. The importance of addressing self-regulation is associated with an understanding of the general principles of task analysis. As we pointed out, Vicente (1999) is a proponent of the constraint-based approach to task analysis. To prove the legitimacy of his approach, he uses the *input/output* analysis method. According to Vicente, the input/output method deliberately ignores the way a task should be performed. The steps used by a person to achieve a result should not be specified; only the adequate result and the performance in the existing constraints should be defined. How a task is actually performed should be ignored. To justify the constraint-based method, Vicente attracts the input/output analysis, which is also known as the black-box approach in cybernetics. By varying an input and obtaining certain results or outputs, we can get some information about what happens inside a black box. However, such data is not always reliable or sufficient.

For example, the black-box concept is used in computer programming when the process needs to be recoded in a new language and the old code is unavailable. Then, the new process is designed and coded based on the input and output of the old program. Therefore, steps that are used to achieve the required output in a new program become known. The input/output approach works for fairly simple processes. For more complex logic, rigorous testing of all possible inputs and exceptions should be done. Otherwise, the new process would not adequately replace the old one. It is not accidental that after a black-box analysis, new programs can sometimes unexpectedly fail.

The reduction of task analysis to the black-box method is essentially a rejection of the real task analysis and, derived from it, ergonomic design in general.

Not only is human behavior flexible but various phenomena of nature also are. We need to know the principles of describing variable human activity, as is done in studying variable phenomena in different fields outside ergonomics and psychology in order to analyze flexible human performance, taking into account the principles of human activity regulation. Units of activity analysis, an understanding of the difference between the goal of a system and a human goal, etc., are vital for such an analysis. The starting point of an analysis of human flexible activity or behavior is the study of activity self-regulation. Without analyzing how self-regulation is understood outside AT, it is difficult to understand this concept in the framework of SSAT. It should be noted that an analysis of the concept of self-regulation outside of AT helps to understand why this concept has not received sufficient recognition in ergonomics. So, we start this book with an analysis of this concept as it is used outside AT. Such an analysis serves as a starting point for further study of the concept of self-regulation in SSAT.

1.2 Self-Regulation from Control Theory Perspectives

We begin our discussion with the analysis of the concepts of self-regulation that derives from control theory. Control theory was one the main source of ideas for self-regulation. According to this theory, our behavior should be viewed as a self-regulative system. One of the founders of these ideas is Wiener (1948). Wiener's model for a kinesthetic feedback control system can be presented as depicted in Figure 1.1. This model presents an input coming from the left to the *subtractor* that is commonly called a *comparator*. The error signal from the comparator actuates the reset of the system to produce an output and required corrections. At the time the concept of *feedback* was only introduced in cybernetics and biology, this model had a considerable scientific interest.

A critical analysis of this model in terms of applying it in psychology was later presented by Powers (1978). Currently, however, this model has only historical interest and we will not dwell on the details and will concentrate on more recent models. The most significant achievement in this field from behavior analysis perspective has been made by Powers (1973). His model in a slightly modified form was used later by various psychologists. Let us consider some of his main ideas.

His model was developed based on the analysis of a tracking task where a subject manipulated a control lever to move a spot of light for tracking a moving target, that is, the subject was trying to keep the spot on the moving target.

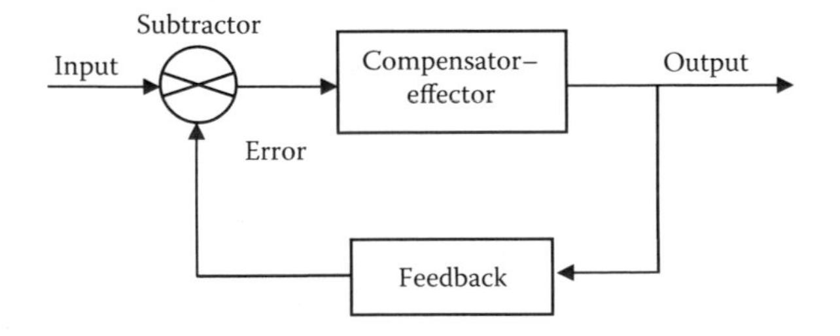

FIGURE 1.1
Wiener's feedback model of neuromuscular control.

Such experiments carry interest for military systems when a subject has to track a target. If a spot deviates from a target, it leads to errors. A subject's task was to eliminate deviations from a target. For such tasks, deviations of the spot from the target are a visual feedback. The fact that any feedback requires reduction of errors demonstrates that such feedback is seen as negative. Not every deviation of the spot from the target is considered an error. A subject develops subjective understanding of what is correct and what is wrong. Powers called the difference between some conditions of situation as the subject sees it a *reference condition*. Errors are always corrected with respect to a reference condition. A reference condition determines where a spot of light will be and not the location of the target itself. Powers defined a reference condition as a goal condition of a variable. A subject controls a variable with respect to a reference condition. If, according to feedback, the deviation is very small, the subject ignores it. When behavioral feedback demonstrates that the deviation reached a particular value, then it should be corrected. There are also disturbances that can complicate corrective actions. Using this analysis, the author infers that the purpose of any given behavior is to prevent controlled perceptions deviating from a reference condition. The second important statement is that purpose implies goal. A goal is defined as a reference condition of a controlled perception. Based on these conclusions, which we present in an abbreviated manner, Powers suggested his general model of a feedback control system. This model is presented in Figure 1.2.

Powers' model depicts a boundary between mental mechanisms of self-regulation and muscle system that interact with the environment. This model consists of the following mental mechanisms: input functions, comparator, and

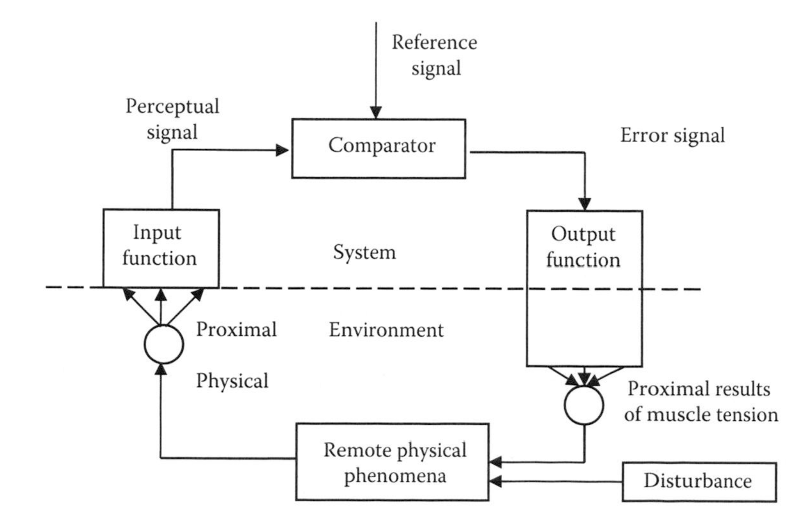

FIGURE 1.2
General model of a feedback control system and its local environment. (From Powers, W.T., *Behavior: The Control of Perception*, Aldine Publishing Company, Chicago, IL, 1973.)

output functions. Input functions are simply mechanisms that translate physical stimulation into a sensory-perceptual process. Comparator compares perceptual data with a reference signal. Goal or reference signal is simply considered as input information for comparator. If there are errors, then output functions are activated. Output functions are reduced to decreasing or increasing muscle tension. Actions directly affect proximal results of muscle tension. The controlled parameters are called *remote physical phenomena*. There are also disturbances that can complicate control of remote physical phenomenon, which in turn affects minute physical variables (*proximal physical stimuli*). The latter influences proximal physical stimuli and therefore input functions. From the model we can see that if there is a condition with zero error signals, then it implies zero output efforts. The starting point for a self-regulation cycle is perceptual process, and external stimulation is transformed into internal perception.

Powers described several levels of control unit organization: higher-order perceptual control systems perceive and control an environment composed of lower-order systems; only the lowest first-order system interacts directly with the external world; higher-order control systems provide awareness of perceptual signals. Control systems at all levels consist of the same mechanisms. However, human behavior cannot be described by the suggested perceptual control system. Input functions are often much more complex and include interpretation, thinking, decision making, etc. It should be noted that the concept of self-regulation has been introduced in works of Anokhin (1969) and Bernshtein (1967). Moreover, Bernshtein has described the levels of regulation of activity. Surprisingly, Powers, when publishing his work in 1973, did not reference these authors.

Powers should be praised for presenting our behavior as a self-regulated system that integrates cognitive and behavioral mechanisms. The attempt was made to present self-regulation as a purposive or goal-directed process. The author demonstrated that an objectively presented situation and its mental representation are not the same. These are important ideas. However, a more detailed examination of his concept of self-regulation shows that it is contrary to modern psychological data. It reduces the process of receiving information to the perceptual processes. However, the process of receiving information also includes memory, thinking, anticipation, and interpretation.

The same external stimulation can be perceived and interpreted differently by a subject depending on individual differences, past experience, etc. Furthermore, a subject voluntarily extracts required information from the same situation depending largely on a conscious goal of activity. Extracting information from a given situation depends not only on cognitive but also on emotional-motivational factors (Bedny and Meister, 1997).

According to Powers, a goal is a reference condition of a controlled perception. A goal performs input functions for a comparator in the described model (Figure 1.2). Such understanding of a goal contradicts with the data obtained in cognitive psychology (Pervin, 1989) and AT (Bedny and Karwowski, 2007; Bedny and Meister, 1997; Leont'ev, 1978; Rubinshtein, 1959). Goal cannot be considered simply as a reference condition of perception or input for the comparator. Goal, as a mechanism of self-regulation, involves all psychic processes and performs much more complex functions in activity regulation. Powers does not distinguish between different types of goals that exist at different levels of behavior regulation. In AT the term *goal* of activity is applied to the highest levels of regulation associated with complete or partial awareness of what should be reached during activity. Goal is the most important anticipatory mechanism of activity regulation that is associated with our consciousness. At the lower unconscious levels of activity regulation, various terms are used in place of the term *goal*. For example, the term *purpose*, suggested by Tolman (1932), is used to designate anticipatory future result of behavior regulation. Sokolov (1963) suggested utilizing such terminology as *neural model of stimuli*. Anokhin (1962) introduced the term *acceptor of an action* and Bernshtein (1966) suggested using *neural model of required future*.

A goal is a special form of anticipatory reflection of reality that represents a required future result of activity. At least part of this reflection must be conscious. This presentation of future results can be clarified or modified during the activity process. However, if the goal is fully modified and changed, then it means that a person is involved in new activity. We discuss the concept of goal in AT in a detailed manner in previous works (Bedny and Karwowski, 2007; Bedny and Meister, 1997). Powers' model is based only on the analysis of tracking tasks. Even such type of task performance involves anticipation at a sensory-perceptive level and conscious goal as the highest level of anticipation.

During skill acquisition, a subject changes her/his strategy of tracking task performance not simply using feedback in the form of deviation from a target, but also forming an image or mental model of the situation that allows anticipating a target's movement. The strategy of activity performance is reconstructed based on anticipatory mechanisms. It becomes possible to forecast a target's movement and change a program of regulation of motor actions based on this forecast. Powers' model of the self-regulation process is possible only if there are errors. According to this model, self-regulation is reduced to elimination of deviations from an externally given goal standard. Correction of behavior is possible only at the final stage of self-regulation when deviation from a goal of the self-regulation process is detected. Powers ignores immediate feedback that allows correcting behavior in the course of its implementation. Immediate feedback prevents errors.

However, self-regulation of behavior or activity cannot be reduced to error elimination based on negative feedback (Bedny and Karwowski, 2006; Bundura, 1989). It includes multiple loops of self-regulation. Powers' model does not take into account the fact that goal can be modified or even changed during a self-regulation process. Described model ignores the emotional-motivational aspects of the regulation of activity, etc.

Regulation of motor actions cannot be reduced to a change in muscle tension. Moreover, it involves not just motor but also mental action. Activity can be fulfilled only in intellectual plane. Therefore, self-regulation of activity also takes place in a mental plane without implementation of external physical actions. For example, a mathematician can mentally manipulate mathematical symbols, find and correct mistakes, and so on. A gymnast can perform a combination mentally by using her/his imagination before the actual performance of the routine. She/he can also correct any mistakes that might occur during the actual performance. Activity performed in a mental form is often not aimed at transforming a situation to achieve practical changes, but to explore a situation, to promote various hypotheses, evaluate their consequences, etc. Such activity is called *orienting activity*, which is also organized based on a self-regulative process (Bedny and Meister, 1997). According to Powers, all levels of regulation have the same components and similar organization. However, the structure of self-regulation at various levels is not identical (Bernshtein, 1996).

Let us consider the basic unit of cybernetic control suggested by personality psychologists Carver and Scheier (2005). Their model is presented in Figure 1.3.

This model of self-regulation contains several elements: an input function, a goal, a comparator, and an output function. An input function is equivalent to a perception process. A goal serves as a standard or reference value for a feedback. A comparator facilitates comparison between an input and a goal or a reference value.

An output function is treated as an external or internal behavior. If a comparator yields a difference between an input function and a goal or standard, the output changes. If a comparator does not detect such differences, the output function or behavior remains the same. There are also two kinds of feedback: negative or discrepancy-reducing loop and positive or

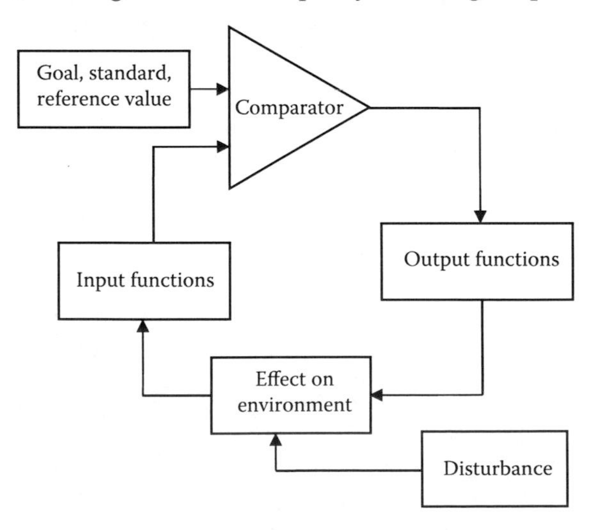

FIGURE 1.3
Feedback loop or the basic unit of cybernetic control. (From Carver, C.S. and Scheier, M.F., *On the Self-Regulation of Behavior*, Cambridge University Press, New York, 1998.)

discrepancy-enlarging loop. In the first case, feedback has a purpose of diminishing or eliminating any detected discrepancy between an input and a reference value or goal. Carver and Scheier (2005, p. 44) explain the second case, the discrepancy-enlarging loop, as follows: "The reference value in this case is not one to approach, but one to avoid. Think of this as an 'anti-goal.' When a subject has an anti-goal she/he attempts to increase discrepancy. Some authors use the term 'avoidance goal' instead of anti-goal" (Meas and Gebhardt, 2005).

We consider such terms as *antigoal* or *avoidance goal* as questionable. For instance, a chess player can formulate a goal of avoiding defeat, try to defend her/his figures' position, and prefer a safe, nonrisky strategy. Avoiding defeat is formed as a final goal of the game in the most general terms. It is specified through a set of intermediate goals that should be achieved. The whole activity is directed to reach the final goal "do not lose the game or avoid defeat." Let us describe another example: Suppose a person is an alcoholic. She/he forms a goal to quit drinking. For this purpose she/he takes medication, avoids parties where people gather to drink, and as a result she/he formulates goals—"go home instead of going to a bar," "go see a movie," "go to a psychiatrist," etc. All in all, a person formulates various goals to oppose the undesirable goal. Such goals can be achieved only by forming a vector *motives* → *goal* when a person tries to achieve the goal.

A schematic depiction of a feedback loop according to Carver and Scheier is presented in Figure 1.3. In this model, a goal plays a role of an objective standard. Such standard is necessary for functioning of a comparator that influences an output.

This model is similar to Powers' model. The main difference lies in the fact that the authors of the feedback loop model consider not only a negative, but also a positive feedback. A goal is seen not simply as an input for a comparator, but also as an independent mechanism of self-regulation. However, a goal in the proposed model is viewed simply as a readymade standard. The authors overlooked the fact that an objectively given goal and a subjectively excepted goal do not always match (Bedny and Karwowski, 2007; Konopkin, 1980; Kotik, 1974). A clear justification of described mechanisms of behavior regulation is not given. An externally given situation is not clearly separated from internal mechanisms.

The self-regulation model includes only one loop that cannot explain a complicated process of activity regulation, there are no emotional-motivational mechanisms of self-regulation, and so on. If we compare this model with the model proposed by Powers, it may be noted that the description of mechanisms of self-regulation is more precise in Powers' model. Carver and Scheier's model does not explain the goal-directed self-regulative process of human activity.

Lord and Levy (1994) also proposed a psychological version of self-regulation of behavior that derives from Wiener's control theory. The authors present their version of the self-regulation model as distinctly different from already existing versions, but their model depicted in Figure 1.4 differs little from the existing ones and in particular from the model proposed by Carver and Scheier (1998).

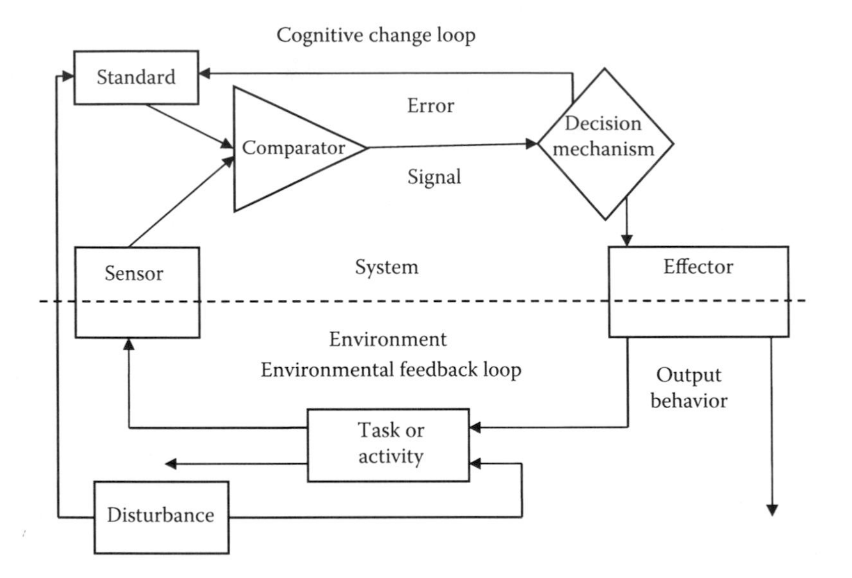

FIGURE 1.4
Self-regulative model of behavior according to Lord and Levy. (From Lord, R.G. and Levy, P.E., *Appl. Psychol. Int. Rev.*, 43(3), 335, 1994.)

It should be noted that the authors of most models of self-regulation do not distinguish between two principles of models development: one based on morphological and another on functional analysis of activity. In morphological analysis, each mechanism of self-regulation represents physiological organ or neural mechanism responsible for a specific stage of self-regulation of activity. The second approach uses the functional principle of self-regulation where each mechanism describes mental or emotional-motivational functions of activity regulation. This approach divides cognition into relatively independent stages that integrate sensation, perception, memory, thinking and decision making in subprocesses responsible for particular stages of self-regulation of activity. Such models do not show connections between parts of nervous system but demonstrate relationship between their functions.

For example, an overall goal of task can be considered as a conscious mechanism that performs integrative functions in activity regulation. Goal is not a part of a body or neural mechanism but rather is formed as a psychological mechanism that ceases to function after a goal is achieved. Functional mechanisms perform specific functions in the regulation of activity but their content and their importance in activity regulation vary. In contrast, morphological mechanisms depict specialized physiological organs that do not disappear or change their constructive features after a particular self-regulative process is completed.

This is incorrect when in the same model one uses functional and morphological principles of description for different mechanisms of self-regulation. As we later demonstrate, self-regulation at the psychological level should be based on functional principles of activity regulation. In Lord and Levy's model, such mechanisms as *sensor* and *effector* are related to morphological description and *standard, comparator,* and *decision making* are examples of functional description. *Task or activity* and *disturbance* are not mechanisms of self-regulation. Any task includes not only human activity but also material components and a sign system.

The function of such a mechanism as *task or activity* is also unclear. What is the relationship between *output or behavior* and *task or activity.* Behavior is an outward manifestation of activity. It is also unclear what is the relationship between *task or activity* and such mechanisms as *standard, comparator,* and *decision mechanism.* The latter are elements of activity and they do not exist outside of activity. All these limitations demonstrate that in Lord and Levy's model, as in the previous models, there is no clear distinction of functions that each mechanism plays in activity regulation. For example, a standard is conceptualized as any type of sensed information that can be stored in long-term memory for future use as a standard for comparison. Authors relate to such standard goals, images, attitudes and values, self-assessment, desired outcomes, affective outcomes, and rate of progress (Lord and Levy, 1994, p. 338). This is a rather broad and not precise definition of a standard. Such an understanding of standard covers a variety of mechanisms. The authors combine completely different functions in the same mechanism and, at the same time, delimit interrelated functions of activity regulation. Goal and standard are not the same because the goal of a task can be clarified and specified during task performance. A goal includes information not only about the past that is stored in long-term memory, but first of all information about the desired future result of a subject's own activity. A goal and an image cannot be contrasted against each other because a goal includes verbally logical and imaginative components of a desired future outcome. Emotional-motivational or energetic components are independent mechanisms of activity regulation. They interact with informational or cognitive mechanisms during the process of self-regulation. In other words, information and energy are interconnected but are not the same thing.

It is not clear how comparator works. Self-regulation is carried out only based on analysis of errors. Nonetheless, humans and even animals can regulate their behavior and activity in general based on prediction of positive or negative results. These examples show that the mechanisms being considered are amorphous, have no clear boundaries; it is not clear how such a model can be used in specific studies and particularly in task analysis.

Following Powers (1973), these authors emphasize the hierarchical organization of the self-regulative process based on understanding self-regulation as a hierarchical, organized control system. A higher-level self-regulative system integrates low-level self-regulative systems. Each level of self-regulation has a similar structure. However, the number of subsystems is restricted by the span of control (Simon, 1999). Activity also has logical organization. Moreover, at each level of hierarchy, the structure of self-regulative system is not the same. Hence, it is necessary to develop multiple models of self-regulation that are specific for each particular level of activity regulation.

It should be noted that the special issue of *Applied Psychology: An International Review* in which Lord and Levy's article has been published was complemented by critical publications on the work under consideration. This greatly facilitates our critical analysis of the concept of self-regulation proposed by these authors. Locke (1994), for instance, titled his article in this issue as "The emperor is naked." For him, all concepts of self-regulation that derive from control theory are too mechanistic and therefore are not applicable to human activity. According to him, the human self-regulative process involves consciousness and this problem is ignored in every self-regulation concept that derives from control theory. Self-regulation concepts reduce activity regulation to negative feedback mechanisms while people deliberately create discrepancy (Bandura, 1989). Human self-regulative process is goal-directed but not homeostatic (Bedny and Karwowski, 2007; Bedny and Meister, 1997). Some control theorists object that their version of control theory posits not only discrepancy reduction, but also discrepancy creation in the process of self-regulation. However, this is not clear from any models of self-regulation control theory offers. We also cannot agree with Lord and Levy that there is discrepancy-based learning. According to SSAT, learning is a self-regulative process that springs out a variety of skill acquisition strategies (Bedny and Karwowski, 2007) because cognition and activity in general are organized as a goal-directed self-regulative process. Such self-regulative processes derive from the laws of regulation of live beings' behaviors, but not from control theory that has been adapted for the description of the self-regulation process of technical systems.

In this same issue, Hacker (1994) also discusses Lord and Levy's work from a critical perspective. He considers another aspect of these authors' publication, paying particular attention to such aspect as "bridging the gap between cognition and action." According to Hacker, control theory will not bridge the gap between cognition and action because there is no such gap. In AT, there is the well-developed principle of "unity of consciousness and behavior" (Rubinshtein, 1959) or "unity of cognition and behavior" (Bedny et al., 2001, 2011). Hence, Hacker rightly addresses these authors to AT.

All these critical comments to a large extent can be attributed to the models of self-regulation discussed so far. A peculiarity of these models is the fact that all of them ignore data obtained in this field by such scientists as Anokhin (1969), Bernshtein (1996), and Sokolov (1969) and others whose work laid the basis for the study of self-regulation not only of technical devises but also of human and animal behavior.

1.3 Self-Regulation in Cognitive Psychology

Let us consider the concepts of self-regulation that derive primarily from psychological analysis of self-regulation in cognitive psychology. The most important work in this field was done by Miller et al. (1960) when behaviorism dominated in American psychology and human behavior was considered as reactive, purposeless, and driven by environmental stimuli. These authors described human behavior as self-regulated, purposive system. We discuss this work in an abbreviated manner because it is well known by scientists. According to these authors, the building blocks of human behavior are represented by the TOTE system (test, operate, test, exit). The *test* phase assesses whether there is a discrepancy between existing and future desired state. If there is such a discrepancy, the necessary behavior is executed at the *operate* phase. A subsequent *test* phase evaluates the result of the second phase. If there is no discrepancy between the actual and desired state, transfer or *exit* to the next hierarchically organized units is possible. The TOTE unit might be controlled at a higher level by means of a plan, which in turn might be a part of a higher-order plan, and so on (Figure 1.5).

It is obvious that authors describe only general principles of behavior self-regulation. Miller et al.'s (1960, 1965) was the first work in the West that showed human behavior not as a reactive but as a self-regulated system. According to this work, human behavior cannot be explained without feedback. This model can be criticized from various viewpoints: It describes self-regulation in a very general manner; it suggests that human behavior is similar to computer functioning; specificity of planning human cognitive and behavioral actions is not discussed. However, the authors did not intend to describe

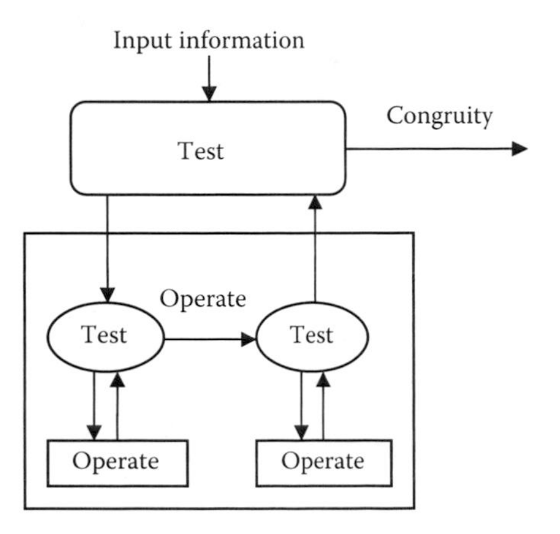

FIGURE 1.5
TOTE as the self-regulative unit of behavior. (From Miller, G.A. et al., *Plans and the Structure of Behavior*, Moscow, Progress, Russian Translation, 1965.)

self-regulation in detail. The main purpose in their work was to show that our behavior is based on the principles of self-regulation, which was especially important because behavioral approach dominated at that time.

The concept of self-regulation proposed by Miller et al. (1960) served as the basis for applying ideas of self-regulation to training research (Welford, 1968), in the development of a general model of the feedback control system (Powers, 1973), in action theory developed by Norman (1986) in the United States, and in the theory of action created in Germany (Frese and Zapf, 1994). Of course, the model presented by these authors is very general, but its theoretical importance is very significant.

Despite the fact that the general model of self-regulation of behavior developed by Miller, Galanter, and Pribram was proposed in 1960, it did not contribute sufficiently to the further development of ideas of self-regulation in the cognitive psychology, but instead replaced self-regulation with feedback. The idea of feedback outside of the concept of self-regulation is not very productive. Let us consider how the concept of feedback is discussed in contemporary cognitive psychology by examining the model of human information processing (Wickens and Hollands, 2000) presented in Figure 1.6.

This model offers some basic ideas for analyzing various psychological processes and their interactions. Information processing is described as a series of stages whose function is to transform or carry out some operation on the information. According to these authors, processing may start from input from the environment, initiated by an operator's voluntary intention to act, or somewhere in between. Hence, there is no fixed starting point in the sequence of stages.

Limitation of this model is obvious. It does not take into account that receiving and selecting information is conducted based on an existing goal, emotional and motivational state of a performer, and his/her system of expectations at a particular time. A performer actively selects information from the environment and from memory. At any given time, all cognitive processes are integrated in a certain way depending on the specifics of the task and the stage of its performance. Some mental processes play a leading role, some subordinate to the others. This suggests that the study of separate mental processes in human performance is necessary but not sufficient. Engineering psychology

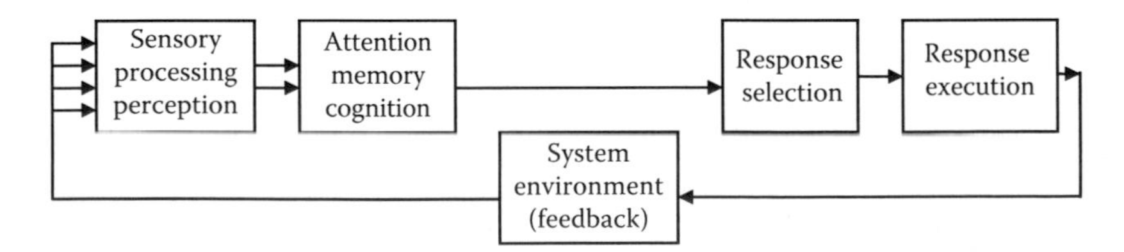

FIGURE 1.6
A modified fragment of the model of human information processing stages. (From Wickens, C.D. and Holands, J.G., *Engineering Psychology and Human Performance*, Harper-Collins, New York, 2000.)

and ergonomics should not just study separate mental processes but rather consider work activity as a whole during task performance.

It is obvious that operators do not choose a readymade response from memory. Further, the response cannot be viewed as a reaction to stimuli. An operator actively forms a program of motor actions, adapts it to a changing situation, and based on it performs required motor actions. Each motor action has its own goal. Motor actions are performed based on self-regulative principles. Feedback can be used at various stages of action execution. Evaluation of actions is performed based on subjective criteria of success. These criteria do not always coincide with goals of actions. All of this will be considered in detail further in the chapter.

Another important point in the analysis of the proposed model is to consider the role of feedback. According to this model a response can be evaluated only after it is completed, an error can be corrected only after it has occurred during performance of a new response, anticipating errors in the course of the response is virtually impossible. An immediate feedback during performance of a motor movement can be used for correction a specific motor response, but such feedback is not sufficient for correction of activity in general. For example, a subject can incorrectly select a motor action and due to immediate feedback correct it. However, this does not prevent errors during task performance. However, in some cases, errors are unacceptable. An operator can forecast an error and adjust the program of response in his/her mental plane, use an immediate feedback during execution of motor response. This means that a motor response can be corrected in the course of its execution. Using feedback at the final stage of response is just one of many possible ways of error correction. This demonstrates that the discussion should focus not on reactions or responses but on goal-directed and voluntarily regulated motor actions. All these issues will be discussed further in Chapter 3.

A model of human information processing stages, similar to any other model of self-regulation outside of SSAT, ignores the fact that a person can perform cognitive actions and evaluate their results using mental feedback. When discussing the presence of feedback, not only after performance of motor actions but also after performance of mental actions, it is very interesting to consider some example from a game of chess. There is a good example from the life of the famous chess player Bob Fisher and his mentor and friend John "Jack" W. Collins. Jack was disabled and he was driven through the streets of New York in a wheelchair. Fisher often accompanied him on these trips and they play chess during such trips. Remarkable was the fact that they were playing a game of chess without a chess board. Such outstanding players can play chess this way because they can mentally imagine a chess board and the pieces on it. Moreover, they could mentally imagine movements of chess pieces during a game and remember their positions. This style of game is known as "blindfold chess." "Pawn to queen bishop four" yelled young Bobby for the whole street to hear, plunging passersby into horror, who thought he was mad. This example demonstrates that both players performed mental actions. They were able to mentally promote various hypotheses, check them, evaluate possible actions of the opponent, and so on. This means that players would not only perform cognitive actions, but also mentally evaluate them using a mental feedback.

1.4 Self-Regulation in Action Theory

There is a set of models developed by a number of authors within the framework of action theory. Let us briefly consider the ideas of self-regulation as viewed in this area. First of all, we would like to stress the fact that the concept of action in AT and the meaning of this term in action theory are not the same. German scientists Frese and Zapf (1994, p. 271) defined *action* as a goal-directed behavior that can be regulated consciously or via routines.

Activity is defined as a goal-directed system, where cognition, behavior, and motivation are integrated and organized by the mechanisms of self-regulation (Bedny and Karwowski, 2007). The concept of goal plays a major role both in action theory and AT.

The analysis of the definition of action suggested by Frese and Zapf demonstrates that *action* in action theory has the same meaning as *activity* in AT.

However, the concept of action in AT has a totally different meaning. It can be cognitive and behavioral and is a main unit of activity analysis. Action is the smallest unit of activity that has a conscious goal (Leont'ev, 1978; Rubinshtein, 1959). When we consider activity from the perspective of morphological analysis, we describe the logical organization of human cognitive and behavioral actions (Bedny and Karwowski, 2007). German psychologists Heckhausen (1991) and Gollwitzer (1996) introduced the motivational concept of action. Frese and Zapf (1994) discussed the behavioral concept of action, suggesting that only behavioral concept of action can be applied to the study human work and the motivational concept of action should not be considered. These two theories are different. Historically, psychologists in Eastern Germany utilize the concept of activity that derived from Soviet psychology. When East and West Germany integrated, AT was labeled as action theory. For example, Hacker (1985), who used to be considered as an AT specialist, is now recognized in Germany as one of the leading specialist in behavioral action theory. So, some basic concepts in behavioral action theory derive from AT. Frese and Zapf (1994, p. 273) wrote, "At the same time it has been influenced by Soviet psychology, particularly Rubinshtein (1959, 1958), Leont'ev (1978, 1981), Vygotsky (1962) and Luria (1959)." These authors also mention Soviet psychologists Oshanin (1976) and Gal'perin (1969). Hence there is no clear-cut border between AT and action theory. At the same time, there is no clear understanding in action theory of basic units of analysis as it is the case in AT. These authors are not familiar with applied and SSAT.

The behavior-oriented action theory, similar to AT, utilizes the concept of self-regulation. However, the process of self-regulation is simplified, stating that an action as a self-regulative system consists of the following steps (Frese and Zapf, 1994): (a) development of goals and decision between competing goals; (b) orientation, including prognosis of future events; (c) generation of plans; (d) decision to select a particular plan from available plans; (execution and monitoring of the plan); and (f) feedback processing. Such a self-regulative process is depicted in Figure 1.7.

This process includes six *steps*. Why Frese and Zapf chose to include these steps in the below diagram is not sufficiently explained. They simply refer to various authors who discussed these steps. Moreover, the role of these steps as mechanisms of self-regulation also is not clearly justified. One the most important steps in this model is goal, which is considered as an anticipative cognitive mechanism (Frese and Zapf, 1994, p. 275; Hacker, 1986, p. 115). Hacker's definition of goal is very similar to the one given in general AT. However, according to Frese and Zapf, goal integrates motivational and cognitive components. Action is *pulled* by the goal and goal is a point of comparison for an action (cognitive aspect of the goal). Hence, a goal in this diagram is a standard for evaluating the result of an action. The second function of the goal is to pull an action toward it. However, we cannot see how a goal pulls an action in the diagram. In AT, there is a concept of the vector *motives* → *goal*. Motives are energetic components and

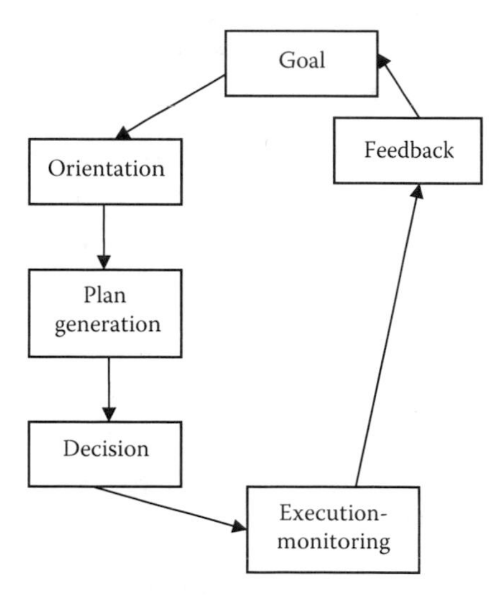

FIGURE 1.7
The self-regulation model of the action process. (From Frese, M. and Zapf, D., Action as a core of work psychology: A German approach, in: Triadis, H.C., Dunnette, M.D., and Hough, L.M. (eds.), *Handbook of Industrial and Organizational Psychology*, Consulting Psychologists Press, Polo Alto, CA, 1994, pp. 271–340.)

goal is a cognitive component in AT (Bedny and Meister, 1997). Independent motivational mechanisms are not presented in this diagram.

Our behavior is always poly-motivated. The inclusion of motivation into the content of goal results in having a lot of goals for the same task at any given time. The question is how a person can focus on fulfilling a task. It is known that during task execution a person has one final goal and a number of intermediate goals of subtasks or individual actions. Due to the confusion that arises when we combine goals and motives, some scientists propose to abandon the concept of goal altogether (Diaper and Stanton, 2004), which is unacceptable from an AT perspective.

We have discussed the first box in Figure 1.7. The next one is about orientation, which is considered as an orienting reflex, as attending a signal. However *orientation step* can be a complex stage of activity or can become an independent, orienting activity that has its own goal, motives, criteria of evaluation, etc. Orienting activity can be considered as a complex self-regulative system. We present a model of self-regulation of orienting activity in Section 3.2. Orienting reflex is only the simplest mechanism of the orienting step and such concepts as conceptual model, mental model, dynamic model, operative image, and situation awareness are much more important. Frese and Zapf consider the operative image concept (Frese and Zapf, 1994, p. 286). This term has been introduced to AT by Oshanin (1977). Works of Frese and Zapf define operative image as an internal long-term representation of condition–action–result interrelationship. However, according to Oshanin, operative image is dynamic and situation specific. It is a component of dynamic mental model, which is also an important concept in cognitive psychology. The long-term representation of condition–action–result interrelationship belongs to a stable mental model or conceptual model (Bedny and Karwowski, 2007).

The next box is about the plan generation step. Relationship between a program formation stage and a program realization stage is not clear in this model. The authors simply describe some general characteristics of a plan as a combination of thoughts and actions. What is the role of thinking in the other steps of action process is also not clear.

Without a clear understanding of the elements of a holistic action as a goal-directed behavior, one cannot understand what is a plan or a program. A plan or a program facilitates consciously or unconsciously keeping smaller units of the whole activity in some order. Planning can be performed at various levels of decomposition. However, in any particular situation, we need to know what elements of the whole are. Without understanding the units of analysis, it is impossible to understand planning or programming of actions.

In the diagram, feedback follows execution, as it is designated by a special box. However, feedback is not a step or mechanism that only follows execution. Feedback in such models is usually designated by lines connected with other mechanisms as well. A one loop model cannot explain functions of feedback with simple progression from goal to execution and from execution to goal. Some general characteristics of feedback cannot substitute the description of feedback mechanisms in a self-regulative model. The authors pay attention to the fact that action execution not always follows in exact order as it is presented in Figure 1.7. However, this possible flexibility is not demonstrated in the diagram. Thus there are discrepancies between the representation of the diagram and the explanation of its functioning. Frese and Zapf's (1994) description of self-regulation is eclectic. On one side they use terminology from AT utilized by Hacker and on the other side they use terminology from action theory.

It is doubtful that the proposed model can be used to describe self-regulation of goal-directed behavior. Explanatory opportunities of suggested diagram are limited and as a result the authors try to use Miller et al.'s (1960) model of self-regulation for their interpretation of action theory. It is worth mentioning that some industrial/organizational psychologists mix motivational action theory suggested by Heckhausen (1991) and behavioral action theory described by Frese and Zapf (1994).

The other concept of self-regulation was suggested by Norman (1986) in his approximate concept of action. According to Norman, we are presently unable to develop the theory of action and we should use approximate theory of action. His model of seven stages of activities described as one loop reflects the main ideas of this theory (see Figure 1.8).

As Norman wrote, these ideas come from studying servomechanisms and cybernetics and psychological works, among which the most important were the works of Miller et al. (1960) and Powers (1973). The approximate theory describes action as a self-regulated process.

It distinguishes the stages of activities depicted in Figure 1.8 but not always used or applied in this order. Utilized terminology raises the question of

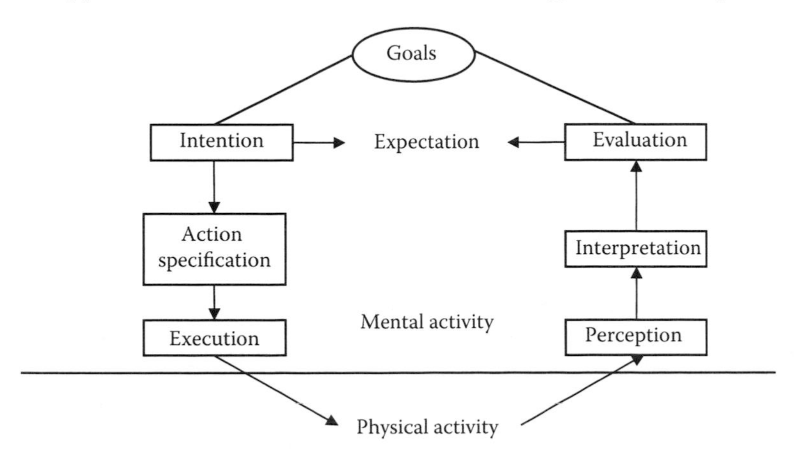

FIGURE 1.8
Norman's model of the seven stages of activity. (From Norman, D.A., Cognitive engineering, in: Norman, D. and Draper, S. (eds.), *User Centered System Design: New Perspectives on Human–Computer Interaction*, Lawrence Erlbaum Associates, Hillsdale, NJ, 1986, pp. 31–61.)

what is the difference between activities, action, physical action, and activity. In psychology, this terminology has different meanings in different theories. This basic question is not discussed. We can infer from this model that forming a goal and an intention and specifying an action sequence are mental events. Execution of an action means doing something or performing a complex sequence of motor actions. Hence, actions are related to motor activities and all other stages are mental processes or cognition. However, a person can perform mental actions—rotate an image to a particular position according to a given goal, conduct mental calculations, make a decision, etc. In AT, these events are considered as cognitive actions (Bedny and Meister, 1997). Each cognitive and behavior action has its own start and end points. Cognitive and behavioral actions are the basic elements of activity and are main units of analysis in AT. In Norman's explanation we cannot distinguish between action and activity.

Let us consider the stages of activities given in Figure 1.8. Norman (1986, p. 41) wrote, "...the process of performing and evaluating an action can be approximated by seven stages." These are: (1) establishing a goal, (2) forming an intention, (3) specifying an action sequence, (4) executing an action, (5) perceiving a system state, (6) interpreting a state, and (7) evaluating a system state with respect to a goals and intentions.

Norman does not believe that there really are clear, separate stages. This is simply approximating activity into stages for practical purposes. These stages correspond with stages of execution. From the cited statement, we can see that *action* is divided into stages. *Activity* is also divided into stages. *Stages* also is synonymous to *activities*. Description of Figure 1.8 also demonstrates that *actions relate only to motor activity*.

We can see that same psychological data have completely different terminology, which is not accidental. Norman is one of the leading cognitive psychologists. However, in cognitive psychology, there are no standardized units of analysis or related terminology. This is unacceptable when developing a self-regulation model. Let us look at the seven stages in more detail.

A goal is a state a person wishes to achieve. An intention is a decision to act so as to achieve the goal. Specification of an action sequence is the psychological process of determining a psychological representation or mental specification of actions that are to be executed. The next stage is the execution of *actions*. At the bottom of Figure 1.8 there is physical activity. The term *action* has two meanings: mental activity with seven stages or simply physical activity. Moreover, physical activity has a number of physical actions. Perceiving a system state is translating a physical state into a psychological state or perception. The last two stages involve interpreting a system state and its evaluation with respect to goals or intentions. However, this sequence of stages and stages themselves are not clearly defined. Some stages are a combination of various cognitive processes. The others are simply separate cognitive processes. Moreover, these stages often are not separate. For instance, perception can be included into all the stages. Goal cannot be considered as a readymade standard for the evaluation of the regulation process, etc.

In explaining this model, Norman uses such terms as action, actions, activity, activities in an interchangeable manner, which is confusing when one wants to understand the specification of actions' sequence without knowing what the term action means. This representation shows that action is a complex self-regulative system that includes seven stages. Specification of an action sequence means that actions are elements of an action system.

This model ignores the emotional-motivational factor that influences cognitive stages of information processing. The term *intention* better suites designation of motivational components. Self-regulation is a multiloop process and cannot be presented by a model with one loop with a linear sequence of cognitive stages, and a simple statement that they can follow in any order does not compensate for its shortcomings. This model can use feedback only after physical activity execution. Therefore, a subject can correct her/his behavior only after producing errors. However, the subject can perform actions mentally and evaluate their result. Vicente (1999, pp. 183–184) considers Norman's model as representing traditional information-processing approaches. According to him, this approach is reductionist in the sense that it tries to break down molar tasks into their constituent elemental information-processing steps. Typically, such steps are organized into a linear sequence that progresses from perception to decision making to actions. He presents as an example such linear steps as Activation → Observation → Identification → Interpretation → Evaluation ··· → Execution →. Vicente wrote that the specifics of such steps are not important. The key point is that the information-processing approach breaks down a task into a many linear elementary information-processing activities. Further, he stated that different models consist of different steps, and as an example for comparison, he presented Norman's model that we discusses earlier. For Vicente, the linear model of human information processing and Norman's model are practically similar. Thus, such terminology as molar tasks, elementary information-processing activities, stages of user activity, and their classification and organization do not matter.

This is not accidental. The terminology that describes these stages, their organization, and validity of choosing these stages is not justified. It is natural that the importance and validity of such models is very limited. At the same time, Norman's model is different from existing linear sequential stages in traditional cognitive psychology. It includes the concept of goal and attempts to present human behavior (including cognition) as self-regulated system. In conclusion, we want to say that in spite of respecting Norman's fundamental achievements in cognitive psychology we still cannot accept his approximate model of action regulation because of its shortcomings.

1.5 Concept of Self-Regulation in I/O Psychology

Analysis of self-regulation in I/O psychology that examines human work is of particular interest. In this regard, let us consider the data presented by Vancouver (2005).

Self-regulation is a critically important concept in I/O psychology. Thanks to self-regulation, a worker adjusts his/her behavior to a situation and selects adequate strategies of task performance. In I/O psychology, self-regulation is described from the control theory perspective (Vancouver, 2005). This author views self-regulation as a homeostatic process. So, we consider this theory to be the most representative theory of self-regulation in I/O psychology.

We already considered how Vancouver understands self-regulation. He defined self-regulation indirectly based on the concept of *regulation*, which, for him, is keeping something regular or maintaining a variable at some value despite disturbances to this variable. If the system regulates a variable based on an internally presented desired state, this is self-regulation. Internally presented desired states are called goals (Austin and Vancouver, 1996; Vancouver, 2005, p. 304). Vancouver did not specify what kind of system is in question: technical system or a human as a system. A technical system's goal is not the same as that of a human. The definition of goal is also incorrect. Internally represented desired states are not always the goal of a human. For example, a teenager can imagine different types of desired situations or states, but they may not have any relation to her/his goals. A child says: "I would like to become an astronaut" but never does anything to accomplish it or understands that for health reasons it could not materialize. Such desire is just a dream. A desired future state can be a goal if a person acts to achieve this future state. A desired future state can also be a potential goal. However, a person should understand that he/she has an opportunity to achieve this future state as a result of his/her own activity. A person should also realize that he/she may use this potential opportunity in the future.

Vancouver's description of the diagram of self-regulation is similar to Powers (1973) and Carver and Scheier (1981) (see Figure 1.9).

Vancouver's (2005) diagram contains similar units to the ones presented in Powers' model. These units contain input function (*I*), comparator function (*C*) and output function (*O*). In this diagram, an input function is reduced to a perceptual process. However, receiving of information should not be reduced to a perceptual process because an operator actively extracts and integrates required information and based on it creates internal mental models of a situation (stable model, conceptual model, dynamic model, etc.), plans her/his activity, and makes decisions (Bedny and Karwowski, 2007; Bedny and Meister, 1997). This means that input functions include not only perceptual process but also memory and thinking. An input function is also interpreting information.

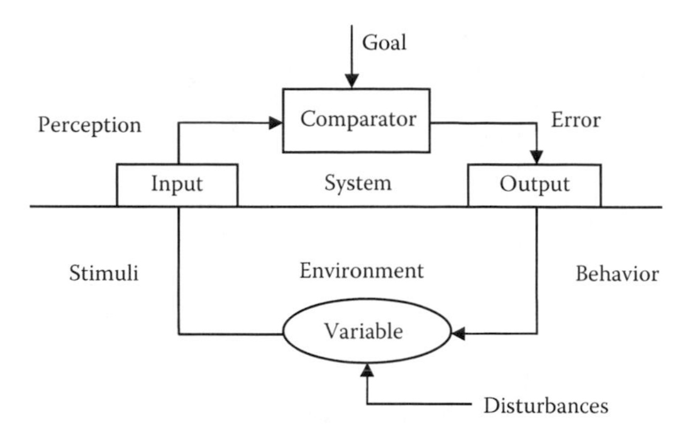

FIGURE 1.9
Vancouver's cybernetic-control system diagram that is similar to that of Powers (1973) and Carver and Scheier (1981). (From Vancouver, J.T., Self-regulation in organizational settings: A tale of two paradigms, in: Boekaerts, M., Pintrich, P.R., and Zeidner, M. (eds.), *Handbook of Self-Regulation*, Academic Press, San Diego, CA, 2005, pp. 303–341.)

In Vancouver's diagram, a goal is simply a reference signal or a standard that performs an input function for a comparator. A goal should not be considered as an externally given in a ready form standard. An objectively given goal can be interpreted in various ways, reformulated, specified, rejected, and so on.

Vancouver interprets the concept of human goal in one instance as an internally presented desired state of a system and in another instance our behavior varies in a certain range around such goal-standard, and then he defines it as a desired level of errors (see Vancouver, 2005, p. 305, 329).

This understanding of goal contradicts to data presented by such scientists as Lee et al. (1989), Kleinbeck and Schmidt (1990), Leont'ev (1978), Bedny and Meister (1997), Bedny and Karwowski (2007), and Rubinshtein (1957). In another work, Austin and Vancouver (1996, p. 345) consider goal as a mechanism that can be conscious or unconscious. As an example of unconscious goal, they present needs for achievement, needs for power, and needs for affiliation. However, according to AT needs belong to motivational factors (Bedny and Karwowski, 2007; Bedny and Meister, 1997).

For instance, if a person has needs for achievement, it is a motivational aspect of activity and in order to satisfy these needs she/he can formulate various goals that she/he wants to achieve, which are cognitive components of activity. Reaching these goals would satisfy a person's needs for achievement. When needs are connected with formulated goals and the vector motive → goal is created, they become motives. Considering a goal as a combination of motives and cognition leads to a situation when it becomes impossible to use a concept of goal in task analysis. As we have mentioned above this is one reason why some scientists insist that a concept of goal should be abandoned when studying human work (Diaper and Stanton, 2004).

Our behavior is really polymotivated. If the goal includes both cognitive and motivational factors, then each task has multiple goals of task. On the contrary, each task has only one overall goal of the task. Of course, there are goals of subtasks, or goals of separate actions that are used by a subject during task performance. We can use the term *potential goal*; however, it is not yet a goal. For example, goal-directed set can be transformed into a conscious goal. However goal-directed set is not yet a goal.

Goal cannot always be clearly defined in advance as a standard. Moreover, a goal and a standard of successful result do not always coincide. Only for the simplest tasks, where a goal is clearly described and presented to a subject in instructions, can a standard of a successful result be the same as a given goal, but even in such cases a subjectively formed standard of successful result may be different from a clearly specified goal because a subjectively accepted goal may be modified and differ from an objectively given one. Moreover, a successful result of activity can also be modified by a subject and distinguished from an objectively given goal. For example, depending on motivation, significance of a task, feelings of fatigue, etc., an employee may make a subjective adoption of a standard of successful result. This standard may be above or below an objectively given or subjectively formulated goal.

There are also complex problem-solving tasks where a goal can be specified or formulated in very general terms at the first stage of task performance. Goals for such problems are corrected and specified during the subject's progress toward a solution. The result of the solution may also be unpredictable. Progress in solving a problem can be represented as the formation of subgoals and attempting to achieve them. Subgoals can also be formulated in general terms and reformulated or clarified during solving subproblems. For such tasks, feedback is changing over time and is dynamic in nature. Therefore we cannot accept situations where a goal, a standard, and a reference value are practically considered as synonyms and unchangeable over time.

Results of activity can be consistent with actually formed goal or act as side effects that may correlate with possible ideal or potential goals. As a result, an actual goal may change. A potential goal can be transformed into an actual one. Formation of goals can be deployed in time as a process. In some cases, formation of a goal could turn into independent activity and goal formation arises as an independent task-problem. This becomes possible when goal-formation process acquires its own motivation. Examples when goal-formation process is transformed into an independent activity can be found in science. Mental reflection of a goal can also change due to changes between verbal and imaginative components of goal. The greater is the role played by the verbal component, the more precise goals become and they are kept in memory long after their achievement. Hence, a goal cannot be considered as a readymade standard that performs a role of input for the comparator, as it is presented in Vancouver's model of self-regulation.

Trying to link goal with self-regulation Vancouver introduced attainment goal and maintenance goal. His notion of *maintenance goal* helps to connect the concept of goal with the concept of self-regulation. The author reduces the process of self-regulation to elimination of deviations from a so-called maintenance goal. Understanding self-regulation as a process of eliminating deviation from *maintenance goal* is a homeostatic principle of self-regulation but human activity is not limited to elimination of errors that deviate from the standard. Such simple tasks are commonly accomplished by technical systems. Typically, if a situation deviates from acceptable limits, a performer formulates a new goal and therefore has a new task that helps to eliminate the deviations. At the next stage, a logical system of actions aimed to achieve this new goal are performed.

For example, an operator performs a task that requires maintaining a technological process according to required technological parameters. However, introducing the concept *maintenance goal* is not reasonable even in such situations. Depending on the time, size, and other specific characteristics of the deviation, an operator formulates the goal to *eliminate deviation*, makes a plan to achieve this goal, performs a sequence of actions, and so on. Moreover, self-regulation cannot be reduced to elimination of errors. According to Vancouver (2005), if a subject makes an error it gives him/her an opportunity to eliminate it and the self-regulation process is activated.

To substantiate his arguments and theorizing, Vancouver utilized the following laughable hypothetical example (Vancouver, 2005, p. 305). In maintenance context, a widget maker has to monitor the state of the shelves in the store. The goal is to keep the shelf full. When customers purchase widgets, a widget maker should replenish them, but only enough to fill the shelf. So the customers are considered to be a source of disturbance of the variable (state of the shelf) that produces an error. In the attainment context, the widget maker has a goal to produce a required number of widgets to fill the shelf and to keep it full. Each workday begins anew with zero widgets made and ends when a goal is reached. Further, the author writes (Vancouver, 2005, p. 315): "as customers purchase widgets, an error is created between the goal and the widget maker's perception of the state of the shelf."

The example demonstrates that the author does not understand the meaning of the production process, work process, task, errors, etc. It sounds like a widget maker performs incompatible functions. When a worker produces a widget, she/he performs a number of production tasks. Transportation also can include a number of tasks and takes time. A customer cannot wait for the completion of the production process. The same person cannot be responsible for production, storing, and selling. Customers do not disturb businesses. The main goal of the production process is to sell a product, not to keep shelves full. In this example, the widget maker has multiple goals that are not compatible in time. Moreover, it is not reasonable to fill the shelves after every sale. Only when the number of widgets approaches the minimal required quantity, a worker formulates a task (therefore goal of task) to go into the production room and bring more widgets to fill the shelf. Usually this is not done by the salesman. When the number of widgets becomes lower than the minimum required and it, for example, results in an inability to serve the customers in a timely manner, it can be considered as an error. In other situations, a low number of widgets can be considered as a permissible level of deviation. Bringing the widgets into shelves is a particular stage of work process that might include a number of tasks. If there are a number of tasks, there are a number of goals for the tasks. Each task includes a number of cognitive and behavioral actions that in turn have their corresponding goals. Hence, self-regulation cannot be considered as elimination of so-called disturbances and errors. The self-regulation process allows not only correction of errors but also their prediction and prevention. A person can forecast future events and based on it regulate his/her activity. Self-regulation happens even when there are no disturbances and/or errors as considered in the described example (Bedny and Karwowski, 2007; Bernshtein, 1947). Our activity is a self-regulative system. Self-regulation is a complex process that regulates the entire activity. Disturbances include danger, unanticipated events, emergencies, etc. (Ponomarenko and Bedny, 2011). Subjects have to improvise and adapt to the contingency of such disturbances. Due to disturbances, the self-regulation process becomes more complex and strategies of task performance change. The self-regulation process continues to function even when there are no disturbances and errors. This was clearly demonstrated by Bernshtein (1967). There are strategies that are utilized in normal work conditions, in dangerous situations or when there are other disturbances, and transitory strategies, that is, when a subject transfers from an existing strategy to a new one. Foundation for all these strategies is the process of self-regulation that involves goal formation. In the following chapters, we will consider some examples of self-regulation in a pilot's activity during performance of a variety of tasks in emergency conditions.

Analysis of Vancouver's publications demonstrates that there is currently no clear understanding of goal, task, self-regulation, errors and other important concepts that are necessary for task analysis. I/O psychologists who study human work cannot use such primitive examples as the one described by Vancouver even in their theoretical discussion.

Vancouver (2005) describes the self-regulation process as the elimination of errors that are emerging as a result of deviation from a maintenance a goal. Hence the process of self-regulation is reduced to an error elimination process, but if deviation from a standard during self-regulation does not exceed some acceptable level of tolerance, it is not an error.

Vancouver treats human behavior as a behavior of rats in a Skinner box without taking into account that people adjusts their behavior, first of all, consciously based on conscious goals. A rat regulates its behavior unconsciously. When a person makes errors and she/he tries to eliminate them consciously, they can understand the consequence of the errors, anticipate their emergence, and based on it correct their own activity. If these errors are irreversible and have serious consequences, they should be qualified as failures. Our behavior or activity varies but these variations can be within the range of tolerance and even remain at the unconscious level. It is important to understand that a subject can consciously or unconsciously vary her/his performance (Zabrodin and Chernishev, 1981), which is not an error but a way of regulating activity. In the study of work activity, errors are understood as deviations that exceed the range of tolerance. Variations in activity performance cannot be totally eliminated. Hence work activity should be designed to keep variations in activity performance within a permitted level of tolerance.

Self-regulation that is based only on eliminating errors can be presented by the following example. Suppose a person drives a car on the road. She/he can control her/his car only based on elimination of errors. In this case she/he needs to hit the car in front of her/him then correct an error by going backward and hitting the car behind her/him and then correct my action again and so on. It is obvious that the self-regulation process not only eliminates the errors but also prevents them. A person regulates position of a car consciously or unconsciously. If a car does not move outside of its lane of motion it is not an error. Any deviation inside that lane is within the range of tolerance. Mostly such deviations are not even recognized and are corrected unconsciously. Moreover, such small deviations are necessary to maintain a feeling of the car's movement. As we have already discussed an immediate feedback about a specific movement is not sufficient for errors' prevention. A subject might erroneously select a wrong motor action and then perform it correctly due to immediate feedback. However, such feedback does not prevent errors.

It should be noted that Vancouver and all previously considered authors have presented models of self-regulation that govern behavior based on errors. The inefficiency in these methods is self-evident. It is well known that when a subject performs motor instrumental actions he/she can quickly regulate his/her performance based not only on deviation of the spatial characteristics of instrument, but also depending on such time derivatives as velocity and acceleration. Nervous system not only detects special deviation of a tool at any given time, but also anticipates further changes in the instrument's position. Principles of regulation based on speed and acceleration are well known in the theory of automatic control.

Self-regulation of biological systems was first described by Anokhin (1962, 1969) and Bernshtein (1966, 1967). We will consider their works in the next section. Here we only want to stress that there are two types of motor actions' regulations: programmed and afferentational. Motor actions of very short duration that are performed with maximum speed can be corrected only during the second attempt of execution. Motor actions that are performed based on feedback derived from velocity and acceleration of movement can be regulated without committing errors.

Cognitive activity also can be regulated by preventing errors. For example, when a ship follows a certain course and an obstacle suddenly appears, a captain can predict consequences of the ship's movements and correct the course of the ship based on this information.

In general, activity can be controlled based on forecast or anticipation, including its cognitive components. A person can mentally evaluate possible consequences of her/his cognitive and motor actions. Hence, the actual performance of cognitive and motor action is carried out only after their mental evaluation, which leads to the conclusion that self-regulation is carried out not only as a result of committing errors, but also based on their prediction and prevention. The term *error* is something that is incorrect and leads to unwanted consequences. Therefore, when we consider models of self-regulation, it is more accurate to speak about possible errors, or possibility of their occurrence. In the latter case, people do not foresee emergence of a particular error, but anticipate a possibility of unacceptable deviations within a controlled process. Self-regulation is primarily in prediction of errors, not only in their elimination.

Deviation from a goal is not always an error. This deviation can be evaluated positively or negatively based on existing standards of successful result. A person might evaluate a result of her/his activity positively or negatively and correct activity accordingly, even if there are no errors, because such an evaluative process involves not only cognitive but also motivational mechanisms (Bedny and Karwowski, 2007). Hence, self-regulation is not a discrepancy reduction process that is based on error signals and their elimination.

In his model of self-regulation, Vancouver instead of such terms as situation or information about a situation utilizes the term *variable*. However term *situation* cannot be reduced to *variable* because it includes many elements, each of which has a certain meaning and sensed differently by different subjects. The elements of a situation interact with each other. The meaning of the elements of a situation and the nature of their interaction should be clarified by a subject during her/his activity. This is why when studying the process of self-regulation, we use terms such as the situation, information about a situation, or simply information. The term *variable* is something that is artificially isolated in an experiment. Subjects might extract totally different information, interpret it differently and formulate a variety of tasks on the basis of the same situation (Ponomarenko and Bedny, 2011).

Cognition interacts with emotional-motivational mechanisms in self-regulative process. Depending on emotional-motivational factors, the cognitive evaluation of discrepancies in performance or the cognitive evaluation of activity result in general can be totally different. Self-regulation cannot be explained without considering cognitive and motivational factors in unity.

Vancouver treats the theory of self-regulation very broadly considering various concepts of decision making as being a part of self-regulation theories. He also analyzes Locke's goal-setting theory or Bandura's self-efficacy concept as theories of self-regulation. These works describe important mechanisms of behavior regulation but not the theory of self-regulation. Locke (1994) has criticized the discrepancy-reduction theory of self-regulation. Bandura and Locke (2003) questioned the meaning of control theory stating that their theories are in opposition with it. A majority of self-regulative theories consider negative feedback as the most important mechanism of self-regulation. However, Bandura (1989) fairly pointed out that people deliberately create discrepancies.

Vancouver wrote that the structure of the control system describes the flow of information between organisms or machines and their environment for the purpose of maintaining stability of some variable in the external environment or organisms' internal environment. This definition is relevant for homeostatic self-regulative process in engineering or physiology, but self-regulation at a psychological level is always goal directed. For example, even in cases when the operator maintains stability of technological parameters, her/his activity is goal-directed, rather than homeostatic. An operator analyzes the situation, forms a goal, evaluates complexity and importance of a task, creates a program of activity performance, makes decisions, performs mental and behavior actions, and evaluates results. Maintaining stability of external variables or the technological process in this example is a result of goal-directed activity that is not regulated in the same way as the technological process. Goal is accepted, or formulated by a subject and achieved by her/him through activity performance.

1.6 Overview of the Concepts of Self-Regulation

In this section, we have described the current state of the field of self-regulation outside of AT. Thus it is necessary to make a brief summary of the obtained data. Currently, the fact that our behavior or activity is a self-regulating system is widely accepted. It was proved by such scholars as Anokhin (1955), Bernshtein (1967), Miller et al. (1960), Powers (1973), Konopkin (1981), Bedny (1987), Bedny and Meister (1997), and others that it is more proper to consider our behavior and activity not as a chain of successive stages of information processing, but as a loop-structured system with various mechanisms and feedback and feedforward connections.

Our behavior varies because it occurs in the constantly changing environment that requires constant monitoring and control of performed actions or behavior. To use the terminology developed in ecological psychology, there are behavior-shaping constraints that define boundaries of actions or behavior variations. Hence the self-regulation process is active even if there are no disturbances and errors. In the process of activity, people deliberately create variations in performance in order to evaluate possible consequences. They carry out such variations on conscious and unconscious levels.

In the AT, it was justified long ago by such scholars as Anokhin and Bernshtein that our behavior or activity is organized based on principles of self-regulation. In the United States, it was first considered by Wiener

(1948) and then more thoroughly substantiated by Miller et al. (1960) and then Powers (1973), but there is no further progress in this area. On the contrary, we can speak of the certain regress. Analysis of the concepts of self-regulation shows that all models of self-regulation are single-circuit or one-loop models that do not reflect reality. However, self-regulation is a multiloop process with feedforward and feedback connections. Self-regulation is largely reduced by some authors to a homeostatic process, resembling fluctuations of the pointer around the neutral position. Wiener pointed out long ago that self-regulation is more like the movement of an arrow in a specified direction. The role of positive feedback actually is not sufficiently justified. Stages of self-regulation also do not have sufficient theoretical justification and there is no clear understanding of the role of separate mechanisms and factors that are involved in the self-regulation process. Based on self-regulation, strategies of human performance are formed and adapted to existing disturbances. Hence the same goal can be achieved in various ways. It is also clear that the goal cannot be viewed as a readymade standard that simply performs the input function for a comparator.

These examples show that self-regulation primarily is viewed as an unconscious, automatic functioning process while in AT self-regulation is considered as a conscious goal-directed process and that unconscious levels of self-regulation perform a subordinate role. Powers introduced levels of self-regulation based on identically structurally organized mechanisms but the structure of self-regulative processes at its various levels is not the same. So, various models should be utilized for their description.

Some authors reduce the self-regulation process to the elimination of deviations from an existing standard. In such analyses, complex human behavior functions as the simplest automatic device. Our behavior and activity should be viewed as a constantly changing goal-directed strategy aimed at not only adapting the subject to external conditions, but also at helping to transform a situation in accordance to the goal of an activity. Moreover, such strategies also depend on past experience and temporal state of the subject. A subject can also use different strategies for achieving the same goal, or create an entirely new goal. In one case, he/she uses discrepancy-reducing strategy and in another case she/he can use discrepancy production strategy or a combination of these strategies.

Strategies of activity or behavior change over time. Carver and Scheier (2005) describe covered in cybernetic literature transformational processes that are of theoretical interest. However, the authors do not have any convincing evidence of the adequacy of their models to psychology.

The negative feedback loop is often viewed simply as a motivational mechanism. Cognitive aspects of self-regulation are considered separately. Some scientists criticize the self-regulation theory of motivation (Locke and Latham, 1990). Bandura (1989) showed that people can deliberately create discrepancies. If discrepancy reduction is the main mechanism of motivation, then people would never act after they achieve a goal standard. In reality, people try to reach a variety of goals that satisfy their needs (Locke and Latham, 1990). In response to such objections, some control theory followers state that in their concept of self-regulation they utilize not only discrepancy reduction but also discrepancy creation feedback. Bandura (1989) also points out that there is not just feedback control but also feedforward control. In response to such critics, some scientists started using a concept of positive feedback or discrepancy enlarging loop in their mechanisms of motivation (Carver and Scheier, 2005, p. 44), which in turn utilizes the concept of *anti-goal*. At the same time these scientists point out that typically a discrepancy-enlarging loop is constrained in some way by a discrepancy-reducing loop.

However, as we will show in the next section, the process of self-regulation involves multiple feedforward and feedback loops, which provide coordination of various mechanisms in the process of self-regulation. Therefore, self-regulation cannot be reduced to the relationship between discrepancy-enlarging loop and discrepancy-reducing loop as it was described by Carver and Scheier. Existing self-regulation theories to a significant degree ignore a subject's consciousness. Locke (1994) even suggested that we abandon control theory altogether. All of such discussions sound a little strange from an AT perspective, where there is no such concept of anti-goal because people create a new goal that can contradict with a prior goal that they try to reach the same way as a regular goal. As we will discuss later, such a self-regulation process is suitable for simple technical systems and some physiological processes. Models of self-regulation that derive from control theory cannot explain complex goal-directed human activity. Moreover, we agree that control theory and its models of self-regulation of behavior look very mechanistic.

Nevertheless, the concept of self-regulation cannot be ignored in psychology. Goal setting theory, self-efficacy, and other motivational theories should not be used as a substitute for such a fundamental concept as self-regulation.

All of our behavior or activity, except for automated involuntary reactions, is organized based on principles of self-regulation. Self-regulation facilitates creation of arbitrarily goals and choosing appropriate strategies to achieve them. We always act differently in different circumstances. Therein lies the of self-regulation process, which cannot be ignored. Nor can we agree that the complex issue of self-regulation is reduced to examining its individual mechanisms—goal, motivation, self-efficacy, etc. The issue is developing correct models of self-regulation of behavior or activity in general.

There are two methods for creating models of object in the systemic-structural approach: object-naturalistic and theoretical-methodological (Shchedrovitsky, 1995). The first one stresses the fact that the description of an object depends on the specificity of the object being studied. The second one claims that the description of an object as a system depends on our ability and purpose to describe the object as a system. Shchedrovitsky praised the second approach. We do not reject the fact that a model of an object depends on the specificity of the object. However, our ability and purpose are critical factors in developing adequate models when we consider an object as a system.

Existing attempts to create models of self-regulation that derive from control theory are not adequate because such models ignore the units of analysis of behavior and the goal-directed character of the self-regulatory process. Often self-regulation is reduced to motivation or its individual components.

All these models are basically reduced to a single-loop self-regulation process, which includes such mechanisms as input, goal as an externally given standard, comparator, and output. Feedback is used at the final stage of activity, when a subject performs motor actions and commits errors. In such models of self-regulation, preventing errors is not possible. Self-regulation is considered as a homeostatic process, the purpose of which is elimination of deviation from a goal.

Very often a situation arises when motor actions or responses do not play a special role in operator's work. For example, there are situations when the main type of operator's work is observation. An operator observers a situation, makes assessment of a situation and decides that nothing needs to be done, or a motor response can be relatively simple. Similarly, a chess player can perform a very complex mental analysis of the situation but a motor response will be very simple. All of the described above models of self-regulation cannot properly account for such activity because mental activity described in the examples has virtually no reliance on motor actions. It is obvious that the model of self-regulation should be able to describe self-regulatory processes of cognitive activity as well. This means that the self-regulation process may be organized as a closed-loop model without any relation to motor responses.

Homeostatic principle of self-regulation is more consistent with the physiological processes of the organism. A lot of physiological processes function based on homeostasis. For example, maintaining body temperature within a few degrees of variation is based on the mechanisms of homeostasis. Exposure to cold constricts blood vessels on the body's surface and produces a shivering reaction. This helps to keep the blood warm. In warm weather, peripheral blood vessels dilate and perspiration permits heat to escape. These automatic mechanisms keep body temperature within the desired range. Numerous physiological states are maintained within relatively narrow limits based on the same principles.

Some psychophysiological functions also can be considered as homeostatic processes.

For example, Hull (1951) introduced the drive-reduction concept of motivation in psychology. His concept can only partially explain the psychological theory of motivation. Hunger and thirst can be considered as examples. A need is considered as a physiological imbalance. Drive is the psychological consequence of a need. Drive does not exactly change in the same way as a need. From a homeostasis perspective, a need is the result of physiological departure from the optimal state of an organism. Its psychological counterpart is drive. When the physiological imbalance is corrected, the drive and motivation to eat are reduced. At the same time, the psychological level of self-regulation can affect the physiological levels of self-regulation and thus the motivation. A person can be hungry but still not motivated to search for food. Conscious processes also influence what a person should do. The feeling of hunger, which is derived from the physiological state of a human organism, can only indirectly influence motivation and behavior. For example, in a dangerous situation, after a long absence of food, a person may not even notice that they are hungry.

The concept of self-regulation is widely used in contemporary psychology. We covered the basic concepts of self-regulation that rely on self-regulation models developed in the framework of control theory. These concepts of self-regulation have been criticized as being too mechanistic. The attempts have been made to describe the process of self-regulation without using the control theory in various fields of psychology. These are multiple commonsense theories that do not have accurate terminology and even the notion of self-regulation is not always used correctly.

For example, Bandura (1977) described the process of self-regulation as having four consecutive stages: observing, judging, rewarding, and regulating oneself. This is not a psychological description of the self-regulation process. Self-regulation is often described similarly in educational psychology (Gibson and Chandler, 1998). Sometimes self-regulation is even used in place of such notions as willpower, volition, or as motivational mechanism. Such understanding of self-regulation is erroneous. The concept of self-regulation becomes meaningful only when self-regulation models are developed. Such models are defined in terms of function blocks or mechanisms and feedforward and feedback connections between them. If there is no clear description of mechanisms of self-regulation and the nature of their interaction, the concept of self-regulation cannot be discussed from a scientific point of view. We agree with the criticism of self-regulation concepts as they are presented outside of the AT and derived from control theory. These critical analyses can be found in the works of Bandura and Locke (2003), Locke (1994), and others. However, we do not agree with the interpretation of the psychological self-regulation process as it is presented in recent psychological publications outside of AT.

Self-regulation has been studied in AT since the 1940s by such scholars as Anokhin (1969) and Bernshtein (1967) whose works were translated into English. Later, several concepts of self-regulation of activity were developed by other scientists that were based on Anokhin and Bernshtein. The more advance theory of self-regulation is suggested in the framework of SSAT. We will discuss this theory in the following chapters.

According to the analysis discussed so far, we can make a general conclusion that models of self-regulation derived from control theory are too mechanistic for analyzing human activity. The process of self-regulation involves consciousness and therefore cannot be based only on the ideas of control theory. To create psychological models of self-regulation, it is necessary to use the SSAT that utilizes such concepts as a conscious goal and cognitive and motor actions as basic elements of the activity, mental feedback, etc. In this chapter, we set ourselves the task of analyzing various concepts of self-regulation that exist outside of AT. The problem of self-regulation from the perspectives of applied AT and SSAT will be discussed in the next section.

2

Concept of Self-Regulation in Activity Theory: Psychophysiology and Psychophysics Perspectives

2.1 General Characteristics of Activity Approach

General activity theory (AT) is considered as one of the most important accomplishments of Soviet psychological science and has an extensive history dating back to the works of Vygotsky (1978), Rubinshtein (1959), Leont'ev (1978), and their followers. Each of these scientists developed their own school of psychology. Subsequent development of the theory took place within these schools of thought, which share some main ideas and at the same time differ from one another. *Activity* in all of them is considered as a goal-oriented system, where a goal is a complex cognitive mechanism that is associated with our consciousness. General AT is a broad approach in psychology or *grand theory or framework* that can be applied in various fields of psychology. This theory plays a particular role in studying human learning and in school psychology. For a long time, it has also been applied in the study of human work until more advanced applied activity theory (AAT) was developed for this purpose. In the 1970s, leading scientists who studied work psychology realized that general AT is a useful philosophical framework but it cannot be directly applied to the study of human work in contemporary industry and started developing AAT. Among those were Bedny (1987), Gordeeva and Zinchenko (1982), Galactionov (1978), Kotik (1974), Konopkin (1980), Landa (1976), Platonov (1970), Pushkin (1978), Zarakovsky et al. (1974), Zavalova et al. (1971), and others.

In general AT, a predecessor of AAT, various scientists carried out studies of voluntary and involuntary memory when subjects performed memorization tasks. These scientists focused on analyzing the effect of motives, goals, individual features of subjects in their memorization process, etc. (see, e.g., Zinchenko, 1961). However, at that time nobody studied short-duration stages of information processing. Only after the emergence of cognitive psychology did scientists in AAT start combining cognitive psychology methods with AT methods.

Zinchenko et al. (1980) demonstrated that short-term memory cannot be presented as rigid, totally involuntary programmed processes because goal, consciousness, motivation, etc., can affect it. Features of short term memory depend not only on the automatic information processing stages but also on the voluntary regulated human activity. Particular attention in AAT has been paid to studies of practical thinking (Pushkin, 1978; Tikhomirov, 1984). One important type of practical thinking is working or operative thinking. It has been demonstrated that emotional components played an important role in this type of thinking.

Scientists in AAT demonstrated that an important aspect of an operator's activity is the development of a mental model of reality. This model influences correct interpretation of situation and efficiency of execution (Konopkin, 1980).

Zincheko and his colleagues conducted similar cognitive psychology studies with short-term memory (Zinchenko et al., 1980). Concepts such as sensory memory, short-term memory, microstages or information processing, etc., were transported into AAT from cognitive psychology. The method of studying short-duration stages of human information processing with some modifications was named microstructural analysis (Zincheno and Vergiles, 1969). Microstructural analysis is explicitly based on the methods and ideas generated in cognitive psychology.

Next, we consider some examples of ergonomic methodology that are based on the principles of AAT and utilized in aviation. The most representative fields in this area of research are aviation, semiautomatic systems in manufacturing, automatic systems associated with remote control of various technological processes and software design. Several works of leading scientists in AAT who work in aviation have been presented for the first time in the special issue of *Theoretical Issues in Ergonomics Science* (TIES) (Bedny, 2004). They give a general idea about studies in this area of research. In this chapter, we consider some new examples that give us a more accurate picture of this important area of study from the perspective of AAT.

Ergonomic evaluation of the aircraft system is based on studying the pilot's work activity. Scientists analyze how pilots strive to achieve the same goal under different conditions and how they change their strategies of performance in emergency conditions (Ponomarenko and Lapa, 1975; Ponomarenko and Zavalova, 1981). Specialists also study how a task is performed by test pilots, regular pilots, and pilot trainees under routine and emergency conditions. Particular attention is paid to the pilot's strategies of gathering information. Strategies of pilots' activity include conscious and nonconscious components. Such methods as interviews, verbal protocol, questionnaires, and other methods that are based on verbal responses are combined with experimental research methods, error analysis, recording of eye movements, chronometrical studies, etc. Comparison of objectively gathered performance data with its subjective interpretation by pilots is an important methodological principle of work analysis.

In experimental studies, psychophysiological criteria such as breathing frequency, electromiograms, galvanic skin response, and other methods are also widely used. All physiological data are interpreted in the context of task performance. Physiological data are specifically important for performance of tasks in emergency conditions. The same physiological reactions can be interpreted differently in a different activity context. Specificity of these methods is that they interpret results based on comparing different data. AAT has formulated the basic principles of the experimental study in aviation (Dobrolensky et al., 1975).

The first important principle of study is the analysis of dynamic features of an aircraft. The main dynamic features of an aircraft are its controllability and stability. Controllability is defined as an aircraft's ability to perform every maneuver the pilot makes. Stability is defined as an aircraft's ability to return to an initially required flight regime during external disturbances, such as weather conditions or others situation that effect the flight's program, without the pilot's involvement.

The second principle of study requires preserving the actual in-flight characteristics of the pilot's activity as much as possible during the experiment. It is important to understand that a pilot does not receive isolated data from various instruments. Information from specific instruments is always received and interpreted in the context of other information. In accordance with AT, a pilot's goal, motives, and strategies of performance should be considered in the experiment. For example, it has been shown that in laboratory conditions, a pilot can discover a failure of an aircraft engine in 3 s. Discovering the same failure during an actual flight takes 280 s.

The third principle states that cognitive processes should be considered in the context of an actual pilot's activity. For example, a goal of a flight is clear, but intermediate goals of the same flight can be often formulated by pilots independently. Interpretation of information often depends on adequacy of a formulated goal. The process of acceptance or formulation of a task's goal by a pilot is an important component of task analysis.

The fourth principle is that an experiment should be performed not only under routine but also under emergency conditions. A pilot's strategies of task performance in these two conditions can be totally different. The introduction of emergency conditions into an experiment is a critical factor for studying potential accidents. It is important to know that signals indicating occurrence of an unusual situation during a flight can be both instrumental (received from an instrument) and noninstrumental (vibration, noise, position of an aircraft, etc.). Recognition of an accident is complicated by the fact that the same signal may have multiple meanings.

The last principle for conducting an experiment involves utilizing multiple criteria as well as analyzing their relationships. The following basic measurements should be considered in every experiment: physical parameters (e.g., flight parameters such as glide slope and altitude), other parameters that may effect system failure, such as precision, time, and reliability of operators responses;

internal psychic processes (cognitive strategies); strategies of gathering information by using, for example, eye movement data; psychophysiological factors such as workload, stress, pulse rate, etc.; and subjective responses of subjects obtained during debriefing. Subjective opinions of pilots should always be compared with objective data. We will not into further details of the studies performed in AAT.

In this book, we pay particular attention to the systemic-structural activity (SSAT) approach. SSAT began as one of the AAT directions (Bedny, 1987, 1981). This theory has been originally presented as an independent one in *The Psychological Foundations of Analyzing and Designing Work Processes* (Bedny, 1987) and has been further developed in the United States and published in several monographs (Bedny and Karwowski, 2007; Bedny and Meister, 1997) and numerous articles. This book does not just integrate different data from various publications in this field but it presents totally new data in the field. At present, this theory provides a unified framework for studying human work.

In general AT, systemic analysis has not gone beyond general philosophical discussion and has not been brought to the level of practical application. In AAT, only some aspects of systemic analysis have been applied. Systemic analysis as an interdisciplinary field should not negate the need for the development of proper systemic psychological methods of activity study. Implementing such an approach becomes possible when activity can be described as a complex structure that evolves over time. The creation of such an approach is possible only when we can develop methods of analysis for describing activity as a systemic-structural entity where cognition, behavior, and motivational processes are considered as a systemic organization. In this case, activity is described as a system consisting of the subsystems and smaller elements that is in specific relation and interaction with other elements of activity. Activity is considered as a logically and hierarchically organized system. Transition from general philosophical discussion about systemic analysis to its practical application is not that easy. Existing methods of systemic analysis of activity are important and useful from theoretical and practical viewpoints. However, they are fragmented and cannot be substituted for a unified and, to some extent, standardized approach to systemic analysis of work activity.

SSAT views activity as a structurally organized self-regulated system, rather than an aggregation of responses to multiple stimuli, or linear sequence of information stages as it is described in behavioral or cognitive psychology. Furthermore, it views activity as a goal-directed rather than a homeostatic self-regulative system. Such a system is considered goal directed and self-regulated if it continues to pursue the same goal under changed environmental conditions and can reformulate or formulate the goal while functioning. Activity is a self-regulated system that integrates cognitive, behavioral, and emotional-motivational components. In SSAT, there are standardized units of activity analysis and their precise description. This makes systemic-structural analysis of activity possible.

For example, in SSAT, cognitive and behavioral actions have precise description and it describes the method of their extraction from holistic activity. There are also other units of analysis such as cognitive and mental operations, functional macro- and micro blocks, members of a human algorithm, which consists of one or several interdependent actions integrated by a high-order goal. Such algorithm describes the logic of activity performance and its members have logical organization. In contrast to AAT, these units of analysis are clearly described and unified in SSAT. In general AT, some units of analysis do not exist at all. There are no principles for their organization into a holistic system. There are no standardized stages and levels of analysis. Therefore, from our point of view, general AT cannot be efficiently used for practical purposes. AAT does not have rigorously developed stages and levels of systemic analysis of activity. At the same time AAT has a number of powerful methods for studying of human work.

Cognitive approach, which is presently dominating, treats cognition and behavior as a process and makes it difficult to study activity and behavior from systemic-structural perspectives. The notion of process does not allow describing activity as a structure. Introducing of standardized and unified units of analysis in SSAT helps to describe activity as a structure that unfolds over time. One can extract from the same activity different structures as independent objects of study depending upon the purposes of a study. Each of these objects of study can be represented as an independent system. Consequently, we may have different representation of the same activity. Dividing activity into distinct elements and components and constructing holistic activity from each component of activity is an important method of the systemic-structural analysis of activity.

SSAT suggests the study of the same activity from different points of view and from distinct aspects, thereby legitimating the use of multiple approaches to the description of a single object of study. This implies that in SSAT research, adequate description of the same object of study requires multiple interrelated and supplemental models and languages. Therefore, activity as a system cannot be described by one best method as it is done in cognitive psychology. It calls for multiple stages and levels for the description of the same activity.

These stages are qualitative analysis, analysis of logical organization of activity (algorithmic description of activity), analysis of activity time structure, and quantitative description of activity (evaluation of task complexity, reliability of task performance, etc.). These stages have a loop structure organization, which means that any stage in the analysis may require reconsideration of a previous stage.

Each stage of analysis can be performed with different levels of decomposition. Macrostructure and microstructural analyses determine the levels of analysis.

At a microstructural analysis, cognitive and behavioral actions are described as a system of operations or functional microblocks. For example, motor actions can be described as a system of motions. Decomposition of activity into logically organized system of actions is related to the morphological analysis of activity. Similarly, cognitive actions can be subdivided into mental operations. The starting point of an action is the initiation of a conscious goal of action (goal acceptance or goal formation) and action is completed when the actual result of action is evaluated.

In morphological analysis, cognition is considered not only as a process but also as a logical organization of cognitive actions and operations. The identification of actions, and particularly cognitive actions, can be complex. For example, during the analysis of computer-based tasks, it may be necessary to use an eye movement registration to extract and classify cognitive actions (see Section 9.2). Depending on strategies of activity performance, the same task can contain various actions and their logical organization can differ.

A contemporary task can be performed by using various strategies of task performance. Therefore, self-regulation of activity during task performance is a critical step in task analysis. From the general analysis of activity during task performance to the analysis of activity self-regulation and then description, it is important to adopt more efficient strategies of task performance for qualitative analysis. When we describe activity as a self-regulative system, the major units of analysis are function blocks. Analysis of activity self-regulation is the subject of this book and we will discuss this problem from various perspectives in of the following chapters. From functional analysis prospective when activity is considered as self-regulative system objectively presented to a subject task and subjective representation of this task are not the same which is important from practical and theoretical points of view. For example, even the simplest sensory task—*detection of signal*—can be perceived subjectively in various ways. The most rigorous instructions cannot prevent subjective interpretation of the task. When designing an experiment, a psychologist should take this factor into consideration. A task includes objective requirements, which are further transferred into a subjectively accepted goal in a variety of ways. Task conditions can also be interpreted subjectively. Such interpretation can result in multiple strategies of task performance. Analysis of self-regulation mechanisms of activity should be combined with morphological activity analysis when the main units of analysis are cognitive and behavioral actions. Actions can be described with more detail using such units of analysis as psychological operations. Algorithmic activity analysis is used for describing logical organization of actions. In algorithmic analysis of activity, in addition such units of analysis as members of algorithm are utilized. A member of an algorithm integrates several actions of the same type by a high-order goal. Therefore, SSAT uses various units of activity analysis. Some of them have a hierarchical relationship. Standardized system of units of analysis and morphological description of activity offered by SSAT allow creating a number of quantitative methods of assessing task performance. In this book, we focused our attention on quantitative assessment of computerized tasks.

2.2 Anokhin's Concept of Functional Self-Regulative System

One important area of AT studies is the psychophysiological approach. This approach is also known as higher nervous activity approach where the first principles of self-regulation of activity were formulated (Anokhin, 1962; Bernshtein, 1967). Consequently, we consider the concepts of self-regulation in this area of study.

The material presented in Section I shows that the concept of self-regulation has become increasingly popular in some areas of psychology. Our behavior constantly varies because it occurs in a variable environment that requires constant monitoring and control of performed actions and activities. As per terminology developed in ecological psychology, there are behavior-shaping constraints that define the boundaries of actions or behavior variations. Hence, the self-regulation process is active even when there are no disturbances and errors. In the process of activity, people deliberately create variations in performance in order to evaluate possible consequences. They can carry out such variations consciously and/or unconsciously.

This leads to the conclusion that our behavior or activity is organized based on principles of self-regulation. It is a well known and scientifically established fact. In the United States, self-regulation was first studied by Wiener in cybernetics (1948) and then more thoroughly substantiated by Miller et al. (1960) and Powers (1973) in psychology. However, further progress in this direction is not observed. Moreover, we can speak of some regress.

Thus, it is not accidental that this concept is not very popular in ergonomics and engineering psychology. Only in one of the leading textbooks in the field of I/O psychology we found a page dedicated to behavior self-regulation (Landy and Conte, 2007, pp. 354–355). In engineering psychology and ergonomics, self-regulation is not discussed at all. The concept of feedback that derives from control theory is widely used instead of self-regulation. However, the concept of feedback outside of the concept of self-regulation is not very productive. The material presented in Chapter 3 demonstrates that the concept of self-regulation is often used incorrectly. Self-regulation is considered as a separate psychological mechanism or as a homeostatic process. Study of some psychological mechanisms in self-regulation can be useful but self-regulation is a systemic integration of various psychological mechanisms and cannot be reduced to isolated psychological mechanisms. Control theory has been developed to analyze the behavior of complex technical systems. Some ideas from this theory can be useful for studying self-regulation of human activity. However, human external behavior, or activity in general, cannot be equated to the functioning of a technical system. Self-regulation is an interdisciplinary concept. In the West, this concept has derived from cybernetics that studies self-regulation of not only living but also nonliving systems.

When studying self-regulation in psychology, it is necessary to consider data obtained by analyzing self-regulation of technical systems and at the same time take into account the specifics of self-regulation of human behavior. The concept of self-regulation was first represented in works of such prominent physiologists as Anokhin (1935, 1955, 1962) and Bernshtein (1935, 1947, 1967). It is important to distinguish the term *regulation* from the term *self-regulation*. Regulation is an external influence on a system that can change behavior of a system in a desired direction. Self-regulation on the other hand is an intrinsic influence on a system that can change its behavior. More precisely, self-regulation can be defined as an influence on a system that derives from another system in order to correct its own behavior.

We want to underline that in this section, we will use terminology that was used by Anokhin and Bernshtein at the time when they created their theory. The terminology in AT and specifically in SSAT has changed since that time. Anokhin and Bernshtein were physiologists, not psychologists. In SSAT that was created much later, the basic terminology was changed based on new data obtained in AT. Therefore, terms such as purpose, goal, action, motion, etc., do not have an exact match in SSAT. However, in our analysis, this discrepancy does not really matter because these terms have similar meaning in the context of our discussion.

In this chapter, we consider Anokhin's concept of self-regulation from a psychophysiology perspective. In the early twentieth century, Pavlov (1927) introduced the concept of conditioned reflex, which in behavior approach is known as conditioned response. Conditioned reflex was then considered to be the basic unit of animal and human behavior.

In Pavlov's studies, conditioned reflex was considered as a reflex arc consisting of three components: the first component is involved in receiving information about the stimulus, the second one is the central component, and the last one is the response. Anokhin (1935) in turn introduced into the study of conditioned reflex the concept of backward afferentation, which according to modern terminology is known as feedback. Due to backward afferentation, living organisms can evaluate the result of conditioned reflex and correct it. Thus, conditioned reflex was now considered as a conditioned reflex loop or as a self-regulative system. This concept was formulated before the idea of feedback in cybernetics was applied to technical systems. With the advance made by Wiener's (1948) works, Anokhin quickly accumulated cybernetics' ideas into his studies at the same time staying true to his physiological theory, and not transforming into the mechanistic ideas of cybernetics. Cybernetics models of self-regulation use feedback and have loop structure organization. They include a number of mechanisms that interact with each other during functioning. Such models are not well adapted for explaining goal-directed self-regulative processes.

Self-regulation plays an important role not only in homeostatic but also in purposive or goal-directed behavior (Ackoff, 1980; Wiener and Rosenblueth, 1950). Anokhin's self-regulative model of condition reflex can explain both homeostatic and goal-directed or purposive self-regulating possesses. It should be noted that this concept of goal in Anokhin's model and in those of the scientists discussed earlier does not have exactly the same meaning as that in AT. It rather corresponds to, what is known in American psychology, as a purpose. Such an understanding of goal is more adequate for studying purposive behavior (Tolman, 1932). According to Anokhin, ideas borrowed from cybernetics are extremely useful for understanding brain functioning. The main idea of Anokhin is that the ability to assess the outcome of an action is one of the main adaptive mechanisms of animal and human behavior. Each result of a performed behavior or activity is evaluated and if it satisfies an existing need, such behavior is stopped. This principle of organism functioning is considered as a self-regulative process. At the beginning of his work, Anokhin used the term *self-regulating condition reflex*. Later, he starts to use the term *self-regulating functional system of behavioral act*. This means that he considered his model as principle of self-regulation of behavior in general. It is considered as a functional system of behavior that can be presented as a dynamic organization that selectively integrates various central and peripheral neural mechanisms. Such selective integration facilitates achieving a specific desired result. A functional self-regulating system is a closed-loop structured system that can receive information from the environment and compare it with the internal state of an organism. Based on continual feedback information, this system can evaluate an obtained result and correct its functioning. When Anokhin described his self-regulating system he did not utilize such mechanistic terms as input, output, standard, comparator, etc., but used terminology that has psychophysiological meaning. According to Anokhin, a functional self-regulating system is organized as a system of interacting mechanisms that allow achieving useful results. Such system has a feedback about the outcome that can be obtained by control center and this center can evaluate the result and perform required corrections if necessary. Any self-regulating system is formed for particular purpose. After achieving the desired result and its positive evaluation, this system disappears. Each functional system selectively integrates central and peripheral apparatuses based on feedforward and feedback connections in order to achieve an adaptive effect.

Functional self-regulating systems may be of various complexities. Some of them facilitate a homeostatic function. Useful results achieved by a system have an adaptive effect on an organism in relatively unchangeable conditions. Other self-regulative systems form adaptive acts in a changing environment to achieve the desired goal. Such functional systems are developed in the course of obtaining new experience. These systems are goal directed. They are not only adapted to the environment but also change it according to the goal of the behavior. They aim at achieving a new adequate result for specific conditions. The basis of homeostasis is a negative or discrepancy-reducing loop that eliminates deviations of a system from its desired state. This is called an adaptive homeostatic system. However, a physiological self-regulative system very often can change the reference value of a desire state. Such systems are adaptive homeostatic systems. For example, blood pressure may have a different valid value in different conditions. The reference value of blood pressure at rest is different from blood pressure during heavy physical work. Such a regulation is due to the adaptive homeostatic self-regulation process. Such systems often utilize not only a negative, discrepancy-reducing loop but also a positive, discrepancy-enlarging loop. A stabilized index or reference value of blood pressure can be transferred to a new stationary level. This adaptive response occurs in an organism when significant changes in functioning conditions of the organism happen. A positive feedback provides transfer to a new functional state of an organism when reference value of blood pressure or other physiological index should be adapted to new conditions. At this level of functioning, deviation from a new reference value is constrained by negative feedback. Hence, self-regulation at a physiological level performs various adaptive homeostatic functions, protecting an organism from negative influences of external environment. Such a system cannot provide an effective interaction of the living system with the changing environment every time a new goal has to be achieved.

According to Anokhin, self-regulation is not limited to maintaining internal constants of the body. For effective interaction with the external world, the self-regulation process should also provide for achievement of certain goals. Such goals arise and change depending on internal conditions of the body and changing external environmental conditions. This is what is known as purposive or goal-directed self-regulation process. Thus, any type of animal or human behavior is an adaptive self-regulative act that has a loop structure with feedback and can be homeostatic or goal directed or purpossive. Anokhin's model of self-regulation of behavioral action includes five stages: afferent synthesis, decision making, program formation, reflex implementation, and action acceptance.

The stage of afferent synthesis permits comparison and synthesis of all data required to perform an adaptive action that is the most adequate in the given circumstances.

Afferent synthesis is the most important stage of a goal-directed self-regulative process.

Simultaneous presence of multiple external influences on an organism requires establishing an adequate relationship between them and comparing such complex stimulation with an internal state of an organism. It has been shown in Anokhin's laboratory that cells of the cerebral cortex have the ability to integrate excitement caused by various stimuli coming from an organism's external and internal environment. Some cortical neurons have the ability to not only combine various stimulations but also select a stimulation relevant to the activity's goal. Such a stage of adaptive act is called afferent synthesis. The following are components of afferent synthesis: (1) creation of dominant motivation, (2) mechanism of memory as a component of afferent synthesis, (3) excitement caused by main initiative stimulus, and (4) excitement caused by the influence of environmental stimulation (situational stimuli that can still influence the main stimulus). This system includes predictive mechanisms that can evaluate a result of a behavioral act.

According to the classical theory of conditioned reflex, only the current main stimulus causes a reaction or response. In contrast in Anokhin's theory, the desired response is elicited not only by the main initiative stimulus but also by some environmental stimuli that can influence the main stimulus.

The other important factor that influences an adequate response is presently the dominated motivational state of an organism. It was discovered that motivation that prevails at a given time affects the cerebral cortex. Under motivational influence, various types of motivation selectively mobilize those synaptic structures in the cortex that are associated with a specific motivational state. Distribution of stimulations in the cerebral cortex produces various levels of energy conditions that provide functional dominance of cortical associations that correspond to dominant motivation. Thus dominant motivation can selectively influence a state of cortical neurons. All these data were obtained through neurophysiological studies. Each type of motivation produces an ascending activating effect from the subcortical area of the brain to the relevant centers on the cerebral cortex. If there are several motivational requirements, then relevant areas of the cerebral cortex are activated based on the dominant type of motivation. After satisfying the dominant motivation it becomes possible to satisfy the next level of motivation and so on. Thus, the dominant motivational state determines *what* an organism must do and the environmental stimulation determines *how* it should do it.

The next component is past experience, when memory mechanisms are critical for afferent synthesis. Anokhin and his colleagues have identified the importance of past experience in the formation of afferent synthesis. Results of afferent synthesis are represented in the frontal lobe in some form.

After a stage of afferent synthesis is completed, decision is made on how to implement a corresponding act. The most important aspects of the decision making process are formed in the frontal lobe. The decision may be carried out on both conscious and unconscious levels. For example, the work of the respiratory system involves decision making that is automatic. In this system—afferent synthesis formed based on the analysis of the oxygen concentration in the blood—the decision "to gain a certain amount of oxygen in the lungs" is made. This process can also be performed partly at the conscious level. Decision making is related also to the formation of goal of a behavioral act. The frontal lobe plays the most important role in this process. The goal concept in Anokhin's model is considered as physiological mechanism for the regulation and correction of inadequate animal and human behavioral acts. A goal is considered as an end state toward which behavior is directed. Therefore, a goal in this model is understood as a physiological mechanism and is not always connected with consciousness.

After the decision-making stage, the program formation process begins. Performance program does not only influence effectiveness of the executive stage but also the evaluative stage of a behavioral act. An acceptor of effect is formed at the stage of program performance. This mechanism is required to correct a behavioral act with its acceptance due to the fact that the generated program is the source of the formation of the assessment mechanism of the behavioral act. When a performance program is formed, efferent stimulation takes place in the efferent collateral axon pathways. It is a *copy of a command* to perform an act that constitutes the acceptor of an effect. This is the system of neural processes in the afferent section of the brain, which reflects all basic attributes of future result.

The acceptor of effect can be considered as neural model of desire future result.

In Anokhin's model, acceptor of effect is equivalent, to some extent, to the psychological concept of purpose or goal. Acceptor of effect includes not only afferent attributes of the final result, which correspond to the final purpose of the act, but also some intermediate performance stages that should be achieved. Therefore, discrepancy between an intermediate result and an acceptor of effect already formed at this stage can be evaluated and corrected.

From a neurophysiological standpoint, an acceptor of effect represents neural cells in the cerebral cortex where continuous circulation of nerve impulses happens. This continuous circulation of nerve impulses is a neural model of a desire result. A command to executive organs leads to performance of a behavioral act according to a developed program of performance.

Through the performance of actions, a specific result is obtained that produces a number of afferent stimulations corresponding to attributes of the performed actions. This stimulation to the brain is the final stage of action execution. Thus, such terms as action or act, result of action, and attributes of action, demonstrate the possibility to describe a final executive stage of the self-regulative process from a neurophysiological perspective.

Signals about the results of an execution of a required act come to an acceptor of effect as a feedback where circulating nerve impulses represent the neural model of the desired result. Emerging discrepancies between obtained result and neural model are immediately corrected in acceptor of effect. If an obtained result matches an existing acceptor of effect (circulation of nerve impulses from a feedback coincides with circulation of nerve impulses in an acceptor of effect) the behavioral act ends and the action acceptance stage is completed.

The model of behavioral act as a self-regulative system is presented in Figure 2.1.

Thus in Anokhin's model, psychic processes with some approximation are correlated with neurophysiological processes. At the same time it should be stressed that there is no one-to-one relationship between neural and psychological functions. The model presents the main initiative stimulus and situational stimuli. These data are integrated and compared with memory data and current motivation (afferent synthesis) and then decision making follows. At the next stage, the acceptor of effect and the performance program of act are formed. Evaluation of performance program of an acceptor of effect is depicted by a feedback loop. After evaluation, an action is executed based on a corrected program of performance.

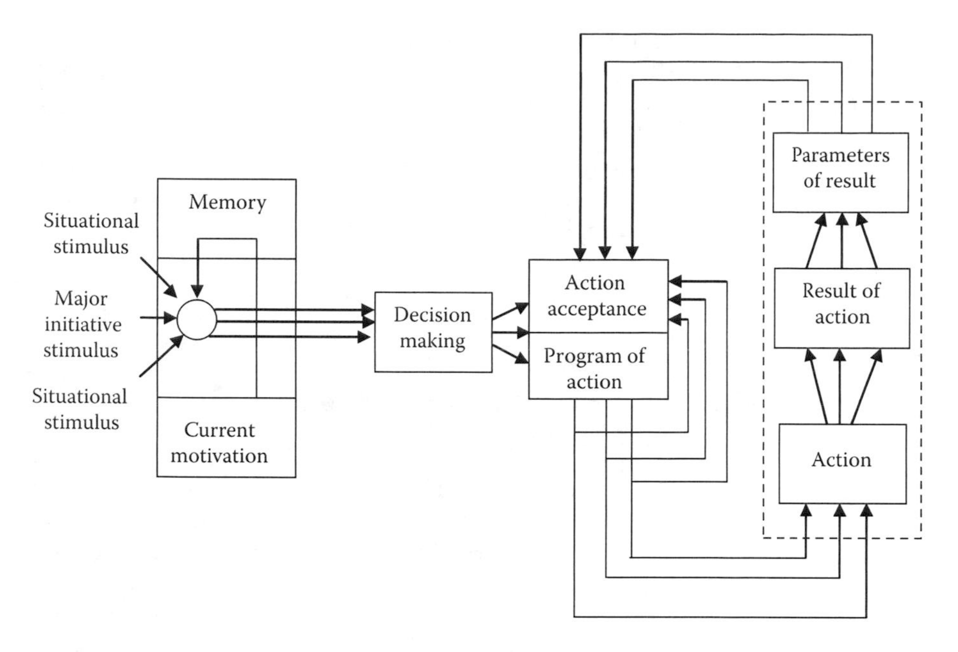

FIGURE 2.1
Functional self-regulative system of activity according to Anokhin.

Actions produce results. The results of actions have different characteristics that are encoded in the nervous system in the form of nerve signals, which are transmitted to the brain where they can be compared with acceptor of effect. Anokhin called these characteristics the parameters of result. Thus, a behavioral act can be corrected during execution based on immediate (ongoing) feedback. There is also a possibility to correct an action during its next execution. The presented model is a simplified schema of the self-regulation of activity. The justification of neurophysiological mechanisms of the model is accompanied by a detailed description of experimental and theoretical data at the neurophysiological level. We have presented a description of the model in a simplified manner, because detailed analysis of the neurophysiological mechanisms of this model is beyond the scope of this book.

In Anokhin's neurophysiological model, a behavioral act is a functional self-regulating system. It demonstrates cybernetics ideas that derived from psychophysiology, but not from cybernetics models of technical systems. At the same time, Anokhin accepted general ideas of cybernetics as evidence of legitimacy of his own basic ideas.

To conclude this section, we would like to acknowledge works of Sokolov (1969) that is related to studying modeling principles of functioning of nervous system. He demonstrated that the central nervous system can be viewed as a modeling system that creates specific changes in its internal structure and such changes are isomorphic to external influences. Sokolov has discovered these modeling principles when studying orienting reflex.

It was demonstrated that orienting reflex does not occur as a direct result of stimulation but rather emerges as a result of the evaluation of discrepancy between trace of recent stimulation and trace in memory of previous stimulation. One important feature of a neural model of stimulus is that it is involved in anticipating significance of afferent impulses based on past flows of impulses. In this study, Sokolov proved that at the neuronal level, orienting reflex functions as a complex self-regulative system.

2.3 Bernshtein's Concept of Self-Regulation and Motor Activity Analysis

A cofounder of the theory of self-regulation, along with Anokhin, was Bernshtein (1935, 1947, 1966, 1996), as has been recognized in AT. Anokhin studied self-regulation of conditioned reflex, which he considered to be a basic behavioral act.

Later Anokhin began to examine this model more broadly and called it the functional self-regulation system of activity. Of course, for this purpose he conducted a number of additional fundamental studies of self-regulation from neurophysiological perspectives that justified his broad view on the concept of self-regulation. Bernshtein restricted himself to studying self-regulation of motor behavior or movements. Both these theories of self-regulation share general principles. These authors consider self-regulation as a process that integrates various mechanisms and includes feedforward and feedback connections and that it is not just homeostatic but also purposeful. The processes of self-regulation described by these authors have similarities but at the same time have some important differences. These theories complement each other and present a general picture of understanding principles of activity self-regulation from psychophysiological perspectives. In this chapter, we consider Bernshtein's work only.

According to Bernshtein, a human organism is an active system that continuously performs various task problems. He considered performing motor actions as the ability of individual to find motor solutions in specific conditions (Bernshtein, 1996). Such tasks include a goal as a future desired result. A person evaluates a situation and creates a plan of actions. When a motor task problem is initiated, an individual determines whether it is necessary to change the ongoing performance program, and the motor task problem is either solved or not solved. Thus, motor actions include a continuous stream of feedback information that is necessary for corrections of motor actions or movements, which means that motor actions should be considered as self-regulative systems. In his theory of movement or motor action regulation, cognitive components play an important role. Bernshtein introduces the concepts of *image of situation* and *image of motor action*, which are important cognitive mechanisms of motor actions' self-regulation. Therefore, in a number of cases, he used psychological terminology because at that time there was no sufficient data to describe some of the mechanisms of regulation of movement in terms of neurophysiology.

The source for constructing such images is a continuous stream of feed-forward and feedback influences based on which images of situation are created and motor actions performed. Motor actions or motions do not just transform an external situation but also have an ability to reflect it.

Movement in humans is mostly facilitated by limbs. A combination of limb movements is a kinematical chain.

Such movements are executed within complex force fields that are constantly changing. There are external and internal force fields. It is very difficult to coordinate movement within such dynamic fields. The external force field acts outside the human body. The internal force field operates within the body and depends on the interaction among segments, muscles, etc. In such a situation, no one can provide precise movement of the body segments in such field in advance developed nerve impulses. In a dynamic force field, the same nerve impulses can cause totally different movements. One-to-one interconnection between central neural impulses that are organized according to the program of performance and movement cannot exist. In this situation, successful performance of movements is the result of feedforward and feedback interconnections between central impulses and peripheral body segments. Feedback enables one to introduce specific corrective impulses during motor actions or movement execution.

It is possible to consider movements of the nexuses of the human body as a result of joined influences of external (outside forces that influence a person) and internal forces (forces of interaction of nexuses of the body, of its muscles and internal organs). Combination of these forces forms a common force field. A person performs movements using his/her limbs each of which is a multinexus cinematic chain. These movements take place in a complex force field that is never constant, which complicates the coordination of movements.

In such conditions, no one even very precise a central nervous system impulse can facilitate achievement of required precession of a given movement. Moreover, the same nervous impulses can produce different effects due to changes in external and internal force fields. Therefore, there is no one-to-one relationship between a central nervous system impulse and a body movement. In such conditions, the coordination of movements is facilitated by constant coordination of central nervous system impulses with peripheral changes. Such coordination is possible only due to feedback that gives information about the current state of the muscles and about the results of the ongoing movements. This feedback allows making necessary corrections during performance of the movement. Bernshtein called these corrections sensory ones because they are based on sensory perception of the movement. According to Bernshtein, the essence of coordination is in overcoming the excessive number of degrees of freedom of the moving part of the body and turning it into a manageable system. The secret of movement coordination is not in wasting nervous impulses on extinction of reactive events but rather in using them for the movement itself. Therefore, in order to achieve a dynamic reliable movement, the central nervous system undergoes the following three stages for biomechanic regulation.

Stage 1. The initiation phase of movement acquisition involves restriction of degrees of freedom of joint body segments, in order to exclude reactive forces that interfere with the desired movements. At this stage of acquisition, movements are very jerky.

Stage 2. In intermediate phase, the degrees of freedom of movement increases because the influence of reactive forces is decreased. Some obstructive forces are overcome by using rapid muscle impulses. Movements are executed more effortlessly and efficiently.

Stage 3. The final phase of movement acquisition is accompanied by the release of the maximum degrees of freedom of various body segments. An individual becomes capable of not only using muscle forces but also reactive forces of body segments. The movements are performed with ease and confidence.

Bernshtein (1947) demonstrated that when a subject attempts to repeat the same set of movements multiple times, it was revealed that each of the movement has some unique characteristics. He called this phenomenon *repetition without repetition*. This demonstrates that each repetitive action is performed not simply by utilizing information from memory. Each repeated action is also constructed and adapted for constantly changeable force fields. We want to draw attention to the fact that terms such as motor actions and goal are not yet clearly defined. For example, goal and purpose are not clearly distinguished. In Anokhin's work, the term *acceptor of an action* and in Bernshtein's (1966) work *neural model of required future* have similar meaning as a goal in AT. However, these terms are closer to Tolman's (1932) concept of purpose. There was no clear distinction between motions, movements, motor actions, etc. However, these ambiguities in terminology are not important for our discussion where the central notion is feedback.

Let us consider feedback from the point of view of movement control. First of all, the source of feedback is the motor sense organ that presents the following information: efforts or muscle tension, positions of parts of the limbs relative to each other, the speed of movement of the joint, acceleration of joint movements, and the direction of movement of the joint. Sense of touch is also used in the movement regulation process.

The second source of information about movements is the vision. The main function of vision is representation of information about the course of changes in situation and perception of the result of movements. With the help of vision one also evaluates correctness of movements.

For example, we can evaluate the correctness of hand movement to the target. Visual information is combined with information from kinesthetic sense organ. The latter informs us about the position of hand during movement even without visual information. One does not precisely realize the significance of the combination of vision and kinesthetic senses. For example, when we are in the garage and the light is on we move our hand to the red garage opener button automatically. However, when the garage door is closed and the light is off and we see only the glowing button, in spite of the target being clearly visible, we experience some inconvenience in moving our hand to the button. This means that not only the visual information about the position of the target and kinesthetic feelings from the muscles are important, but visual tracking of the movement of the hand toward the target is important. Combined information about a visual position of the target, visual information about movement of a hand toward a target, and kinesthetic feelings of muscle facilitate effectiveness of this motor action. Therefore feedback is provided by a combination of different types of information and the person is not aware of some of that information.

The sense of touch also plays some role in movements' self-regulation. The feeling of pain can in some cases be used in movement regulation. The sense of pain can be used as information about a range of amplitude or effort limits. Acoustical information plays a specific role in the acquisition of the rhythm of movement.

Not all information received by the sensory organs is related to feedback but only the one that is used for the assessment of appropriateness of movements and their correction can be considered as a feedback. Therefore, we have to know the specifics of the motor task in question to predict feedback of motor movements.

The role of various sense organs in movement regulation changes during the skill acquisition process. At the first stage, vision plays a significant role in movement regulation, whereas during the later stage kinesthetic sense plays the leading role—it is essential for the internal cycle of movement regulation. The external cycle utilizes information from external receptors and the internal cycle utilizes information from internal or kinesthetic sense organs. These two cycles interact, and their relationship depends not only on the stages of motor skills acquisition but also on the level of complexity of the skills. Relationships between these two levels determine relationship between conscious and unconscious processes during performance of motor movement. The internal cycle is predominantly involved in the unconscious regulation of movements while the external cycle in the conscious regulation of movements. However, the internal cycle of self-regulation can also be involved in conscious performance at precise movements. In the course of development of motor skills, a certain violation of relationship between the external and internal contours of the regulation of movements can lead to loss of motor skill. This is commonly observed among gymnasts. For example, if a gymnast performs a motor movement automatically and tries to control the movement consciously he/she can sometimes loose the skill.

Time of the movement performance based on the inner cycle regulation is shorter than when it is based on the external cycle of regulation. So if quick motor response is required, it is better to use muscular rather than visual feedback. In other words, regulatory functions of movements should switch from external receptors, specifically from vision to kinesthetic receptors and sense by touch. From this it follows that the internal cycle of movement regulation sometimes is more preferable. Sometimes combinations of external and internal cycles of movement regulation can be adopted, which would increase not only the speed but also reliability of motor response. Studies have shown that it is often useful to utilize not only the final feedback at the end of movements, but also the immediate feedback during the execution of movements. It is also useful to set specific indicators showing results of the manipulation of controls during the implementation of actions and in the final stage, when the information on the results of motor actions is given.

Relationship between unconscious physiological feedback and conscious psychological feedback is of particular interest. It has been shown in various studies that it is possible to control the biological processes to a certain extent. For example, involuntary and unconscious reactions of a vascular system were combined with visual data of their states can become conscious. As a result, a person could, within certain limits, regulate their vascular responses. In other words, the combination of unconscious or not fully conscious biofeedback with the consciously perceived psychological feedback can in some cases help to control biological reactions (Jack et al., 1971; Lisina, 1957).

Similar examples were carried out using myograms. A person with an amputated hand can acquire skills to control a prosthetic hand better when he/she can visually perceive an electromyogram on the screen (Person, 1965). All such methods are based on the combination of different contours of movements' regulation.

Such methods were widely tested in sports by Farfel (1969). In many of his studies, he used the method of switching from a proprioceptor channel to a visual channel.

According to Bernshtein (1947), self-regulation has a hierarchical organization. He described four nervous system levels of the self-regulation of motor actions. Each level has appropriate mechanisms of movement regulation. The concept of hierarchy is based on the idea that the nervous system has subsystems and high-order superordinate components. Higher-order subsystems provide guidance for the subsystems that are below them in the hierarchy. The higher-order nervous system is also responsible for the regulation of verbal logical thinking. This level of self-regulation may be involved in the regulation of complex motor actions. It is also important in the regulation of motor skills at the beginning of their development. Bernshtein (1996) considered skill development as a constructive, problem solving process with a number of stages. The first level of movement regulation involves meaningful symbolic aspects of movement regulation and is particularly important at the first stage of skill acquisition. This level of regulation is specifically important when motor activity is combined with complex cognitive activity. Suppose I need to write a letter. It is a conscious level of movement regulation. This level is also associated with the sense components of movements. In performing this task, I need to write the sequence of words that includes corresponding letters in specific order. In this situation, the highest level of self-regulation is involved in the symbolical coordination during speech and writing. The second level of movement regulation is involved in special aspects of movement performance. This level provides the body movements in space. For example, it regulates movement of a hand in space during writing. The third level is associated with kinesthetic sensitivity of movements. Kinesthetic level utilizes information about our body parts from receptors in joints and ligaments and in the muscle fibers. In our example, this level determines some handwriting features. The lowest level controls the muscle group tonus. This level works in cooperation with equilibrator senses, which deal with body position. The relationship between levels of regulation and their importance can be changed during the skill acquisition process. The first level is responsible for the conscious regulation of motor actions. The lower levels, which are developed during the training process, are responsible for the unconscious level of regulation. The conscious level of motor action regulation performs auxiliary functions as a kind of *scaffold* at the first stage of skill acquisition. At the stage of motor skill acquisition, this level of motor regulation is abbreviated or even becomes redundant.

Bernshtein (1966) developed a general model of movement regulation. This model includes various mechanisms that have feedforward and feedback connections. The coordination of movements is provided by the coordination of central neural impulses with peripheral body segments (see Figure 2.2).

Let us consider this figure briefly.

1. The initial stage of motor action is the formation of an ordered apparatus. This mechanism introduces the meaning of control parameters (*neural mode of required futures*).

   According to Bernshtein, an image of motor action serves as a leading factor that determines order apparatus of movement. Using psychological concepts in his physiological theory of the regulation of motor actions or acts, Bernshtein's explanation is as follows. "Utilizing the concept of image of the action result, which belonging to the field of psychology, to describe the leading mechanism of a motor act, emphasizes the fact that we cannot yet pinpoint its underlying physiological mechanisms. This does not mean that we do not recognize its existence and exclude it from our consideration." Furthermore, he wrote, "...because physiology is not yet sufficiently developed in the study of movements it cannot describe the physiological mechanisms of the image" (Bernshtein, 1966, p. 241).

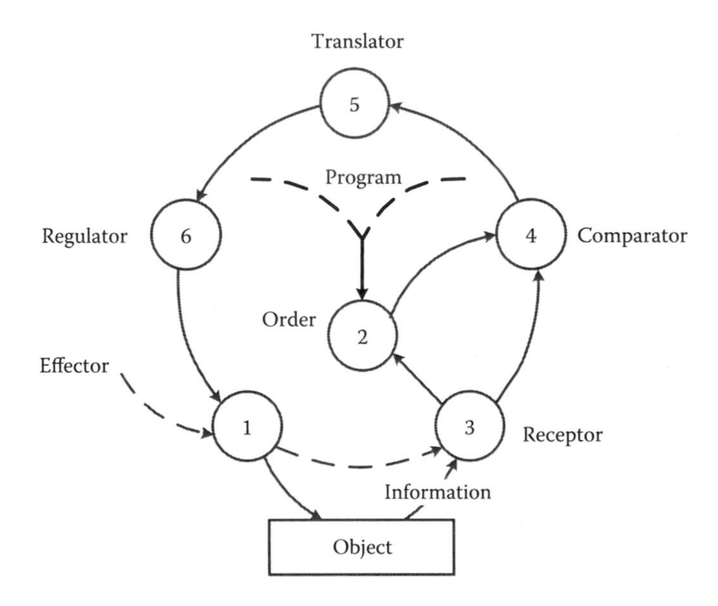

FIGURE 2.2
Self-regulation model of human movements according to Bernshtein.

2. The next mechanism is program of performance, which is formed based on interaction order apparatus and comparative apparatus. This mechanism provides probabilistic prognosis or probabilistic extrapolation of the course of events in the environment, based on the information received about the current situation. The program of performance can be formed on conscious and unconscious levels. That is why the person is not fully aware of how he/she carries out his/her motor actions. The program can be modified during motor action execution because of complex interactions with other mechanisms of self-regulation. Effectiveness of usage of feedback and an ability to correct a motor action depends to a great extent on the degree of precision and details of a program of performance, as well as its compliance with a motor task.
3. Translating apparatus, which translates data from comparison apparatus into corrective impulses that are used by the regulator. Here we can talk about the corrective action program.
4. Comparative apparatus is involved in the evaluation of any discrepancies between the desired and actual parameters of the movement. This mechanism is also involved in the program formation process.
5. Regulator apparatus that governs the effectors. This mechanism is involved in the conversion of data received from a performance program into the system of execution commands. It forms a system

of nerve impulses or commands that regulate the functioning of effectors. This mechanism can also consciously regulate functioning of effectors.

6. Effectors or motor apparatus, whose work is governed by specific parameters. Efficiency of muscle functioning, their strength, endurance, movement time, etc., can be considered as characteristics of effectors.
7. Sensory synthesis provides selective election of information from the external environment using various receptors.

Based on the material presented, we can make the following basic conclusion. A person performs movements by limbs, which are multilink kinematic chains. These movements are performed in a very complex and dynamic force field, and they are never permanent. In these circumstances, only self-regulative systems can provide the required precision of movements. Coordination of movements can be viewed as overcoming of the degrees of freedom. The essence of coordination is overcoming superfluous degrees of freedom in joint movement. The purpose of coordination is not expending neural impulses to depress reactive forces, but using them for performing actions.

The analysis material presented in Sections 2.1 and 2.2 demonstrates that there are two basic sources for the psychological theory of self-regulation in AT. One source derived from cybernetics and associated with the work of Wiener (1948) and his colleagues, and the other derived from works of physiologists such as Anokhin (1935, 1955) and Bernshtein (1935, 1947). The basis for the concept of self-regulation in AT is derived not so much from cybernetics control theory but from works of Anokhin and Bernshtein. These authors were the first who introduced the concept of feedback in physiology. Their work has important meaning in AT as well. However, the fundamental synthesis and formulation of the principle of feedback control for any complex systems including living organisms and machines has been made by Wiener. American scientists are not sufficiently familiar with Anokhin's theory of self-regulation. They are more familiar with the works of Bernshtein. His work was important in the development of the proposition that offers the idea of integration of American ecological psychology developed by Gibson and the Russian motor control theory (theory of self-regulation of motor activity) proposed by Bernshtein (Dainoff, 2008).

We reviewed the Anokhin and Bernshtein model of self-regulation. Their detailed analysis involves the description of fundamental physiological and neurophysiological data. These issues are out of scope of this book, the purpose of which is to present psychological and related ergonomic aspects of self-regulation. Detail description of these theories and their neurophysiological basis were given in various works of Anokhin and Bernshtein.

2.4 Applications that Derived from the Psychophysiological Study of Self-Regulation

The findings of Anokhin, and specifically Bernshtein, had profound implication for the development of physiology and psychology of work. Their work also had important influence in some areas of cognitive psychology. We will now discuss some of the data obtained from AT, cognitive psychology, and motor learning.

One of the important issues of motor movements' regulation is the possibility of controlling motor movements by using feedback not only after completing motor movements, but also in the process of their performance. According to some authors, motor actions that are performed with high speed or have short duration cannot be regulated with feedback during their execution. In the latter case, immediate feedback is important for the regulation of motor actions.

Let us consider as an example interesting data presented by Novikov (1986), who dealt with this issue. He was researching the process of development of motor skills of blue collar workers. In his studies, he used the data obtained by Bernshtein, according to which motor actions can be regulated not only based on spatial characteristic of movement, but also on speed and acceleration characteristics. As an object of study, he chose benchwork and specifically the operation of filing. This author's biomechanical calculation shows that without immediate feedback, deviation of file from the horizontal surface would reach 7°–8°. However, if a worker has already developed skills, this deviation will not exceed 0.5°, which is 14–16 times less. Hence, immediate feedback is vital. Using the terminology of automatic regulation, we can say that the execution of motor actions can be seen as a regulation based on deviations from a predetermined trajectory. It is known that any system of automatic control operates with a certain lag time. In order to improve performance of control systems, various methods of regulation are used in engineering. Regulation can be performed not only based on deviation in time, but also its derivative in time. In other words, regulation can be performed based on speed and acceleration. Using specific combinations of deviations and their time derivatives, we can not only evaluate the amount of deviation occurring at any given time but also accurately predict future changes in a system state and prevent system output deviation beyond permissible limits. Novikov's study of filing has shown that physical actions are governed not only by deviation in time, but also by speed and acceleration. It was discovered that feedback during regulation of movements based on speed and acceleration may be performed with a time delay equivalent to 0.06–0.12 s. The study of Yarovoj (1966) showed that when correcting position of a file, a worker needs much more time. In his experiment, various positions of an instrument were presented to subjects and they had to correct its position as quickly as possible. It was found that the total time of correction was 1.44–1.63 s, and latency time was 0.45–0.57 s. This significant adjustment time was due to the deviations approaching the operative threshold. Correction of an instrument's position when performing actual filing also approaches an operative threshold. Hence, precision of the file's position can be achieved not only due to its position deviation but also to the velocity and acceleration of its movement.

Correction of file based on velocity and acceleration was considered as an immediate kinesthesis feedback that is used at an unconscious level. That's why correction time is less than the latent time of simple sensory-motor reaction. There are also motor movements that are carried out without feedback. For example Schmidt and Russell (1972) discovered that very quick movements that have duration no more than 160 ms are regulated without feedback. If movements have greater duration they include feedback. It was discovered that ballistic actions can be evaluated after their completion. A result of this evaluation can be used for the correction of a following ballistic action.

In most cases, motions are performed with a visual feedback. Novikov et al. (1980) studied discrete tracking, demonstrating that the introduction of visual feedback delay split a movement into several stages. The first stage is a ballistic motion; the second stage is a series of corrective motions.

A motor programming stage has been discovered in cognitive psychology by means of the additive factor method (Sternberg, 1969a; Sternberg et al., 1980). Later, Sanders (1980) discovered motor programming and a motor adjustment stages based on Sternberg studies. In AT, Gordeeva and Zinchenko (1982) performed a microstructural analysis of motor actions and motions. Three stages of movement were discovered: program formation stage (latent stage), executive stage (motor stage), and evaluative stage. As we have already discussed, the first and third stages are cognitive components of movements. New interesting data has been obtained in these studies where Sternberg's ideas played an important role. This is an example of interaction of cognitive psychology and AAT. Gordeeva and Zinchenko (1982) discovered that motor components of actions (excluding the program formation stage) can be divided into a ballistic stage (acceleration stage) and a slow down stage. The ballistic stage is not sensitive to visual feedback, but the slowing down stage is.

Adams (1971) used Anokhin's and Bernshtein's ideas about the acceptor of actions, ordered and comparison apparatus, and also the neural model of stimulus (Sokolov, 1960) in development his concept of movements' regulation in motor learning.

In his work, Adams made some critical comments regarding Anokhin's model, noting lack of differentiation of some mechanisms. The weakest component of Anokhin's model according to Adams is the fact that a single mechanism not only generates and initiates a reaction, but also performs function that involves confirming correctness of a reaction. It should be noted that Anokhin, Bernshtein, and Sokolov did not just create psychological models of self-regulation of our activity or behavior. They are the first who demonstrated that principles of self-regulation derived from certain physiological mechanisms of the nervous system. This allows us to construct a flexible psychological model of self-regulation of motor actions and activity in general. Based on ideas of these authors, Adams introduced the concept of memory trace and perceptual trace. Performance of movements generates a perceptual trace that is developed based on such feedback characteristics of a movement as visual and proprioceptive information. Execution of movements is accompanied by comparison of ongoing feedback and perceptual trace of the movement. He also hypothesized that memory trace is important for movement control as well. Considering that the feedback is not available in the beginning of a movement, a second construct such as a memory trace is needed to explain the selection and initiation of the movement. According to Adams (1968, 1987), some motions are carried out based only on a memory trace, particularly ballistic motions. Correction of such motions is usually carried out on the bases of examining preceding action results, involving an open-loop cycle. This concept suggests that motor actions and movements can be regulated based on open- and closed-loop cycles. Based on physiological data, Adams created a psychological model of self-regulation of motor movements. Adams' theory has some drawbacks. It postulates that every movement has its own perceptual or memory trace. Such movement regulation can result in an overload of memory.

Schmidt (1975) tried to overcome this drawback. He introduced the concept of schema, originally utilized by Bartlett (1932). A schema is a structure in our brain that organizes experiences we gain when interacting with the environment. Schmidt suggests that during motor skill acquisition we also develop a motor response schema. This schema includes the following elements: (a) initial conditions: information about body and limb positions and the state of the environment in which movements take place; (b) response specifications: requirements for force, direction, speed, and other dimensions (it is assumed that a general motor program exists for generating movement); (c) sensory consequences: feedback generated by movements; and (d) response outcome. Thus, Adams' and Schmidt's concepts of movement regulation are, to some extent, similar to Anokhin's, Bernshtein's, and Sokolov's ideas.

Let us consider, as an example, some studies of self-regulations in cognitive psychology, in field of motor movement regulation and their relationship with data that has been obtained in physiology. In general AT, the study of self-regulation also was limited to physiological data for a long time. However, studies of labor demanded proper psychological development of the principles of self-regulation. Anokhin's and Bernshtein's works served as the basis for studying psychological principles of self-regulation in AAT and SSAT. Psychological aspects of self-regulation have been connected with the analysis of real strategies of tasks performance and identifying the role of self-regulation mechanisms in the formation of such strategies.

Here, as an example, we consider some data that has been obtained in AAT.

Let us consider tracking tasks. Tracking tasks are a specific type of tasks that are often performed in dynamic systems. Tracking involves adjusting responses to a given set of dynamic conditions. Such tasks involve analysis of relationship between input signals and an operator's output reaction. Tracking performance usually is measured in terms of errors. Relationship between input and output of a system is considered as transfer functions that have been used in models of tracking behavior to describe human performance. From a self-regulation standpoint, the transfer function reflects a behaviorist approach that reduces the analysis of human behavior to an S–R relationship. From a self-regulation point of view, the transfer function approach ignores the significance of errors to a subject, human motivation, and the importance of explorative strategies in tracking task performance. Zabrodin and Chernishov (1981) conducted a study that demonstrates the low efficiency of tracking models. Visual harmonic stimuli were tracked, with the experiment starting with a 0.05 Hz frequency and gradually moving to higher frequency levels, until the subject could no longer perform the task. The experimenters noticed that the subject's responses contained micro motions with additional harmonics not anticipated by a tracking goal. It was evident that subjects did not perform a linear transformation of input signals. Additional micromotions produced by subjects that were not related to the subjects tracking goal provided additional information that helped them increase precision of tracking performance. Decreasing or distortion of information from such micromotions reduced quality of a tracking task performance. These additional micromotions from the perspective of self-regulation and the efficiency of utilized strategies produced useful information that was used in tracking. Such findings contradict with existing mathematical models that describe transfer functions. According to these models micromotions are unnecessary and should be considered as errors. Zabrodin and Cherneshov (1981) did not consider these important findings from a self-regulation theory perspective. However, this study was very interesting and useful. We interpret this study from self-regulation theory perspectives. According to Zabrodin and Cherneshov, additional movements can be viewed as explorative motions that perform cognitive functions. Therefore, when studying tracking tasks, attention should be paid to such mechanisms and components of activity regulation as goal, strategy, feedback, and significance of errors for subjects.

Usually, cognitive psychology concentrates on studying human information processing and ignores energetic components of activity. The last involves emotional-motivational components of activity. According to AT, cognition and emotional-motivational factors function in unity. This relationship influences strategies of task performance. Next we consider, as an example, some studies of the factor of significance in AT. Study of self-regulation demonstrates that cognitive and emotional-motivational factors are tightly interconnected. In AAT and SSAT, significance reflects how information or some factors are personally important for a particular subject. It has been demonstrated that such factors influence the subjects' performance. For instance, the speed of information processing changes depending on motivational factors and significance of information for a subject. These factors are not taken into consideration by information theory. In engineering psychology, the human operator is often thought of as a communication channel. There were attempts to use this theory for determining the speed of human information processing but they were not very successful. This theory has more theoretical than practical value. Due to self-regulation, the pace of human information processing can change. It has been discovered by Krinchik and Risakov (1965) that the significance of information can influence the reaction time. In their experiment, the information value of the stimuli was changed. For this purpose, experimenters used monetary rewards, specific instructions, and electrical shock. The reaction time for significant information was lower. This is explained by the fact that the speed of information processing depends not only on quantity of information but also on motivational factors. Interesting data in the study of significance were obtained by Kotik (1978). He demonstrated that significance influences an operator's sensitivity to various stimuli and selective features of attention. He presented a red light (significant) signal and a white light (insignificant) signal to subjects. Instructions emphasized that delayed reaction to the significant red signal will be punished by electrical shock. In addition, a 100 dB sound was presented to the subjects along with either red or white signal. Psychophysiological reactions to the noise under these two conditions were measured using pulse rate, breathing rate, and galvanic skin response. When the significant red light signal was presented along with noise, there was no psychophysiological reaction to noise. When a nonsignificant white light signal was presented with noise, psychophysiological reactions significantly increased. Thus when signals were significant, sensitivity to stressful conditions was reduces and the speed of reaction increased.

Konopkin (1980) developed the model of the self-regulation of sensory-motor activity. Interesting data were obtained in his laboratory. An important aspect of an operator's performance is the speed of information processing. Konopkin discovered that this depends on various factors that cannot be explained by information theory. In his study, it was shown that if an operator has preliminary information about the pace of incoming stimuli, he/she can program various components of activity in advance, and therefore perform tasks more effectively.

In the first experiment, eight subjects performed a complex choice reaction task. A vertical screen with a dimmed glass square subdivided into 16 equal squares was been placed in front of a subject. The stimuli were the 16 squares individually illuminated in a programmed manner. Subjects had to press corresponding buttons positioned radiantly in relation to the start position on the panel. The experimenter calculated average reaction time (RT) for any specific condition and, at the next stage, the speed of processing information (V) according to the existing formula ($V = H/T$). In this well-known formula,

H is information conveyed by a stimuli according to $H = \log_2 N$, N is the number of equally likely presented stimuli, and T is the average duration of a correct reaction determined as the total time of reactions divided by the number of correct reactions.

In the first set of experiments, two conditions were used: 3 and 1 s intervals between the visual stimuli. For any pace, two through eight signals of equal probability independent of each other were presented. Prior to the set of experiments, the subjects were informed of how many signals would be presented and at what pace. The obtained data revealed that in all cases, when subjects worked with a 1 s interval, their reaction time was shorter than when they worked with a 3 s interval. The difference was statistically significant.

In another set of experiments, subjects were given false instructions about transition from one pace to another. The real pace was 1.5 s, but two false paces given were 1 and 2 s. Four signals were used in this experiment. When the false 1 s interval instruction was given, the speed of information processing was higher than that of the false 2 s interval instructions. The results were statistically significant.

In the first experiment, the subjects received verbal instruction about pace. In the second experiment, the researcher gave nonverbal instructions. In the preliminary briefing, sound stimuli were presented to the subjects at a pace equal to the pace of the visual stimuli to be presented later. The study revealed similar results. This study demonstrated that the subjects created mental model of satiation and regulated their pace accordingly. The subjective understanding of activity conditions can be considered as a specific mechanism of self-regulation. At the same time, a subject's mental representation of situation cannot determine the programming components of activity directly. Subjects with the same understanding about work pace may use different programs of performance. It has also been discovered that the speed of a simple reaction that has been considered as a stable individual characteristic can change when instructions presented to subjects change (Nojivin, 1974) and that the speed of information processing depends on the information about the task's duration. In all experimental conditions, subjects have to react with maximum speed. In various experiments, they received different information about the task's duration. If subjects perceive duration of task as significant, reaction time increased. It is important to note that subjects were not aware that they changed their strategy.

We have considered only a few examples of studying self-regulation in the framework of AAT. They give some general idea of research in this area. Data presented in this section demonstrate that Anokhin's and Bernshtein's works played an important role in the development of the psychological concept of self-regulation in AT. It should be noted that the more general theory of self-regulation is currently represented in SSAT. This theory is based on data obtained in AT and reflects the principles of self-regulation in goal-directed activities. This theory of self-regulation will be discussed in the next section.

2.5 Analysis of Activity Strategies in Signal Detection Tasks

Modern research in psychophysics shows that even for relatively simple tasks as the detection of a signal in noisy conditions, people demonstrate sophisticated strategy of task performance. Such strategies include not only the sensory processes, but also complex cognitive processes. The decision-making process on the sensory-perceptual level is of particular importance for psychophysics tasks.

The concept of strategy is closely associated with the concept of self-regulation. Beyond the concept of self-regulation, the concept of strategy is not particularly productive. Any strategy is the result of voluntary or involuntary processes of self-regulation. In the latter case, strategy of activity is not clearly understood by a subject. Oddly enough, the concept of self-regulation is rarely used in psychophysics. The reason is that the theory of self-regulation appeared much later than the underlying research in psychophysics. What is interesting is that studies in psychophysics confirm the fact according which human activity during performance of sensory-perceptual tasks is based on the principles of self-regulation. This section analyzes the results of research in psychophysics that confirms this point of view. In the following, we consider some examples of such interconnections and influences.

There are two types of thresholds in psychophysics: absolute and difference thresholds. AAT additionally introduced an absolute operative (working) threshold and a difference operative threshold (Dmitrieva, 1964). Absolute and difference thresholds are not sufficient for the study of human work because they require the subjects to make maximum efforts to detect stimulus or difference between stimuli. Operative thresholds, on the other hand, have intensity or differences 15 times higher than is used by psychophysics for such stimuli. Operative or working thresholds eliminate extreme requirements for the detection of stimulus or discrimination of two stimuli.

Operative or working threshold has been determined based on experimental studies. Let us consider an experiment that demonstrates the value of an operative difference threshold. In this experiment, the difference between two stimuli is viewed as an independent variable and the accuracy or speed of the discrimination process as a dependent variable. Assume that an operator needs to discriminate between the brightness of two visual stimuli. The greater the brightness difference between two stimuli is, the more rapid and precise the discrimination of these stimuli should be. However, it has been discovered that the relationship is not a linear function. Up to a certain level, an increased number of differences enhances accuracy and speed of discrimination but if the differences between stimuli continue to increase up to a certain level, such increase no longer enhances the discriminative process. This level represents a region (interval) of optimal discrimination that has beginning and ending points, where the beginning point is the minimum difference in brightness and the ending point is the maximum difference in brightness. If we continue to increase the difference between brightness of two stimuli and exceed the ending point of the considered region (critical level of difference in brightness), the accuracy or speed of discrimination deteriorates due to the need to readapt to this more extreme difference. Therefore, there is a region where an increase stimuli differences does not improve the discrimination process according to the criteria discussed. Therefore, we can select the beginning point of the considered region as the value of operative threshold where we can observe the termination of the improvement of the discrimination process for the first time. It has been discovered that the beginning point of region is 15 times greater in brightness than the existing difference threshold. Similar data was obtained for an absolute threshold. Differential or bottom threshold values can be found in standard tables of thresholds and then value of operative threshold can be calculated by multiplying obtained data by the coefficient of 15.

Another important aspect within the frame of AAT was an attempt to study cognitive processes from systemic analysis perspectives. For example, even a simplest psychic process, such as sensation, is considered as an element of complex activity system, which includes goal, motivation, and strategies. Sensation is considered as a process that interacts with other psychic processes. The psychophysical task, like any other tasks, requires problem solving. It means that even simple psychophysical tasks can be interpreted by subjects in a variety of ways. Based on these ideas, Zabrodin (1985) introduced the notions of *sensory space* and *space of decision*. Sensory space includes possible alternative images of the situation. The space of decision includes alternative possible responses that can lead to accomplishing the goal of a psychophysical task. Thus, the simplest task of extracting weak signals from a background noise is considered from the perspectives of problem solving and strategies of performance. These recommendations can be used most effectively in the self-regulation of activity. This is particularly relevant in the application of the signal detection theory. Even in the simplest situation tied to the extraction of weak signals, a task can be treated from the perspective of problem solution when the subject uses complex strategies derived from the principles of self-regulation of activity. Despite the fact that the concept of self-regulation has not been used in psychophysical studies, the results of these studies indicate that the activity is formed based on the principles of self-regulation. This explains the need to consider some of the research in psychophysics.

In many work situations, meaningful signals may occur in the presence of *noise* that interferes with receiving meaningful information. Signals should be detected by an observer and two responses can be produced: "yes, there is a signal," or "no, there is no signal." This important psychophysical task is called *signal detection in noise*. The signal detection theory is used in psychology and ergonomics (Green and Swets, 1966; Swets, 1964) for the analysis of such tasks. This theory is applied in situations when warning signals occasionally occur in the background of random noise of varying intensity

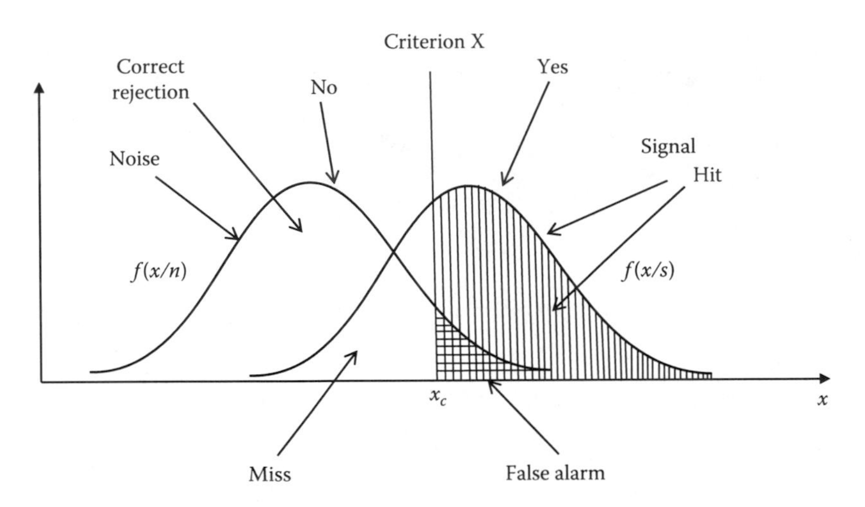

FIGURE 2.3
Model of detection of signal from noise; X—neural activity of sense organ under the influence of *noise* or *signal-plus-noise*; P—probability of *yes* or *no* responses. (From Swets, J.A., *Signal Detection and Recognition by Human Observers*, Wiley, New York, 1964.)

that can be depicted by normal distribution (see left-hand side of Figure 2.3). In turn, the varying intensity of signal-plus-noise might also be depicted by normal distribution (see right-hand side of Figure 2.3).

Horizontal axis demonstrates neural activity of sense organ under the influence of *noise* or *signal plus noise*. Vertical axis demonstrates the probability of *yes* or *no* sensory responses. The location of the curve *signal-plus-noise* on the right is due to the fact that the intensity of the signal-plus-noise is higher than the intensity of the noise itself. If the signal is not very well discriminated from the noise, these two distributions can partly overlap each other. The stronger the noise and the weaker the signal, the more these distributions overlap.

The distance d' between the curves is the difference determined between *signal-plus-noise* and *noise* only by mathematical means divided by the standard deviation of these distributions:

$$d' = \frac{M_{SN} - M_N}{\sigma}$$

where
M_{SN} is the mathematical mean of *signal plus noise*
M_N is the mathematical mean of *noise*

The smaller the d' the harder is the signal detection. The overlapping area is the source of confusion. Signal detection in the overlapping area can be associated with the situation when an observer detects a signal when only noise is presented, which would be a *false alarm*. The other situation is when an observer detects a signal and there are both a noise and a signal, which is the *correct detection*. There is also a *correct rejection* that occurs when an observer correctly determines that there is no signal. The fourth situation is when a signal and a noise are presented but a subject *misses* the signal.

It is important to note that a simple sensory-perceptual task that requires detection of signal includes decision making. Therefore the sensory-perceptual process also involves the cognitive mechanism. An observer has to develop a *criterion* for making a yes-or-no decision and such decision affects the frequency and type of errors. The position of such criterion is presented by point *Xc* and designated by a vertical line in Figure 2.3.

The movement of this line or criterion to the left or right along the horizontal axis affects accuracy of signal detection and type of error. Typically, the shifting of the criterion to the left or right is explained by the individual characteristics of a subject and associated with the cognitive mechanisms of decision making. From an SSAT point of view, shifting of decision criterion along the horizontal axis depends not only on cognitive factors but also on the observer's emotional-evaluative mechanisms or significance of signal detection and personal significance of errors. Position of criterion *Xc* on the horizontal axis depends not only on individual features of an observer, but also on the significance of a task for an observer, goal, motivation, etc., that are important mechanisms of activity self-regulation. In spite of its significance, the concept of self-regulation was not utilized in the signal detection theory. For example, if subjective significance of a correct detection is very high and the significance of committing errors is low for an observer, then the observer's criterion shits to the left. Therefore not only sensory-perceptual and cognitive (decision-making) mechanisms but also emotional-evaluative mechanisms are involved in the performance of psychophysical tasks. The point *Xc* is a critical value for a positive or negative response. When the sensory effect is greater than the *Xc*, an observer produces a positive response (there is a signal). If the sensory effect is less than *Xc* (on the left of the criterion *Xc*) an observer gives a negative response. Since criterion *Xc* moves along the abscissa (horizontal axis), this axis may also be considered as an axis of solutions.

In the works of Swets (1964) and Green and Swets (1966), an observer's response is presented as a point in sensory space or region where this region criterion has a one-dimensional value. In other words, the criterion for decision can change its value only along a horizontal axis. On the contrary, in AAT, the idea of a multidimensional sensory space has been applied for signal detection tasks (Bardin, 1982; Bardin and Voytenko, 1985). These authors assumed that sensory space could be evaluated in terms of multiple criteria. Each criterion can be represented by its own axis. The set of such axes represent a multidimensional space. Bardin and his colleagues demonstrated that an observer, while performing a signal detection task in noisy conditions, used additional subjective criteria that were not given by an instructor. To simplify their analysis, Bardin and his colleagues restricted themselves to considering only of two axes' criteria in the multidimensional space (see Figure 2.4).

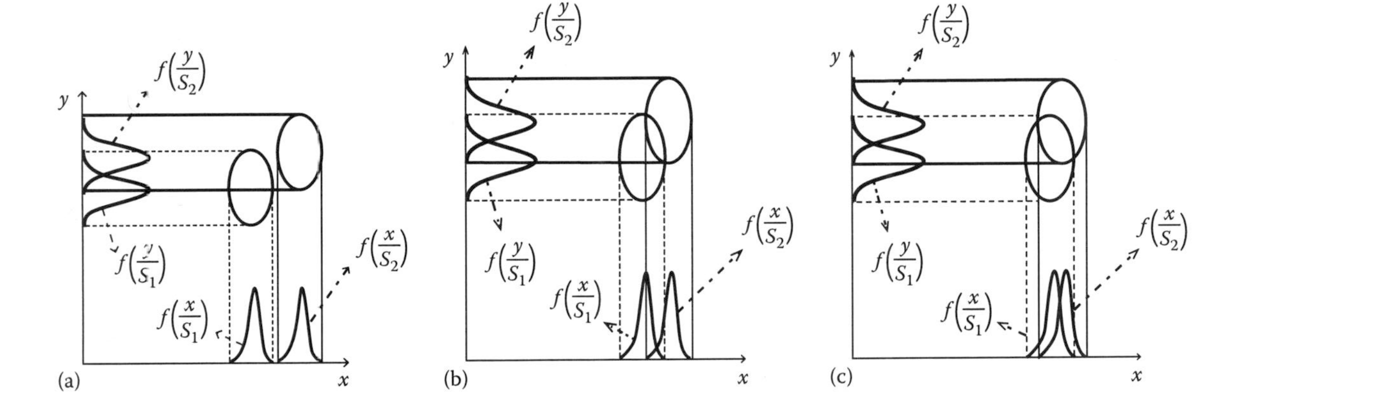

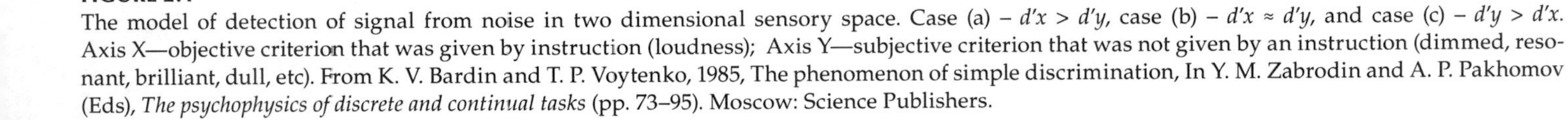

FIGURE 2.4
The model of detection of signal from noise in two dimensional sensory space. Case (a) – $d'x > d'y$, case (b) – $d'x \approx d'y$, and case (c) – $d'y > d'x$. Axis X—objective criterion that was given by instruction (loudness); Axis Y—subjective criterion that was not given by an instruction (dimmed, resonant, brilliant, dull, etc). From K. V. Bardin and T. P. Voytenko, 1985, The phenomenon of simple discrimination, In Y. M. Zabrodin and A. P. Pakhomov (Eds), *The psychophysics of discrete and continual tasks* (pp. 73–95). Moscow: Science Publishers.

In one of their experiments, subjects had to discriminate two stimuli according to their loudness. It has been discovered that when two stimuli are very similar in loudness and discrimination according to this criterion is very difficult, the subjects independently started to use the other criterion. They began to perceive unnoticed qualities in the acoustical stimuli and use them as the discrimination criteria. For example, they reported that the sound seemed to become dimmed, resonant, brilliant, dull, and so on. These qualities were named *additional sensory features*. Such features may be acoustical or may also possess other modality qualities. Some of these qualities are difficult for subjects to verbalize; they can vary in their intensity and specificity, which causes them to have their own axis of measurement in sensory pace. The experiment where the subjects use additional qualities in the discrimination of two acoustical stimuli is presented in Figure 2.4.

Figure 2.4 presents three situations that demonstrate relationship between a signal and a noise. There are two distributions of signal loudness, $f(x/s_1)$ and $f(x/s_2)$, along the x-axis and additional sensory features distribution for these signals along the y-axis, $f(y/s_1)$ and $f(y/s_2)$. The sets of sensory attributes for signals s_1 and s_2 are depicted by circles. In case A, distributions $f(x/s_1)$ and $f(x/s_2)$ along the x-axis do not overlap while additional sensory features distribution $f(y/s_1)$ and $f(y/s_2)$ along the y-axis do overlap. Here we see that $d'x > d'y$. In such a situation, an observer can easily discriminate signals according to loudness (the difference between signals is shown along the x-axis).

However, the difference in the level of loudness of the compared two stimuli s_1 and s_2 begins to decrease. After a certain period of time, the level of loudness of the two stimuli become approximately similar and $d'x \approx d'y$ (see Figure 2.4b). If the difference between two stimuli along the x-axis continues decreasing while the differences along the y-axis is constant (see Figure 2.4c), an observer can start using the second criterion (along the y-axis) because this criterion is easier to use.

Bardin and Voytenko (1985) demonstrated that the use of these additional criterion depends first of all on physical characteristics of the main and additional criteria. We would like to emphasize the idiosyncratic factors here. For example, in the case when $d'x \approx d'y$, an observer sometimes cannot notice some of the discriminative features of the stimuli and totally ignores the difference according to additional criterion because the instruction was to discriminate two stimuli according to loudness. When an observer changes the criterion for discrimination, this means that he/she changes his/her strategy of task performance. Such strategy is formed due to mechanisms of self-regulation such as the vector *motive* → *goal*, mental model of task, positive significance of correct discrimination and negative significance of committing specific errors, subjective criteria of success, etc. Sensory-perceptual and cognitive processes are not the only components involved in the performance of psychophysical tasks. Emotional-evaluative mechanisms of activity regulation associated with the factor of significance are also critically important.

From an SSAT perspective, the analysis of the psychophysical tasks clearly demonstrates that even a simple task to detect a signal in noisy conditions requires involvement of complex mechanisms of activity self-regulation. Psychophysical tasks require problem solving and involve emotional-evaluative mechanisms of activity regulation. A task that is objectively presented to a subject in a psychophysical experiment and a subjective representation of a task in experiment are not the same. A subject forms his/her own mental model of the task. During task performance, the subject develops strategies of task performance, which can be modified and corrected. Zabrodin (1985) introduced the notion of *sensory space* and *space of decision*. The sensory space includes possible psychic images of the tasks and the space of decisions includes alternative possible responses for goal accomplishment. These terms are very similar to the terminology in field of self-regulation in SSAT. The analysis of activity self-regulation during performance of psychophysical tasks allows discovering preferable strategies of task performance. Such qualitative systemic analysis should be performed before formalized analysis of psychophysical tasks. At the second stage, a formalized description of considered task should be made. These two stages of analysis have mutual influence on each other. The fundamental assumption of the systemic-structural activity concept of self-regulation is that even the simplest task that requires detection of a signal in noisy conditions should be treated as complex self-regulative process, which is the basis for the development of adequate strategies of goal attainment.

The analysis of the presented material shows that subjects use sophisticated strategies of activity in the performance of the simple psychophysical tasks. At the heart of the formation of strategies is the process of self-regulation. Bringing the theory of self-regulation to the analysis of psychophysical tasks significantly increases the efficiency of research associated with operator work, which involves detection of weak signals in noise.

3

Concept of Self-Regulation in Systemic-Structural Activity Theory and Strategies of Task Performance

3.1 Concept of Self-Regulation and SA: Comparative Analysis

In the previous chapters of this section of the book, we described the current state of the area of research known as self-regulation. We presented critical analyses of various concepts of self-regulation beyond activity theory (AT). It was demonstrated that the concept of self-regulation in AT was derived from works in physiology, particularly from the works of Anokhin and Bernshtein. Later, with the advance of Wiener's works, cybernetics ideas were accumulated into the study of self-regulation in AT.

The main difference in studying self-regulation inside and outside of AT lies in the theoretical foundation and principles of theoretical interpretation of obtained data. The study of self-regulation in AT always had a ground in physiological and psychological theoretical data, which allowed to avoid some mechanistic ideas from cybernetics that equated the concept of self-regulation in technical system and self-regulation of the conscious goal-directed human activity. Moreover even physiological ideas of self-regulation were not totally adequate for studying the principles of self-regulation of goal-directed and conscious human activity. That is why the psychological concept of self-regulation has been developed in AT (Bedny, 1987; Bedny and Meister, 1997; Konopkin, 1980; Kotik, 1978). Presently, the most developed theoretical approach to studying self-regulation in AT has been suggested by Bedny in the framework of SSAT (Bedny and Karwowski, 2007; Bedny and Meister, 1997).

Human activity is characterized by such features as goal-directedness, consciousness, ability to voluntarily regulate and change this activity, adapt to the environment, and to overcome adverse conditions and to convert them. The basis of such characteristics of activity is the process of self-regulation. Self-regulation of activity is mainly a voluntary and conscious process, in the sense that this process has a conscious goal and a person can recognize certain aspects of goal attainment, to correct the activity or change the goal of the activity. Here, we suggest that not all aspects of activity regulation are accountable. The process of self-regulation includes unconscious components. However, the goal of an activity must be conscious in the self-regulation process.

In the process of self-regulation, a subject has the ability to formulate goals, the ability to change ways of achieving goals, and so on. The main point of analyzing activity self-regulation is the application of the concept of self-regulation to task analysis. Suggested in the framework of SSAT approach to the study of activity self-regulation is well adapted for task analysis. In ergonomics, the study of activity self-regulation is utilized for discovering the most effective strategies of task performance. A subject can not just modify strategies of task performance but even change the goal of an activity when internal conditions of the performer and external conditions of the situation change. The term *strategy* is one of the basic concepts in the analysis of activity self-regulation. Strategy can be viewed as a plan or program of activity performance that is responsive to external contingencies, as well as to the internal state of the system. Strategy is dynamic and adaptive, enabling changes in goal attainment as a function of external and internal conditions of the self-regulative system. The analysis of activity strategies during task performance is the main purpose of utilizing the concept of self-regulation in the study of human work. Strategies of activity are tools for achieving goals of activity.

Due to self-regulation, activity can be seen as a situated system that is constructed or adapted to situations according to the mechanisms of self-regulation. It includes flexible reconstructive strategies (situated components) and preplanned and preprogrammed (prespecified) components. The self-regulative system of activity is at the same time a functional system that is mobilizes, forms, and disappears upon achieving the goal of an activity. In this regard, the analysis of self-regulation of activity is also known as its functional analysis that is performed based on two functional models. One model is called self-regulative model of orienting activity and the other one is general model of activity self-regulation. In this section and in Section 3.2, we consider only the self-regulation of orienting activity. The main units of analysis in self-regulative models of activity are functional mechanisms or function blocks. Such blocks are subsystems with specific regulatory functions within the structure of a self-regulation system. Functional analysis of activity is the description of the process of self-regulation by means of related functional mechanisms.

Extracting the functions that provide realization of the process of self-regulation enables its unified description. In any situation when functional mechanisms are described in relation to other functional mechanisms by using feedforward and feedback connections we use the term *function block*. Each function block represents integration of cognitive processes that are involved in a certain stage of activity regulation. The need for such an approach to the study of activity is due to the fact that division between mental processes is conditional because memory is connected with perception; thinking is not possible without memory and so on. For instance, noise signal detection involves decision making.

Any function may be realized by various mental operations or actions. The content of the function block or mechanism can be changed but the purpose of each function block in a self-regulation process remains the same. The meaning of function blocks in any specific activity can be understood only in relation to other function blocks. Function blocks can be compared, to some extent, with the concept of modules in cognitive psychology, where it is described as parts that are independent in some sense and have different functions (Sternberg, 2008a,b). Each function block in a model of activity self-regulation has rigorous psychological justification.

Typically cognitive psychology focuses on studying individual cognitive processes.

However, work psychology and ergonomics are not so much involved in the study of individual cognitive processes as they are involved in the study of their combinations in a specific type of work activity. Thus, at certain stages of the analysis of work, it is not so much necessary to study isolated cognitive processes as it is to study their combination.

Moreover, the combination of cognition and activity, in general, is not a linear system. It has a loop structure organization with feedforward and feedback connections to ensure a constant possibility of correction and change strategies of performance. The concept of feedback is not sufficient for the analysis of activity self-regulation. Outside of the concept of self-regulation, this concept is not very productive.

The psychological type of self-regulation is a goal-directed process with systemic principles of organization. It can change its structure based on experience. Such a system can form its own goals and subgoals and its own criteria for activity evaluation. A psychological type of self-regulation provides integration of cognitive, executive, evaluative, and emotional-motivational aspects of activity.

It is necessary to distinguish between concepts of self-regulation and self-control. Self-regulation is a broader concept. Self-control is a component of self-regulation which is a goal-directed and largely planned system of mental and motor actions aimed at preventing errors. The formation of self-control can be seen as a process of forming specialized skills needed to perform a certain class of tasks or problems. These skills include thinking as an important component. The ability to pick out signs by which one can consciously perform self-control is in some cases quite a complex skill. Individual features of personality, particularly attention and motivation, perform important functions in the process of self-control. One should distinguish between ongoing and final self-control. Ongoing self-control may be carried out in the uninterrupted process of activity. Final self-control is carried out upon completion of some stage(s) of a task or the whole task. Sometimes, we can use the concept of preventive self-control. A performer should anticipate critical points of controlled processes and utilize preventive self-control. Not just cognitive but also emotional and motivational components are important in the process of self-control. Self-control is associated with prevention of not only motor but also mental errors. The development of skills of self-control is an important component of training.

In the analysis of work activity from the perspective of self-regulation, all cognitive processes are considered in terms of their value in the different mechanisms of self-regulation, thereby defining the functional role of cognitive processes in the structure of work activity regulation. As a result, a model of self-regulation becomes a means of an unified analysis of cognitive processes for various types of work. Therefore, a self-regulation model is at the same time a human information processing model.

As has been demonstrated before, models of self-regulation beyond AT do not differ much from each other (see, e.g., Boekaerts et al., 2005). They include several control-free regulatory mechanisms such as goal standard, input, comparator, and output (see, e.g., Carver and Scheier, 1998, 2005; Vancouver, 2005). These models are based on the homeostatic principle, the main purpose of which is the elimination of deviation from the specified and readymade standard and reaction to disturbances. Self-regulation is presented as a cycle with the feedback from prior performance (Zimmerman, 2005). Moreover, according to Carver, Scheier, and Vancouver, a self-regulative system can function only after receiving feedback from externally performed behavior or appearance of deviations of the controlled variable as a result of interference of disturbances. Such regulation is based on the analysis of the errors that were already made in the prior manipulation with external variables. However, a subject cannot just respond to the unacceptable in his view deviations of variables in the external environment. According to SSAT, he/she can also regulate his/her behavior (external and internal) and can prevent unwanted deviations before executing material actions. This means that self-regulation of activity can be conducted in the inner mental plane. People can often form their own goals, make hypotheses, test the effectiveness of hypothetical strategies, and assess the possible outcomes of cognitive and future motor actions without resorting to a real transformation of the external situation. Therefore, feedback on performance does not always come from the external environment. Subjects can also utilize mental feedback for the evaluation of cognitive actions. Blindfold chess is a great example that demonstrates the importance of cognitive feedback in the regulation of mental activity (see Section 1.1). Chess players perform various cognitive actions and evaluate their result mentally. They can not only evaluate their own actions but evaluate possible actions of their opponents, predicted results, and so on. This fact practically is ignored in all models of self-regulation outside of SSAT.

This example clearly demonstrates that information about the possible result of activity and separate actions can be predicted and real strategies of performance can be formulated and corrected mentally. Thus, a person may act in mental plane and use mental feedback. This allows him not only to correct errors that he made but also to prevent them. This aspect of self-regulation is omitted in the studies of self-regulation outside of SSAT.

In SSAT, feedback can be of the following types: instantaneous or ongoing (implemented in the course of activity or actions), final (performed after completion of same stage of activity), external (information is received from external receptors), internal (information comes from internal receptors; e.g., kinesthesia) or from internal mental actions—negative (discrepancy-reducing) and positive (discrepancy-enlarging). In AT, feedback also can be artificially created or natural for existing situation. One of the most important distinguishing features of this classification is the presence of feedback for the results of cognitive actions that are performed in the mental plane. For example, a person can manipulate mental images of the situation, assess results of mental actions, and decide whether to perform real actions. Mental feedback has an important preventive function.

Models of self-regulation beyond AT include content-free regulatory mechanisms, which do not have sufficient theoretical justification in psychology. Such mechanisms of self-regulation are borrowed from the technical disciplines. All considered models of self-regulation outside AT suggest a feedback only after execution of external motor actions. In this short introduction, we reviewed some important aspects of self-regulation in SSAT and demonstrated that self-regulation models that are utilized outside of AT are not adapted for task analysis. In this section, we consider the concept of orienting activity and compare it with the concept of situation awareness (SA). This will allow us in the next stage to consider a model of self-regulation of orienting activity and its relationship to SA.

Dynamic reflection of the situation is the main purpose of orienting activity and SA is considered as one of the mechanisms of orienting activity (Bedny and Meister, 1999). Based on orienting activity, a subject creates a dynamic mental model of a situation, interprets it, and predicts near future events. We cannot agree with the statement that a dynamic mental model of a situation is constructed based only on perception. Various cognitive processes that interact with each other are involved in the creation of such a model. One can speak about the dominance of one cognitive process or another. Thinking and memory play a special role in the creation of a mental model. Imagination is also important in some situations. Creating a dynamic mental model of a situation is impossible without the interpretation of a situation and the understanding of its possible near future development. It involves an understanding of how elements of a situation interact with each other (Bedny et al., 2004). The nature of interaction between elements of a situation is not a perceptual property but a result of thinking. Imaginative and verbally logical components interact in this process.

Activity has four stages: goal formation, orientation, execution, and evaluation. The main purpose of orienting activity is the creation of dynamic mental model of situation in accordance to the goal of activity, interpretation of it, and prediction of the near future. Sometimes, activity, which involves the comprehension of a situation, is called gnostic activity. One should distinguish between orientation as a stage of activity and orienting activity as an independent type of activity.

Orienting activity, similar to SA, includes perception of a situation, its comprehension, and prediction of the near future. The concept of reflection that derives from philosophy has an important meaning for AT. Cognitive processes perform reflective, regulative, and evaluative functions. Reflection can be considered as a mental representation of reality. The concept of reflection also plays an important role in physiology. The reflection of reality includes predictive functions. Such functions of the nervous system were described by Anokhin (1969), Bernshtein, (1969) and Sokolov (1969). A process of reflection in physiology is considered as a mechanism of the nervous system that models the external world by specific changes that occur in the internal structure of this system. Sokolov (1969) demonstrated that neural mechanisms model the external world by specific changes in the internal the structure of the nervous system. Due to reflection, the nervous system creates a neural model of the environment. Studies in psychology and physiology showed that a reflection of the situation can be committed at conscious and unconscious levels. In the latter case, the external environment is reflected by the nervous system, but such a reflection does not reach the conscious level. However, the studies conducted by Sokolov demonstrated that reflection at an unconscious level affects the behavior of a human organism. So, we have to take into account not only conscious but also unconscious aspects of reflection when studying human activity. Thus, orienting activity usually includes unconscious components.

Human ability to consciously and unconsciously reflect a state of the environment has been studied in psychology. Interesting results were obtained in this area by Konopkin and Zhujkov (1973). They considered the ability of humans to reflect statistical characteristics of the environment at conscious and unconscious levels. They also studied the ability of humans to use information about statistical characteristics of the environment that has been obtained at the unconscious level of reflection, at the conscious level of activity performance.

It should be noted that a number of classical works in cognitive psychology (Edwards, 1961; Kahneman and Tversky, 1984) have been devoted to examining human ability to reflect statistical characteristics of the environment at a conscious level. We do not consider them here, since we are interested in a person's ability to use such information at an unconscious level. In this context, we are interested in the works of Konopkin and Zhujkov (1973).

Their method of study included a presentation of special tables consisting of the sequence of numbers "6" and "9." The task for the subjects was to strikethrough these two figures in different ways. Each subject worked with only one type of table and were informed that the purpose of the experiment was to test their attention. This was done to ensure that the subjects did not know the real purpose of the experiment, which was to ascertain the ability of the subjects to unconsciously reflect the probabilistic structure of the signals. After filling the table, they were asked an unexpected question about the relationship between the numbers. The experiment used 16 versions of the tables. Therefore, each subject had an opportunity to work only with one version of the table. Subjects worked at their preferred pace, which helped to eliminate errors. During the experiments, each subject was given 1000 numbers.

In the first version of the presented table, the ratio between numbers "6" and "9" was 50%/50%. In each following version of the table, number "9" increased by 1%, and the number "6" was reduced by 1%. In all variants of the tables, the numbers were presented in a random manner.

We will not go into the experimental procedures in detail. Here, we present only some results obtained from this study.

Using a special device, the same two numbers were presented to subjects randomly. Their probability of presentation varied in the threshold region and then in the area where subjects can clearly distinguish the frequency of two stimuli presented. The reaction times for these two stimuli were measured. After the experiment, a discussion was conducted with each subject. Based on the comparison of experimental data, observation, and the discussions, interesting results were obtained.

In this experimental study, subjects started to notice the difference in relationship between the numbers when the ratio approached 58%/42% with some small variations.

It had been discovered that even in the cases where the subjects were not aware of the difference in the frequency of stimuli presentation, they could account for this frequency and choose strategies that improved reaction speed and reduced errors. The difference in the probabilistic structure of the signals can be reflected even on an unconscious level and therefore can have an impact on performance. For example, although in the initial stages of the experiment the subjects were not aware of the difference in the probability of the signals, the reaction time varied for different signals. When the difference in the frequency of stimuli presentation became conscious, subjects changed their strategies of reactions performance consciously. When the subjects could consciously reflect the frequency of signals, they most often guessed the appearance of the stimulus and responded adequately with great confidence. Some strategies were specific for particular subjects; for example, several subjects utilized strategies when more attention was focused on the rare stimuli, and so on.

The basic conclusion of this study was that when people are dealing with signals of different probabilities, they are capable of unconscious reflection of the probabilistic relation between signals. Based on unconscious reflection, probabilistic relationship between signals generated unconscious psychophysiological adjustment to the relative frequency of individual signals, which in the future determined reaction time. Unconscious reflection of probabilistic features of a situation is a real psychophysiological mechanism of tuning reactions to probabilistic features of the situation.

During transition from unconscious to conscious processes of reflection of a situation, the strategy of activity may change. This experiment clearly demonstrated that the stage of activity that precedes execution includes not only conscious but also unconscious components. Therefore, SA does not determine the total execution of activity. Sometimes SA can be ineffective, but execution turns out to be effective. Wickens and Hollands' (2000) wrote that driving can be effective but the level of situation awareness can be low—a driver unconsciously reflects the situation and such reflection is effective only at particular moments. Such unconscious reflection of situation can provide effective execution without SA. In our further experiments, we will demonstrate that reflection of a situation and the following execution involves complex strategies of performance with conscious and unconscious components.

The term *conscious reflection of a situation* has exactly the same meaning in AT as *situation awareness* in cognitive psychology. Reflection can be considered as a mental representation of reality. In our further studies, we follow Endsley's terminology and utilize the latter term. Conscious reflection of situation or SA does not function independently. It exists in human activity as one of a number of others mechanisms that are involved in activity regulation. One drawback of the concept of SA is that it ignores unconscious components of reflection.

One of the important means of orienting activity is exploratory actions. They can be internal or external and mental or motor. Their purpose is manipulating with the elements of the situation in order to analyze their effects and interpretation of the situation. For example, a chess player can mentally move the figures on a chess board in order to analyze the situation. The results of these actions are evaluated using mental feedback. In reality, however, there are situations when a person can manipulate material objects to a certain extent and based on practical outcome get an idea about the real situation. Such manipulation is not directed to transform the situation for achieving the goal of activity. Its purpose is interpretation of the situation and the possibility of understanding future events.

Orienting activity, similar to SA, is a dynamic and changeable phenomenon (Bedny et al., 2004). Even in those cases when the external situation remains unchanged, the reflection of the situation is modified, becomes more or less detailed. In any particular moment, a subject can extract different structure from a situation depending on the goal of the orienting activity. We can call it a dynamics reflection of a dynamic situation. Reflection is more complex when a situation is dynamic. Such complex reflections can be provided by a self-regulating or self-tuning system, but not by an isolated, even very interesting, psychological mechanism. That is why in SSAT orienting activity is described as a self-regulated system. SA or conscious reflection of a situation is only a mechanism in such self-regulated system. In other words, orienting activity including SA should be studied from a systemic perspective.

The concept of goal is important for SA analysis and for orienting activity but in AT this concept is understood in a totally different way. Endsley and Jones (2012, pp. 68–69) in their section "Goals versus tasks" wrote:

> The GDTA goals seek to document cognitive demands rather than physical tasks. Physical tasks are things the operator must physically accomplish such as filling out a report or calling a coworker. Cognitive demands are the activities that require expenditure of higher-order cognitive resources, such as predicting the enemy's course of action (COA) or determining the effects of an enemy's (COA) on battle outcome. Performance of a task is not a goal, because tasks are technology dependent. A particular goal may be accomplished by means of different tasks depending on the system involved.

The title of their section and their view suggest that *goal* has a different meaning for these authors compared to that in AT. Here, we only want to mention that goal is not a cognitive demand but a desired future result of a subject's own activity that is connected with motives. It includes cognitive, imaginative, and conscious components. There is a goal of a task, a goal of a subtask, a goal of an action, etc. Goal is dynamic and may have various personal interpretations. Goal can be reformulated in a more or less precise manner by a subject independently. Goal cannot be considered a cognitive demand because only a task, complex or simple, presents some cognitive demands for its performance. Goal can be presented by instructions, formulated and reformulated, accepted and rejected by a subject independently. There are also voluntarily or involutedly formulated goals. Zarakovsky et al. (1974) introduced the term *intention engrams of the memory*. They are interpreted as traces of memory, which represent a combination of a potential goal with some motivational factors. Such engrams can be activated in some situations and a goal would be triggered almost automatically. Zarakovsky called these mechanisms potential goals that can be transformed into actual goals. A potential goal is not yet a real goal. It becomes one through the activation of information in memory. There are requirements and conditions in every task. Anything that is presented to an operator or known by her/him about the task are task conditions. Requirements are what should be achieved during task performance. When requirements are accepted by subjects, they become the subjects' personal goals.

According to Endsley and Jones (see their first sentence in selected paragraph), a physical task does not have cognitive demands and goals. We cannot agree with such a statement. If one performs physical actions, one has to understand that she/he wants to achieve and this is the goal of the motor task. In some cases, exploratory motor activity might be carried out without a clear idea of what result can be obtained. In these circumstances, a goal may be verbally expressed as follows: What will happen with this object if I perform this physical action with it? In some cases, an obtained result can be a consequence of an accidental influence on a situation. Therefore, such a result is not the consequence of a goal-directed activity. However, work activity, including physical activity, is goal directed. Goal-directed behavioral actions have their goals and are performed in combination with cognitive actions. Otherwise, we can speak about automatic reactions or motor unconscious operations that are included in motor, goal-directed actions. Task according to AT is a situation requiring achievement of a goal in specific conditions (Leont'ev, 1978; Rubinshtein, 1940). The task determines the method of goal achievement in particular conditions.

Let us look at an example of a physical task execution. A mother says to her son, "It is about to rain and two chairs are in the backyard. Please bring them inside." Here, the statements "it is about to rain" and "two chairs are in the backyard" are conditions of the task. The statement "please bring them inside" is a requirement. If the son politely accepts his mother's requirement, it is converted into his personal goal. The son goes to the backyard and brings the chairs. The goal is achieved and the task is completed. According to Endsley and Jones (2012, p. 68), a task is a physical act and there are no goals of tasks, which "seek to document cognitive demands." However, as can be seen, a physical task still has a cognitive goal. Moreover, a task also includes some simple cognitive components without which, in our example, the son cannot perform the physical task. He can interpret the goal incorrectly: the mother asks for the chairs to be brought into the living room, but the son brings them into the kitchen. The son can also reformulate the goal: if the chairs are heavy and there is a heavy downfall, he might decide to bring only one chair. The son's goal is always associated with motives. For example, if the son evaluates the task as very demanding (chairs are heavy, heavy rain, he is reading a very interesting book, or watching a movie at the time, etc.) he can reject his mother's request. Endsley and Jones ignore the concept of motivation when considering such concepts as goal and task. The authors overlook the fact that a goal does not exist without motives. There is no unmotivated goal (Leont'ev, 1981; Rubinshtein, 1957; etc.). Motivation influences a goal formation process. Let us consider another example presented by Endsley and Jones (2012, p. 25): "For instance, if a driver sees an accident about to happen, his goal changes from navigating to avoiding the accident." According to AT, this is the formulation of a new goal in emergency conditions. This is a highly motivated goal formation process because the situation is associated with a threat to life and/or material losses. Depending on the significance of a situation and the motivational factors, the subject can select different goals associated with avoiding the accident. Thus the task can be complex or simple. The more complex the cognitive or physical task is the more cognitive demands are required for task execution.

Endsley and Jones's stated: "Performance of a task is not a goal, because tasks are technology dependent." From AT perspectives, when a person performs a task this means that he/she is trying to achieve a goal. Further, in a work environment, every task is technology or system dependent. Task can be considered from technological and psychological perspectives

(Bedny and Harris, 2005). We can keep the same technological process and to some extent change work process. In such situations, we talk about changes in the method of human performance. If equipment configuration changes, method of human performance or human activity also changes as a result. Based on the analysis of the human activity structure, we can evaluate the efficiency of a design solution of technological components of a system (evaluate usability of technological components of a system). Thus, we can study task and its goal from technological or psychological perspectives. These two methods are interdependent, but still are different methods of task analysis. If a task is analyzed in a production process from a technological perspective then technologists or process engineers would be involved in it. If a task is studied from a work activity analysis perspective, then the main players here would be psychologists and physiologists.

In SSAT, goal should be considered as a system concept and as a psychological concept.

These two understandings of goal are not the same. Human goal is an integrative mechanism of human activity, which determines selection and interpretation of information. SA includes human consciousness and therefore is tightly connected with a human conscious goal. Therefore, SSAT can be extremely helpful for studying SA.

Task in a production situation from a psychological perspective is a stage of work process or work activity directed to achieve a goal, where it is a psychological concept. In AT, a goal has always played an important role in the analysis of tasks. A goal is always conscious to some degree. A task, according to Leont'ev (1978), is a goal that is given in certain conditions. Thus, a task always includes a goal. Endsley and Jones (2012, p. 68) present a section titled "Goals versus tasks." However, such opposition from activity perspectives is unacceptable. The goal of a task is a requirement that is accepted and interpreted in a specific way and should be achieved during task performance. The same situation can produce a totally different situation awareness depending on the goal accepted or formulated by a subject.

From an SSAT perspective, task analysis always includes the analysis of a task's goal and associated with it other goals that are formed during task performance. AT distinguishes task goals, a goal of subtasks, and a goal of separate cognitive and behavioral actions. The goal of a task can be presented to a subject in a readymade form as an objectively presented requirement. Then, the key point is its correct interpretation by the subject and how this goal is accepted by him/her. It is well known that a person can reformulate a goal based on her/his subjective preferences. The goal of a task can be formulated by a subject independently. The question arises as to how adequate such goal is to a specific situation. All of these factors also depend on the emotional-evaluative and emotional-motivational components of activity. A goal does not exist without motives. A goal is a cognitive component, and motivation is an energetic component of activity. The goal of a task is the desired future result that should be achieved through a task performance.

A goal and a motive create the vector *motive → goal*, which gives direction to activity. It is natural that this vector cannot be seen in the strict sense as it is used in physics. It is a psychological vector that has certain properties that allow comparing it with a vector in physics. Emotional-motivational factors are critically important in the analysis of goal and task in general. These factors are not considered by the SA method at all. Thus, the concept of goal in AT is different from the way it is described in SA. In SA, the goal of a system is mixed with a human goal during the performance of various tasks. Situation awareness as a psychological concept can be understood only in relation to a human goal. Relationship between these concepts can be presented in the following manner: system's goals → human goals → various types of situation awareness.

The model of orienting activity that will be discussed in the next chapter examines the process of conscious and unconscious goal formation and reflection of a situation in general. These aspects of analysis of a goal and reflection of a situation as well as a decision making mechanism are not addressed by the SA method described by Endsley and Jones. Nonetheless, decision-making mechanisms are always involved in the final evaluative stage of SA. One should distinguish between decision making involved in orienting activity and decision making involved in executive activity.

The model proposed by Endsley is not logically consistent. In her model, SA is treated simply as another box in a flowchart of the human information processing system. Boxes such as SA, decision making, and performance of actions as stages of information processing in such models suggest involvement of various psychic processes in the functioning of each box. Therefore, the box labeled *information processing mechanism* in Endsley's model cannot be described as an independent mechanism (box) of information processing. This model pays more attention to the functioning of perception, memory, and attention, but does not concentrate on the functioning of thinking mechanisms that are the main components in the creation of a dynamic model of a situation.

According to SSAT any functional mechanism of self-regulation, including SA, is a specific integration of cognitive processes organized as self-regulative system. However, this does not exclude the possibility of a special analysis of the role of individual cognitive processes in the efficient functioning of SA or other mechanisms of orienting activity.

Endsley and Jones (2012) studied SA requirement for various types of unmanned vehicle operations. As an example, they present a special table for the SA requirements for unmanned ground vehicles. These tables have approximately 180 factors related to equipment and environment: (1) Level 1 SA: vehicle status, speed of vehicle, heading of vehicle, past vehicle location, etc.; (2) Level 2 SA: vehicle operations, distance traveled, area coverage, deviation between aperture size and robot size, etc.; (3) Level 3: projected location of robot *relative to operator, relative to stating position, relative to other systems; projected destination of vehicle; projected control actions*; etc. (Endsley and

Jones, 2012, pp. 224–225). However, this is not yet task analysis because such a list of factors tells us very little about SA in task performance. SA analysis cannot be reduced to the compilation of such a list of unrelated factors, which make sense only in the context of specific tasks.

According to SSAT, situation awareness is associated with certain stages of task performance. First, tasks that should be performed by an operator need to be listed and then factors that can influence SA for a particular task should be identified. Some factors can be unimportant because their awareness during task performance is unquestionable. Other factors related to equipment and environment can be important for SA for a specific task and can be taken into account differently in a context of specific tasks. So the first step of task analysis is identifying a list of tasks, but if the system already exists, then a list of tasks is known. Ergonomists and those who designed the system, or people who already use it, can develop a list of tasks for an existing system. If the system is in the development stage, identification of a list of all possible tasks can turn into a complex phase of analysis and design of such system. This issue is considered in Bedny and Meister (1997, pp. 211–219).

In works of Bedny and Meister (1997) and Bedny and Karwowski (2008a), we have presented an example of design of unmanned remotely controlled underwater vehicle (UUV). One of the most complex tasks is to ensure adequate movement of UUV under water. Situation awareness of position and direction of the vehicle's movements is difficult because of distorted correlation between the axes of the operator's body and that of the vehicle. We consider this example briefly in the context of SA. In SSAT, there are various qualitative methods of task analysis: objectively logical analysis, sociocultural analysis, individually psychological analysis, and functional analysis (Bedny and Karwowski, 2007). We consider only two of them here. The simplest qualitative method is objectively logical analysis. This method includes observation, analysis of documentation, utilizing questioners, experts' analysis, etc. Functional analysis is more complex. This method studies activity as a self-regulative system. It is qualitative systemic analysis. The concept of SA or the possibility of conscious reflection at all stages of UUV movements was considered in detail by this method of study. In our study, we used a simplified functional analysis of orienting activity as a preliminary qualitative stage. At the following stages, we utilize an algorithmic description of task performance, design time structure of activity, and quantitative assessment of task complexity, which determines cognitive demands on task performance. It should be noted that qualitative, formalized, and quantitative stages of analysis have a loop-structure organization.

Position, orientation, and direction of UUV movements on seabed with respect to operator's body and beacons are one of the most complex and frequently utilized tasks.

Such a task is not only connected with a high probability of errors, but also requires significant mental effort to control its execution and creates major problems for SA and for orienting activity in general, which becomes rather complex during performance of such tasks.

In our studies, we used analytical methods of analyses. We prepared drawings that demonstrated various options for remote control of a UUV. There was a defined hypothetical trajectory of a UUV movement on seabed. It required the execution of various rotations of a UUV, turns at various angles, forward and backward (reverse) movements, movements on an uneven surface leading to skidding of a UUV, etc. Qualitative analysis of orienting activity reveals that there is a high probability to lose conscious reflection of the ongoing situation or of SA. Maintaining SA requires considerable mental efforts because the task is very complex.

We have compared three versions of remote controlled UUV. Based on these methods of analysis, the best version of a remote-controlled UUV was selected.

As we have mentioned, the main difficulties in the remote control of a UUV are associated with the inability of an operator to quickly and accurately reflect the position of the UUV in relation to his/her body axis. An operator might have difficulty choosing to turn right or left, or backward or forward and feel inclining and reclining of the UUV. SA or conscious reflection of position and movements of a UUV is a critical factor in the remote control of such vehicles.

In order to reduce complexity of task and decrease cognitive demands for its performance, we suggest the following:

1. Rotate a panel: An operator can press the left button by the left hand to turn the UUV to the left. Simultaneously, a control board with an operator which is entirely independent of the UUV starts turning to the left. If an operator stops pressing the left button, the rotation of UUV and the control board with an operator stops as well. If an operator presses the right button with his/her right hand, a control board and UUV turn to the right.
2. Rotating a control board indicator's image in the opposite direction: This indicator is mounted on a horizontal panel independent of the UUV. It has coordinate lines and the UUV is depicted by a moving point on the screen. When the UUV and the control board are rotated to the left, the control board indicator is rotated to the right at the same angle and vice versa.

   As a result, the operator's body axis and indicator axis are exactly in the same position as the axis of the UUV during its rotation.
3. Inclining and reclining of control board should be coordinated with the feedback from the UUV.

Quantitative assessment of the complexity of the considered task has been performed for various versions of UUV. Complexity of a task was considered as a multidimensional system with a number of measures that reflect task complexity. It has been concluded that the suggested method of remote control of UUV significantly reduced the complexity of the considered task and therefore reduced cognitive demands for its performance due to improvement of SA and orientation in the situation in general.

An analysis of this study brings us to the following conclusions:

1. Improvement of SA should be carried out in the context of specific tasks due to the fact that the same information is used differently in the context of different tasks.
2. SSAT views SA as one of a number of other mechanisms of activity self-regulation, which should be considered in the context of the functional analysis of activity. This issue will be discussed in detail in subsequent chapters of this section of the book.
3. Justification of the principle of remote control of UUV and improvement of SA in particular can be done not only based on qualitative analysis of tasks under consideration, but also utilizing formalized and quantitative analyses of the complexity of various tasks where SA is a critical factor.

It is interesting to note that a rotated image is now used in some modern cars. The essence of the principle is that an image of the situation on an electronic display can be rotated in a certain direction depending on the position of the car. This principle was first suggested by Bedny (1987), Bedny and Meister (1997) and described briefly in Bedny and Karwowski (2007) for remote control of UUVs for more complex tasks. These tasks are much more complex than controlling a car movement because an operator remote-controlling a UUV does not have any visual contact with the situation, while a car driver is located inside the car and has direct visual contact with the situation. Thus, the principles of design and the complexity evaluation of task performance can be very useful in finding solutions for complex ergonomic design problems and for SA improvement.

The material presented in this chapter brings us to the following conclusion.

The concept of situation awareness (SA) is currently receiving increasing attention in work psychology and ergonomics. There is also interesting data in the study of distributed situation awareness (Salmon et al., 2008). Despite the popularity of this concept, an analysis of literature reveals that this construct is not clearly defined. There are different understandings of this construct and some authors are even questioning its existence (Dekker et al., 2010). We consider SA as one out of a number of other functional mechanisms of activity regulation that integrate in various ways different cognitive processes that are important for a particular stage of activity regulation. In AT, there is a concept of orienting activity, a component of activity that is involved in the reflection of a situation in accordance with the goal of activity. Dynamic reflection of the situation and its interpretation is the major purpose of orienting activity. The concept of orienting activity existed long before SA was introduced in ergonomics. Orienting activity is an important psychological concept that can be used in applied studies.

3.2 Self-Regulation Model of Orienting Activity

In the previous chapter, we considered general principles of self-regulation from SSAT standpoint. We also made a brief comparative analysis of such concepts as orienting activity and SA. The objective of this chapter is to describe the model of self-regulation of orienting activity and show the role of SA in this model. SSAT functional analysis views activity as a goal-directed self-regulative system.

Before considering the model of self-regulation, let us review the main characteristics of this approach. The key problem in studying self-regulation of activity is continuing reconsideration of activity strategies when internal or external conditions of performance change. Sometimes the result of such self-regulation includes not only changes of performance strategies but a change of the goal itself.

When studying work activity, it is often not productive to separate mental processes such as sensation, perception, memory, and reasoning because these processes are interrelated. The process of perception involves memory; working memory is tightly connected with operative thinking, and so on. This is why functional analysis suggests that activity during task performance be studied not only in terms of psychological processes, but also in terms of function mechanisms or blocks.

When we study activity from a self-regulation perspective a functional mechanism or a function block is main unit of analysis. The stages of activity that involve particular functions in activity regulation are called functional mechanisms. When functional mechanisms are presented as components of the model of activity self-regulation with their feedforward and feedback interconnections, they are defined as function blocks. The functional model of self-regulation consists of various function blocks. The same function can be achieved by utilizing different cognitive processes. Hence, any function block includes various combinations of cognitive processes. However, their integration can be carried out in different ways depending on the specificity of the task at hand.

The content of a function block can be described in terms of mental operations or behavioral actions. Each function block has the same purpose in structure of activity self-regulation, but its content varies depending on specificity of a performed task. A function block represents a coordinate system of subfunctions that has a specific purpose in activity regulation. For example, there is a function block that is responsible for the creation of the goal of an activity; the other one is responsible for the creation of the conceptual model of a situation; there is a function block responsible for the evaluation of the difficulty of an activity or the formation of a level of motivation, etc. The key characteristics of a functional model are summarized as follows: main components of a self-regulative system are function blocks; each block is directed to achieve a specific purpose in activity regulation; all blocks have feedforward and feedback connections. Each function block in the model has a rigorous scientific justification. The division of activity system into functional blocks of self-regulation has a similar ideology, which was considered by Sternberg. He wrote, "One approach to understanding a complex process or system begins with an attempt to divide it into modules: parts that are independent in some sense and have different functions (Sternberg, 2008a, p. 112)." This method is one of the basic principles of systemic analysis. Naturally, the particular method which is used by Sternberg is different from our method of extraction of various functional blocks. However, the general idea of systemic analysis in both cases is the same.

The functional model is adapted for the analysis of various tasks where activity during task performance is considered a goal-directed functional system. Each functional system is formed to achieve the goal of a task. When the goal of a task is achieved such functional system disappears and a new system is formed in order to achieve the goal of the next task.

The proposed model can be considered an information processing model. However, in this model, the human information processing system has a goal directed, loop-structured organization. Attention should be paid not only to studying separate cognitive processes but to how they are integrated at various stages of activity regulation. It is assumed that, if necessary, a specialist would proceed with analyzing individual cognitive processes that are also seen as a self-regulatory system. Therefore, cognitive analysis is not rejected but is regarded as a stage in activity analysis.

During task analysis, each functional block determines the range of issues that are connected with this block and should be considered when examining the role of this block in activity regulation. The nature of these issues is also associated with specificity of activity during the performance of a particular task. The most important functional blocks for each particular case should be considered. It is also important to pay attention to the relationship between and the mutual influence of the functional blocks. Hence, meaning of a function block in any specific activity can be understood only in light of its relationship with other function blocks. The model (Figure 3.1) describes a self-regulation process that precedes decision making and follows performance of executive actions, the purpose of which is to transform a situation according to the goal of an activity.

The purpose of this model is to analyze and describe possible strategies of performing orienting activity that includes perceiving a situation,

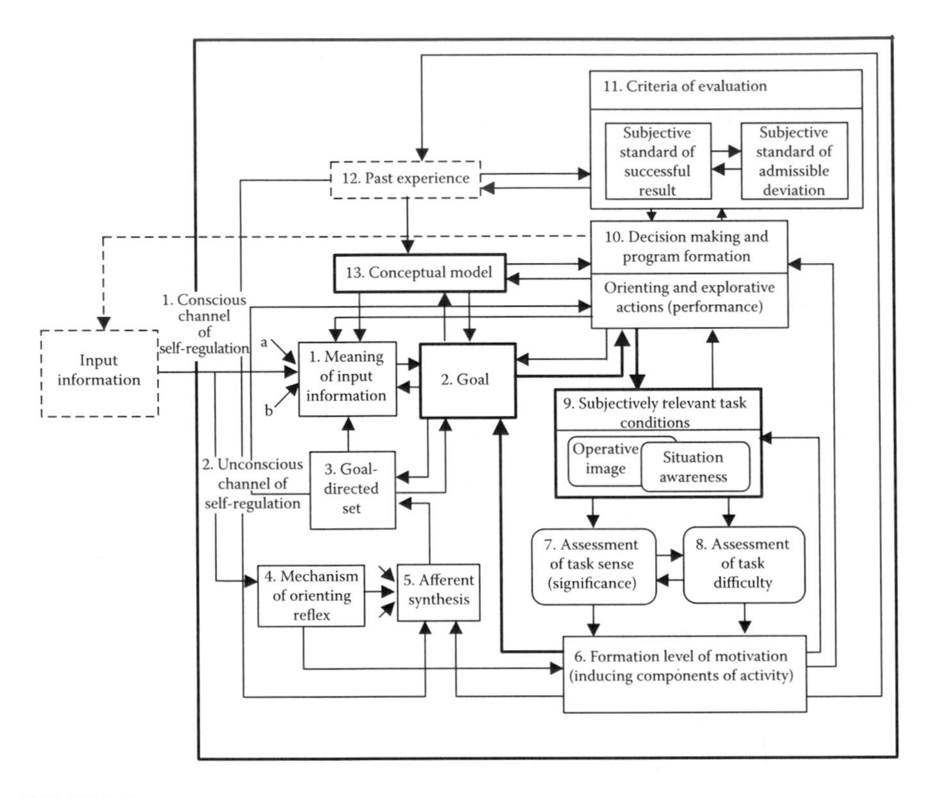

FIGURE 3.1
Model of self-regulation of orienting activity.

its comprehension, and projection of near future events. This is similar to what Endsley and Jones (2012) described as situation awareness. However, here we consider orienting activity, which performs similar however not totally identical functions. Orienting activity is a much broader concept than situation awareness. It includes a conscious goal, which is connected with motives, conscious and unconscious components, subjective criteria of success, etc. Major purpose of orienting activity is acceptance or formulation of goal of activity, the formation of a dynamic mental model of the situation, its interpretation, evaluation of the significance of the goal and elements of the situation and the development of adequate motivation needed to perform orienting activity. In orienting activity there is conceptual model. This is relatively stable model that is not situation specific.

However, this model plays an important role in the formation of dynamic and situation-specific mental models.

Stable and dynamic mental models cannot be fully verbally described by a subject.

They include verbalized and imaginative components. The latter cannot always be verbalized or conscious. Explorative actions and operations play an important role in the mental representation of reality. If we know how a mental representation of the situation is developed, we can understand how the execution part of the activity is carried out. For example, a chess player performs a very complex mental activity before moving a figure, while the motor action is very simple; he/she takes a certain figure and moves it to a particular position. Thus with our model, we describe how a person creates a goal and a subjective mental model of a situation, including SA, what types of exploratory actions and operations are utilized, what types of possible mental models are developed, how a subject selects preferable mental models, and so on. This means that the product of orienting the self-regulation activity is not just transformation of a real situation, but formation of mental representation of a situation that precedes execution. A complicated process of self-regulation is performed mainly mentally by using cognitive actions and operations. Self-regulation is performed not only based on the analysis of already committed behavioral errors, but also based on their mental forecasting. We would like to emphasize that the model of orienting activity involves first of all exploratory mental actions. This allows a subject to create various mental pictures of the same externally unchanged situation and choose the most appropriate decision. The decision making mechanism is excluded from SA in spite of it being involved in the final stage of SA. Decision making as a component of orienting activity should be distinguished from decision making as a component of executive activity. Before a subject chooses a response she/he needs to choose the most adequate mental model of a situation.

This chapter presents a modified version of the self-regulation model of orienting activity. The model presented in Figure 3.1 includes conscious and unconscious levels of self-regulation that interact with each other. The dashed box *input information* is not a function block. It depicts the information that has been presented to a performer.

Let us consider the unconscious level of self-regulation (channel 2). The incoming, usually unexpected information activates the orienting reflex (block 4). The orienting reflex is conveyed by such responses as moving eyes or a head to the stimulus, altering sensitivity of different sense organs, a change in blood pressure or heart rate, and changes in breathing rate. At the same time, some electrophysiological change occurs in brain activity. Orienting reflex appears to play an important role in the functioning mechanisms of involuntary attention. Specific kinds of novel neuron detectors were discovered in the brain cortex (Sokolov, 1963). It was determined that in healthy people electric activity of the frontal lobe of the brain significantly increased when they concentrated on some objects.

Orienting reflex provides automated tuning to external influences and effects general activation and motivation of a subject (see connection between blocks 4 and 6). Orienting reflex (block 4) also influences afferent synthesis (see horizontal arrow between blocks 4 and 5). This arrow depicts the main information that initiates response or orienting reflex. For example, in aviation, the instrumental information (main stimulus) determines the orienting reflex of a pilot but such a stimulus never exists in isolation. There is always some source of background, additional (situational) information. These situational stimuli have some influence on the response to the main stimulus. Hence, afferent synthesis block 5 also can receive irrelevant stimuli (diagonal arrows). For example, in aviation, this would be some noninstrumental information or irrelevant environmental stimuli (noise, vibration, etc.). Therefore, instrumental information (horizontal arrow) and noninstrumental (diagonal arrows) when combined can influence afferent synthesis. Block 5 (afferent synthesis) is also affected by motivational blocks 6 and 12 (past experience). Changes in the motivational block can in turn influence afferent synthesis. Therefore, the block labeled *afferent syntheses* performs integrative function by combining temporal needs and motivations, mechanisms of memory that represent relevant past experience, effect of irrelevant (non-instrumental) information, and effect of the most significant (instrumental) information. Hence, a person does not react to some isolated stimulus as described in classical conditioning, but reacts to a combination of influences such as main stimulus, irrelevant stimulation, temporal motivational state, needs, and past experience. Afferent synthesis allows determining and selecting with a certain degree of accuracy the main stimulus from an infinite variety of influences of external environment in accordance with a person's temporal need, past experience, and specificity of the situation. Special neurons in the brain that are involved in performing such integrative functions were discovered by Anokhin (1969).

Influences from afferent synthesis promote formation of block 3 (goal-directed set). There are different kinds of sets. A general stable system of sets that a person forms as a result of her/his life experience can become an important personal characteristic (Uznadze, 1961). A goal-directed set formed by instructions and specific situations also exists. This *set* is defined as an internal state of human organism that is close to the concept of a goal, but is not sufficiently conscious or completely unconscious. Such set gives a constant and goal-directed character to an unconscious component of activity. A set can be characterized as readiness or predisposition of a subject to process incoming information she/he is not well aware of, which is provided in a specific situation. A set is, to a large extent, an unconscious regulator of activity that allows activity to retain its goal-directed tendency in constantly changing situations without a subject's awareness of a conscious goal. A goal-directed set manifests itself as a dynamic tendency to complete interruptive goal-directed activity. In our model of self-regulation, these kinds of sets (3) influence meaningful interpretation of information (block 1) and goal (block 2) by interacting with a conscious channel of information processing.

If activity regulation is predominantly unconscious then a *set* has a direct impact on block 10 (making decision about situation or strategy of explorative actions). If a *set* is inadequate, further interpretation of a situation can also be incorrect. As a result, explorative actions associated with function block 10 are not purposeful or goal directed, and become chaotic. For example, a pilot can lose reserve time in emergency situations as a result of such undesirable explorative activity. Explorative actions are often performed as unconscious mental or motor operations.

If block 3 (set) affects block 1 (meaning) with insufficient activation of block 2 (image-goal), meaning in block 1 is primarily nonverbalized. The concept of nonverbalized meaning has been studied by Tikhomirov (1984). Such nonverbalized meaning is sometimes called *situation concept of thinking* (Pushkin, 1978). According to the functional model (Figure 3.1), the nonverbalized meaning is formed under the influence of a *set* and not through conscious explorative actions, but through unconscious explorative operations from block 10. Such meaning helps in creating nonverbalized hypotheses in emergency situations.

Orienting reflex (block 4) and afferent synthesis (block 5) can lead not only to the formation of a goal-directed set, but also to involuntary formation of new goal of activity. Blocks 2 (goal), 4 (orienting reflex), 5 (afferent synthesis), 6 (motivation), and 10 (making decision and program formation) are involved in this process of urgent goal formation.

When we consider function block 2 (goal) in a self-regulative process, it is important to know that objectively presented requirements should be interpreted and accepted by a subject and transformed into an individual goal. Different individuals may have an entirely different understanding of a goal, even if objectively identical requirements or instructions are given. A goal performs an integrative function in self-regulation. It integrates all other function blocks into a holistic self-regulative system. This understanding of goal is totally different from its interpretation in cognitive psychology. In AT, a goal is always associated with some stage of activity (interpretation, acceptance, formation, etc.).

An adequately developed goal-directed set contributes to the rapid goal formation process. If the goal-directed set does not match the situation, it slows down the goal formation process. Sometimes, an inadequate goal-directed set could cause incorrect goal formation.

On the proposed model, block 3 (goal-directed set) is directly related to block 2 (goal).

This demonstrates the possibility of rapid formation of a goal at a sometimes insufficient conscious level. In such a situation, a goal-directed set can simply be transformed into a goal without a conscious information processing channel (function block 1 *meaning* is not involved in this process). Influence of motivational factors (interaction of blocks 6 $\rightarrow$ 2) is a necessary condition for urgent transformation of *afferent synthesis* into a conscious *goal*.

Interaction between these blocks provides fast automatic switching to a new goal. It is a way of forming goals by quick involuntarily shifting of attention to a new stimulus, which is a predominantly involuntary goal formation process. This process of goal formation can be achieved by forming an involuntary goal or activating potential goals that are stored in memory as engrams. The key features of this process are the time limit and emotional-motivational activation. However, the formation of a goal in an emergency situation is mostly performed at a conscious level.

An unconscious set can be transferred into a conscious goal and vice versa (see interaction between blocks 2 and 3). For instance, when a subject drives home and discusses some issue with a passenger, the goal *to drive home* is transformed into an unconscious set. The driver shifts her/his attention and formulates various goals that are associated with the ongoing conversation. At the same time, the goal *to drive home* does not disappear but is transformed into a not entirely conscious set. At certain times, when it is required to take an exit, this set is transferred back into a conscious goal. This demonstrates the ability of the subject to switch from one task to another or from unconscious to conscious level of self-regulation and vice versa.

During unconscious information processing (channel 2), goal (block 2) is not yet activated. Furthermore, block 1, which actualizes mostly nonverbalize information, is responsible for preliminary interpretation of information. At this point, signals from various elements of situation are not yet integrated into a holistic system or dynamic mental model. Their integration and interpretation as a holistic mental model of situation becomes possible only after the activation of blocks image-goal (2) and subjectively relevant task conditions (9). Block 9 is responsible for the formation of a mental model of a situation. We use the term *mental model* when not just verbally logical but also imaginative components are critically important during information processing and these components are integrated into a holistic mental picture of a situation. Function blocks 2, 9, and 13 are outlined in bold to indicate that they include imaginative components and can be considered as mental models.

Conceptual model (block 13) is relatively stable and changes slowly in a time model. It reflects various scenarios of possible situations that are relevant to particular tasks. For example, a pilot's conceptual model in contrast to past experience is more specific to what kind of work activity he/she has to perform according to his/her duties. We can talk about conceptual model of the flight from Tokyo to New York that has been developed during training and is part of past experience.

In contrast to the relatively stable conceptual model (block 13), the dynamic mental model is adequate to a particular situation. Function block 9 (subjectively relevant task conditions) is responsible for the creation of such a model. It provides reflection not only of the current situation, but also anticipation of the near future and infers what took place in the past. Block 9 includes two subblocks, i.e., *operative image* and *situation awareness*. Therefore not only logical or conceptual components but also imaginative components of activity provide a dynamic reflection of reality. The imaginative reflection of a situation can be largely unconscious and is easily forgotten due to difficulty of its verbalization. Imaginative and conceptual subblocks partially overlap. An operator is conscious of the information being processed by an overlapping part of an imaginative subsystem. The nonoverlapping part of the subblock *operative image* to a large degree provides an unconscious dynamic reflection of a situation. Relationship between these two subsystems changes constantly. Conscious and unconscious components of dynamic reflection can to some degree transform into each other (Bedny and Meister, 1999). Hence, a person can mentally manipulate inner images and symbols to create an internal model of events progressing in time. This dynamic reflection can be enriched with additional data from internal and external sources that are necessary for each particular period of time. Imaginative manipulation of a situation can be, to a large extent, unconscious and easily forgotten due to difficulty of its verbalization. SA as a component of function block 9 includes a logical and conceptual subsystem of dynamic reflection in which an operator is very conscious of information processing.

A nonoverlapping part of imaginative reflection can also be considered as containing a preconscious reflection. With the shifting of attention, increased will, and a change in the situation, a preconscious reflection can become conscious, or vice versa—what was conscious earlier can become unconscious. All this can be reflected for an individual as *vague feelings* that can also affect conscious components. Therefore, function block 9 (subjectively relevant task conditions) is involved in the dynamic reflection of a situation and the creation of a dynamic model of a situation. It also provides a constant transformation of information on conscious and unconscious levels according to goals that an operator faces. Usually, subjectively significant elements of a situation to an operator are presented in a dynamic reflection of a situation but they are not always objectively important. This can lead to erroneous orientation in a situation and distortion of the internal model of reality. All data that are contained in the SA subblocks can be verbalized. Data in the subblock *operative image* can be verbalized partially or cannot be verbalized at all because some of its aspects are associated with unconscious processing of information (there are no verbal equivalents) and others can be very quickly forgotten. Therefore a dynamic reflection of a situation cannot be reduced to an interview, a questionnaire, or other verbal methods of study.

In cognitive psychology experts are turning to the study of knowledge representation in memory when analyzing mental models. Mental model is considered as activation of stored schemata or knowledge in our memory. This reduces the formation of mental models to internal associations that operate automatically. However, a mental model is not just a function of memory. It provides adequate interaction with an outside world through cognitive actions and operations. Internal cognitive actions transform idealized objects such as signs, numbers, and icons, etc. in accordance to the goal of these actions. Thanks to this cognitive activity becomes object oriented. Cognitive and behavioral actions connect a subject with the external world. Actions may be perceptual, thinking, imaginative, and mnemonic and so on (see Section 6.1).

This means that in the formation of mental models not only mechanisms of memory are involved, particularly the activation processes, but various other cognitive processes as well.

Actions are conscious units of activity that have a conscious goal. Operations are unconscious units of activity and do not have conscious goals. They are components of actions. Thanks to these actions and operations people actively manipulate with operative units of information extracted from memory and units of information allocated in the external environment.

Operative units of activity (operational units of information) are perceived in an outside environment or extracted from memory as a simultaneous holistic entity. A structure of such units may include elements of information that can be unconscious. Thus, operative units of activity include signals from the external environment or information that is stored in memory (engrams). A person actively manipulates these units in accordance with goals of actions and the goal of an activity in general. Therefore, mental models are constructed by a subject in the process of a goal-directed activity.

An individual may create a mental model of reality by performing a sequence of mental actions or operations. Mental operations are associated with unconscious aspects of creating such a model. In situations when only unconscious mental operations are involved in the creation of a model, this process is often perceived as being simultaneous. In more complicated situations, direct recognition and interpretation may be impossible, and Gnostic activity may involve a system of explorative conscious actions. All cognitive processes are involved in the creation of a mental model of reality. However, thinking and specifically operative thinking plays a leading role in this process. From this analysis we can see that a mental model includes imaginative and verbal components and conscious and unconscious elements.

Without emotional-motivational mechanisms, a goal achievement process cannot take place. A mental model is a result of a complex self-regulative process. Cognitive psychology does not consider the concept of cognitive actions as goal-directed elements of activity or goal as a conscious desired result of activity connected with motives, and a mental model is reduced to a conscious verbalized process. Emotional-motivational mechanisms of activity regulation, thinking processes, or the concept of a Gnostic dynamic process are not considered either in cognitive psychology when discussing the development of a mental model. The fact that a mental model is a result of a complex self-regulative process is also not discussed in cognitive psychology. All this is covered further in this section.

Let us consider an example where conscious and unconscious levels of self-regulation interact with each other. This is a real-life situation of a person driving his car. On a mountain road his car's brakes failed. It was impossible to stop to avoid an oncoming car. He saw a pile of gravel on the roadside and decided to drive his car into this pile. This helped to stop his care and avoid collision with the oncoming car. The driver had to change his goal from driving to stopping the car. In this case, it is a quick formation of a new goal that is performed not only on a conscious but also on an unconscious level. In this case, the orienting reflex is activated, which increases motivation. The interaction of conscious channel 1 (blocks 1 and 2) and unconscious channel 2 (blocks 4 and 6) creates the vector *motive* → *goal*. This in turn affects the decision-making block (see blocks of 4, 6, 10) and a dynamic mental model is developed. Thus, interaction of conscious and unconscious channels of information processing increases the speed of the formation of a goal and a dynamic mental model.

Let us consider some aspects of the interaction of various function blocks in a self-regulative process in a more detailed manner. This will allow us to use the model of self-regulation in task analysis more effectively. Conscious and unconscious processing of information involves not only cognitive but also emotional-motivational mechanisms. Therefore, the interaction of blocks 6 and 2 becomes especially important at this stage. This interaction facilitates the formation of the vector *motives* → *goal* and activity becomes conscious and goal directed. Thus SA and dynamic mental model in general cannot be studied without analyzing emotional-motivational factors.

When a main channel is conscious, information goes directly from channel 1 and through block 1 to block 2. Goal (2) influences blocks 13 and 10 and activates conscious explorative actions and conscious decisions. Under the influence of blocks 10 and 6, function block 9 (subjectively relevant task conditions) is activated. As a result, a dynamic model of the situation is developed at a conscious level of activity regulation. In contrast to cognitive psychology where a dynamic mental model is a result of purely cognitive functions, in AT a dynamic mental model of a situation is developed during the interaction of cognitive mechanisms of blocks 10 and 9 and motivational mechanisms of block 6.

It is necessary to considered how a dynamic mental model of the situation interacts with goal. Without a goal there is no goal-directed activity and a dynamic mental model cannot be developed. Integration and interpretation of information and the formation of a dynamic mental model of a situation become possible only after the formation of the goal of the activity (block 2). In AT, goal determines the specifics of selection of information and therefore is a key factor in developing a dynamic mental model, including SA. This is critically important for further conscious orientation in a situation and specifically for SA.

Despite the fact that externally presented situation can be the same, a dynamic mental model can be developed in different ways if the goal of the orienting activity changes. Thus at the first stage, the subject accepts or formulates the goal of activity. When the formation of the goal is completed, an opportunity to the development of a dynamic mental model of the situation (block 9) becomes possible. The dynamic mental model is developed in accordance with the formulated goal of orienting activity. Goal becomes a part of the dynamic mental model of the situation.

Motivational mechanisms of activity are included in block 6 and are considered to be inducing components of activity. This block influences conscious and unconscious aspects of information processing. Without motivation, there can be no *afferent synthesis* or *set*. This block is involved in the creation of the vector *motives* → *goal* and the formation of a dynamic model of a situation (block 9).

Block 6 is tightly connected with block 7 (assessment of sense of task's significance) and block 8 (assessment of task's difficulty). Block 7 is responsible for the evaluation of the significance of a situation or its elements (emotional-evaluative mechanism). The relationship of blocks 7 and 6 is designated by a bold arrow. Block 6 (motivation) reflects inducing components of activity. Block *sense* (emotionally evaluative mechanism) influences logical and meaningful interpretation of a situation. The more significant a situation is for a subject, the higher is the level of motivation. Block 6 is involved in a goal formation process and it switches attention from one feature of a situation to another in a dynamic model. In other words, a factor of significance is involved in extracting adequate features of a dynamic model of a situation (see interaction between blocks 7, 6, and 9). Function block 11 is involved in the formation of subjective criteria of success and the evaluation of the activity result. Here it is worth noting that objective requirements to a result of activity and subjective criteria of success are not the same. Moreover, the goal of an activity and criteria of evaluation are often not the same.

Emotional-motivational aspects of orienting activity regulation are critically important for the reflection of a situation and for SA in particular. Hence, these aspects of self-regulation should be considered in more detail. The specificity of human information processing depends not only on cognitive, but also on emotional-motivational mechanisms of activity. This factor is not taken into account in the models of self-regulation created outside of SSAT and in domain which is known as situation awareness. Hence, we consider these aspects of self-regulation briefly.

When studying motivation, such functional mechanisms as *goal* (block 2), *assessment of task difficulty* (block 8), *assessment of task sense* (*significance*) (block 7), *formation of level of motivation* (block 6), and *decision making and program formation* (block 10) are particularly important. Motivation is considered as an energetic process of dynamic interaction with these blocks. Interaction of these function blocks demonstrates close relationship between informational and energetic components of the activity.

Let us briefly consider function block *difficulty* (block 8) and its interaction with the other blocks.

There are situations when the main function of an operator's work is the ongoing analysis of various situations and periodical intervention in the control process when necessary. Control parameters are dynamic. At any given point of time, they present to an operator multiple interrelated data that should be perceived, compared, evaluated, etc.

Analysis of such situations, even in the case where there is no need to intervene in an ongoing control process, should be considered as a task that is performed by an operator in accordance with the goal of observation formulated by him/her. In other situations, after analyzing all the data, an operator might conclude that there is a need to intervene in the control process. Tasks involved in the continuous observation of a control process can be very complex in some cases. The more complex such tasks are, the more cognitive efforts they require. Tasks complexity can be considered from functional and morphological analyses perspectives (Bedny and Karwowski, 2007; Bedny and Meister, 1997). Here we consider task complexity, and associated with it task difficulty, from a functional analysis perspective.

Functional analysis distinguishes between objective complexity of a task and subjective evaluation of its difficulty. A subject might evaluate the same task as more or less difficult depending on its complexity, her/his past experience, individual differences, and even temporal state. The higher the task complexity, the higher is the probability that this task will be evaluated as difficult. Cognitive task demands during task performance depend on the task's complexity. A subject experiences not the complexity of a task but its difficulty. It is important to find out how a subject evaluates task difficulty. An individual might under- or overestimate an objective complexity of a task. For example, a subject can overestimate a task difficulty and task can be rejected in spite of the fact that objectively the subject would be able to perform it. Moreover, overestimation of task difficulty produces emotional stress and even if a task is accepted, the quality of performance may be affected. On the other hand, if a subject underestimates task difficulty, he/she can fail to perform it. Such psychological concepts as subject's evaluation of her/his own abilities in comparison to task requirements, self-efficacy (Bandura, 1997), self-esteem, etc., are useful for analyzing the functional block *difficulty*, which depicts a cognitive mechanism of self-regulation that influences motivation. The function block *difficulty* is task specific. A subject can estimate a task difficulty correctly, overestimate it, or underestimate it. Reasons for overestimation or underestimation of a task difficulty can depend on stable personal features or on a purely task specific situation. The evaluation of task difficulty can also be a function of past experience. This cognitive mechanism (function block 8) is a critical one for the motivational process. Block *difficulty* interacts with a number of other blocks in the self-regulation process. In this work, we will consider the interaction of block *difficulty* with the block *sense* (understood here and further as a person's evaluation of subjective significance of a task or situation). Hence, the block *sense* is tightly connected with the motivational block (this connection between block 7 and 6 is designated by a bold line). These two blocks include emotional-evaluative and inducing mechanisms of activity. The individual sense creates a predilection of human consciousness (Leont'ev, 1978). The function block *sense* (block 7) predetermines the significance of a task and its elements, the situation, and a subjective value of obtaining a desired result of the task elements and of the task as a whole that provides a sense of achievement. Blocks *difficulty* and *sense* have a complex relationship, and the interaction between them influences the process of motivation. The interaction of these blocks can explain motivation in a totally different light in comparison with existing theories. Let us consider the construct of self-efficacy developed by Bandura (1997), according to whom the stronger the belief in self-efficacy is, the stronger a person will pursue the desired result. He suggests that people with high personal efficacy set more difficult goals and show greater persistence in their pursuit, while people with low efficacy set lower goals, often resulting in a negative influence on motivation that in turn increases the probability of abandoning a goal in the face of adversity. It implies that all motivational manipulations are effected through self-efficacy. From self-regulation or functional activity analysis point of view, if a person evaluates a goal as a very difficult one due to her/his low self-efficacy, the resulting negative influence on motivation (inducing component of motivation) increases the probability of a goal being avoided or abandoned. On the other hand, if a particular goal of task is significant or has a high level of positive subjective value, those with low self-efficacy can nevertheless be motivated to strive to achieve the goal.

The basic postulate of goal-setting theory is that difficult goals, if accepted, lead to greater job performance than easier goals do (Locke and Latham, 1990). We would say that goal-setting theory merely substitutes rather complex motivational issues with the simple statement *if the goal has been accepted*. The model of self-regulation of orienting activity shows that there is a complex relationship between difficulty and motivation. Increasing difficulty of a task does not always lead to an increase in the level of motivation, as it is stated in the goal-setting theory suggested by Lee et al. (1989). However, the level of motivation depends on the complex relationship between the function blocks *assessment of task difficulty* and *assessment of the sense of task* (significance). If a task is evaluated as a highly difficult one and its significance as very low (attainment of a task's goal is not subjectively important) a subject is not motivated to perform a task. In such situations, a subject does not have any reason to spend a lot of her/his efforts on the task. On the other hand, if a task is evaluated as very difficult and at the same time very significant for a subject, he/she is motivated to complete it even with a risk of failure. There can be other scenarios. For example, difficulty and significance of a task are low. In this case, work is very boring and a subject has a low level of motivation.

Sense (significance) is one of the function blocks of self-regulation that interacts with the goal of a task. The goal of a task can have not only positive value for a subject, but can also include some negative aspects. If the goal of a task has only positive value for the subject then the function block *sense* has a homogeneous structure and has only positive significance. However, if the goal of task includes attributes that have positive and negative personal values, then the function block *sense* has a heterogeneous structure and has positive and negative significance. The proportion of these two types of significance determines the integrative character of the evaluation of task significance and plays an important role in the formation of motivation.

There are other functional mechanisms which also influence motivation. Motivation, among other things, depends on a subjective criterion of successful result (see function block 11 *criteria of evaluation*). This subjective standard can deviate from objective requirements, so the satisfaction of goal attainment depends on this criterion.

Function block 11 *criteria of evaluation* has two subblocks. One is called *subjective standard of successful result* and the other is *subjective standard of admissible deviation*. A subjective standard of successful result can significantly deviate from the objective standard presented through instructions. This standard can be modified during performance. Modification can be done through feedback from function block 10 *making a decision and program formation*. If a subject achieves a required goal but her/his level of aspiration exceeds a goal, a subject would not be satisfied with an obtained result. The concept of a subjective standard of successful result is also deemed important in social learning theory (Bandura, 1989). However, in this theory, there is no clear understanding of what is the difference between a goal and a subjective standard of success. Subjective standard of success can deviate from a goal, particularly at the final stage of task performance. A subjectively accepted goal can be used as a subjective standard of success but a goal itself might not contain enough information to evaluate the result of a task performance. This standard has a dynamic relationship with a goal and past experience (Bedny and Meister, 1997) and can be modified during goal acceptance and task performance. For instance, a person might be tired during the second part of the shift but does not want to decrease productivity. As a result, the qualitative criteria of success could be lowered and quantitative criteria could increase.

In orienting activity, the criterion of success is not related to the executive components of activity. It is mechanism that is involved in creation of adequate mental reflection of the situation. An individual can develop different interpretations of the same situation. The decision making process is a part of the formation of criteria of success, which are used for the final acceptance of the mental representation of a situation or a way of its interpretation. Based on data from block 10, a subject can modify criteria that are used to interpret a situation.

Subblock *subjective standard of admissible deviation* is also an important evaluative mechanism of orienting activity. Criteria of evaluation may vary within certain limits. Such variations are determined by the subjective standard of admissible deviation and can be seen as a range of tolerance for the subjective standard of successful result.

So, here an objectively established goal deviates from subjective criteria of success.

Functional analysis of activity describes human cognition during work more comprehensively than cognitive task analysis. Functional models of activity allow a more detailed analysis of performance strategies. Explanatory and predictive features of these models can be explained by the fact that goal, motive, meaning, sense, etc., are considered as parts of a system of interconnected mechanisms that have certain functions in the self-regulation model of activity, which allows describing activity as a self-regulative system. While cognitive psychology assumes the existence of a fixed goal, the functional model emphasizes the process of goal acceptance or formulation and its specific functions in the formation of strategies of activity performance. Further, the model of self-regulation explains why an operator can neglect safety requirements when she/he has high aspirations of reaching a goal based on her/his subjective criteria of success. A subjective standard of success can also influence precision of an operator's performance. As an example, the accuracy with which a pilot can read aviation instruments often depends more on the significance of an instrument than on its visual features. It has been discovered that depending on the goal created by a pilot, the same display can perform different functions and the same apparatus can be used for the evaluation of flight parameters in one situation and for the evaluation of functioning of another apparatus in another situation. Further, the concept of situation awareness (SA) that has been developed in cognitive psychology (Endsley, 2000) is understood in SSAT as a functional mechanism of activity regulation. Finally, we want to stress that ergonomics and applied psychology do not study separate cognitive processes but work as a whole, where cognitive processes are integrated into a system. Functional analysis of activity helps to study cognition as a system where all cognitive processes are considered in their entirety.

SSAT considers situation awareness as a functional subblock, which, together with an operative image, is involved in constructing a dynamic mental model of a situation. SA integrates various cognitive processes with operative thinking and mechanisms of working memory playing a leading role in this mechanism.

Operative thinking performs a complex analytical-synthetic activity based on which various versions of a mental model of a situation and forecasting of its development in the near future are constructed. At the final stage, a subject selects the most appropriate mental model. Even if only one version of the mental model of a situation is constructed, a decision mechanism is important. Such a decision would be called sanctioning and may include an act of will that is related to a decision to authorize an acceptance of developing a mental model.

The function block 9 (subjectively relevant task conditions) is a mechanism that plays an important role in the prediction of future events. For example, this mechanism is important in such issues as anticipation, extrapolation, and prediction of possible errors. In the homeostatic model of self-regulation, the possibility to predict future events is not discussed at all. In engineering psychology, these problems are discussed from a narrow perspective when specialists consider specific examples that demonstrate the possibility of a person to predict some events. In some situations, some interesting data in this area were obtained. However, this problem should be discussed in a broader context of self-regulation of activity. The possibility of prediction depends on the formation of appropriate strategies of task performance in specific circumstances.

SA functions under the influence of an emotional-motivational mechanism that changes sensitivity to elements of an external stimulation and information retrieved from memory.

Emotional-motivational processes facilitate reconfiguration of cognitive processes and their tuning to an external situation. Functions of attention and perception change and a subject begins to notice substantial and, in some cases, subtle features of an external situation. SA is one of the most important mechanisms of orienting activity and its results. However, it is formed by human activity. Formation of a conceptual model, goal of orienting activity and its dynamic model (subjectively relevant task conditions) are the main results of orienting activity. Dynamic reflection of a situation is the main purpose of orienting activity. The model of orienting activity discussed demonstrates that cognition is not a linear sequence of the information processing steps but rather a self-regulative system. Moreover, we have studied not just separate cognitive processes but also their specific integration at the various stages of activity regulation.

In each individual case, depending on the nature of the task, adequate function blocks should be utilized for a specific analysis.

The simplest experimental method of studying self-regulation could be limited to using just function block *"goal"* and the next step of analysis could involve a variation of conditions of task performance and following analysis of the possible strategies for achieving the goal of the task.

3.3 General Model of Activity Self-Regulation

Orienting activity provides conscious and unconscious reflection of a situation in accordance with developed goal of activity. We can say that orienting activity performs a diagnostic function about a situation and promotes hypothesis about a current and future state of a situation. However, execution of various tasks includes not only orientation in a situation but also its interpretation or diagnosis. Most often, there is a necessity to identify the most efficient way of transforming a situation in existing conditions to achieve a particular goal. For practical purposes, it is often sufficient to use the self-regulative model of orienting activity. If the transformation of a situation is sufficiently complex, it is necessary to use the general model of activity self-regulation.

Endsley and Jones (2012, p. 11) wrote, "situation awareness is the engine that drives the train for decision making and performance in complex dynamic systems." They utilize the following schema to illustrate this idea:

Situation awareness → decision making → performance

Based on this, we can present another schema in a simplified manner:

Reflection of situation → final decision about situation → decision about performance → performance.

The first stage "reflection of situation" includes conscious and unconscious components. Reflection also includes goal and stable or dynamic mental models of a situation or both. Several competing mental models about the same situation can be developed and therefore the second step includes the final decision making about reflection of a situation. Only after that stage a subject can make a decision about performance and then a subject starts performance to achieve a desired goal. Of course this is a very simplified schema but it demonstrates the difference between SSAT approach and the approach that derives from the concept of SA.

A general model of self-regulation includes all functional blocks that exist in the already discussed model of self-regulation of orienting activity. However, this model additionally includes functional blocks that are associated with executive components of activity aimed at the transformation of a situation and achievement of the goal of a task.

Figure 3.2 represents the general model of self-regulation of activity.

We will analyze only the new blocks of self-regulation that are added in this model.

In the general model of self-regulation, we have tried to keep the numbering of blocks the same as in the previous model, so blocks that are found in both models and perform the same functions have the same numbers. Therefore, block *making a decision about corrections* is number 11, and the next block *program of performance* is number 14.

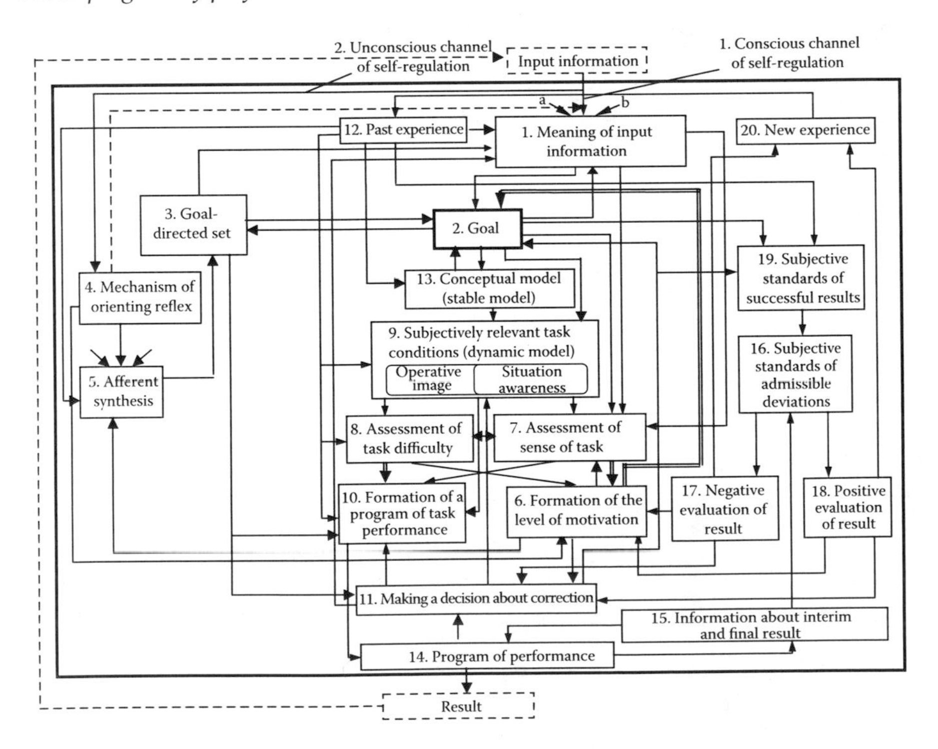

FIGURE 3.2
General model of activity self-regulation.

The model of orienting activity has 13 blocks. The general model of self-regulation includes 20 function blocks (Figure 3.2). The need to include additional blocks is due to the fact that here we are talking about activity that includes decision about performance and the performance itself. Let us first look at these new function blocks.

In this model, we describe stages that include the formation of a program of task performance (block 10), making a decision about correction (block 11), and program of performance (block 14) in more detail. The evaluative stage of activity that includes function blocks 15–19 also deserves attention. In the self-regulative model of orienting activity, for simplification, we did not consider function block *new experience*. Here it is included as block 20.

The interaction between past experience (function block 12) and new input information is an important stage of meaningful interpretation of input information. When input information correspond to to the past experience, it is easy for a subject to interpret the meaning of the input information. At this stage, objectively presented information is transformed into a subjective input for the subject. When input information does not match past experience or is very complex, then variation in meaningful interpretation of input information or even its misinterpretation is possible. Past experience includes not only cognitive components but also data about emotional-motivational components and experience about subjective evaluation of task difficulty. However, more precise interpretation of meaning of input information is possible only after the formation of the goal of a task.

Goal (block 2) and motivation (block 6) in this model, similar to the previous model, create the vector *motive* → *goal*, which gives a goal-directed character to the self-regulative process. This vector is depicted by a double line. A goal performs integrative functions in the process of self-regulation and integrates all mechanisms of self-regulation into a systemic entity. A goal is a system-formation factor in the process of self-regulation because it is a mechanism that brings together all other mechanisms into a coherent self-regulation system. At this stage, the goal of task becomes an integrative mechanism of activity regulation. Goal also performs important functions in the selection of information. Conceptual model or stable model (block 13) and subjectively relevant task conditions or dynamic model (block 9) were considered earlier when we described the self-regulative model of orienting activity.

We also briefly recall functions of some blocks that have been considered in orienting activity because they are also mechanisms of the general model of self-regulation. The other mechanism that is important for self-regulation is assessment of task difficulty (block 8). The more objectively complex a task is, the more is the probability that a subject will evaluate it as subjectively difficult for him/her. This is an important aspect of activity regulation. An individual may under- or overestimate the objective complexity of a task. If a performer overestimates complexity of a task, she/he can reject it or lower the subjective standard of successful result (block 19, see Figure 3.2).

FIGURE 3.3
(a) Subjective evaluation of difficulty (very high) and (b) subjective evaluation of significance (very high).

Function blocks influence each other. For example, block 6 (formation of a level of motivation) also depends on interactions between blocks 8 (assessment of task difficulty) and block 7 (assessment of task sense) or the subjective significance of a task for a subject. Due to the complex relationship between function blocks 7 and 8, no simple conclusions about the motivation processes can be made. The relationship between assessment of the sense or significance of a task for a subject (block 7) and assessment of task difficulty (block 8) is specifically important for block 6 (formation of level of motivation). This complex, dynamic relationship between these blocks can be better understood if we consider Figure 3.3.

In Figure 3.3, there are two vertical bars with a pointer next to each. The higher the position of the pointer the higher is the difficulty or significance of the task. In our example, both pointers occupy a high position meaning that a subject evaluates a task's difficulty or its subjective significance as very high. We can assume that in such situations a subject will have a high enough motivation to perform the task. But suppose that a subject evaluates this task significance for him/her as low and the difficulty remained high, as shown in Figure 3.4.

In this case, the subject's motivation is very low. The subject may even reject performing the task. These figures depict positions of these two pointers that can change in various ways, reflecting the wide range of changes in motivation.

Let us briefly consider the stages of motivation that derive from our concept of self-regulation (Bedny and Karwowski, 2007). According to the model of self-regulation, there are five stages of motivation. The first one involves

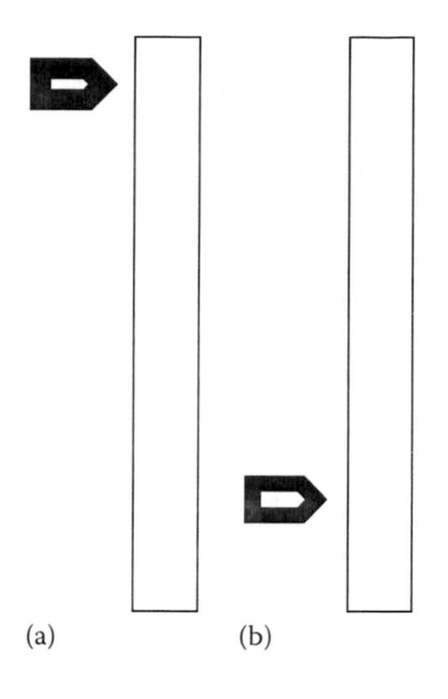

FIGURE 3.4
(a) Subjective evaluation of difficulty (very high) and (b) subjective evaluation of significance (very low).

preconscious motivational stage. The main function blocks that are involved in motivational regulation of activity are blocks 3 (set), 4 (mechanism of orienting reflex), 5 (afferent synthesis) and 6 (formation of the level of motivation). At the unconscious stage, information about external situation interacts with needs and a goal-directed set. This emotional-motivational state precedes a meaningful interpretation of the situation and formation of a conscious goal and is triggered by external stimuli that are not sufficiently conscious. Input stimuli activate related mechanisms of nervous system and motivational processes through the orienting reflex. Afferent synthesis integrates this motivational state with information from main stimuli, information from environmental stimuli that is relevant to the situation, and information from memory, and this is the basis for forming a goal-directed activity. As a result, the motivational tendency, which is not well understood by the subject, makes activity to be goal directed. This preconscious motivational tendency can trigger a conscious goal-formation process or an executive stage of activity.

The second stage of motivation is involved in the formation or acceptance of a conscious goal (*goal related motivational stage*). This stage of motivation has two possible ways of development. One way is transformation of set into conscious goal (see relationship between blocks 3 and 2). Set can sustain unconscious motivational tendency over time. In emergency situations, the interaction of conscious and unconscious channels of information processing can increase the speed of the goal formation process, described in the previous section. Specifics of relationship between these two channels determine precision of the goal formation process.

The third stage is called the *task-evaluative stage of motivation*. The main blocks in this stage are block 7 (assessment of sense of task), block 8 (assessment of task difficulty), and block 6 (formation of the level of motivation). We have also discussed the possible ways of their interaction and influence on motivation in Section 3.2. A produced motivational state influences blocks 9 (dynamic model) or block 13 (stable model) through block 11 (making a decision about corrections).

The next stage is called the *process-related stage of motivation*. Blocks 7, 8, and 6 interact with each other in the same manner as described earlier and produce a motivational effect on blocks 10 (formation of a program of task performance), 11 (making a decision about corrections), and 14 (program performance).

The fifth stage of motivation is the *result-evaluative stage of motivation*. At this motivational, stage blocks 17 (negative evaluation of result) and block 18 (positive evaluation of result) and their interaction with motivational block 12 influence the motivational process.

An analysis of the motivational stages of activity regulation demonstrates that motivation is dynamic and can be changed during task performance or during work performance in general. All motivational stages are interconnected and can be in agreement or in conflict. For example, there can be positive, goal-related and negative, process-related stages of motivation. These stages of self-regulation can even be in conflict. For example, a subject can accept a given goal positively but work performance can produce a feeling of monotony or a negative motivational state. If the negative, process-related motivational state exceeds a value of the positive, goal-related motivational state, the subject can stop working or drastically reduce the quality of performance directing his/her attention to events not related to work. There are various ways to overcome this negative relationship between stages of motivation. For example, if there is a feeling of monotony, it is useful to provide information about ongoing productivity. Information about approaching the goal can reduce the feeling of monotony. Other methods involve increasing a goal's significance and precise formulation of a goal. Precise formulation of the temporal and quantitative characteristics of a goal helps to overcome negative factors of process-related motivation. The following example demonstrates relationship between goal-related motivation and result-related motivation. If the goal of a task is not significant for a subject, he/she can be indifferent to positive and negative evaluations of task performance.

Numerous approaches have been suggested in psychology literature to assess work motivation with the existence of separate, unrelated mechanisms. In contrast to these theoretical approaches, SSAT considers motivation as a dynamic component of the process of self-regulation of activity. Motivation is determined by the interaction of different mechanisms of self-regulation. At various stages of self-regulation, there are qualitatively distinct stages of motivation. The analysis of interaction between different functional blocks of self-regulation demonstrates close unity of cognitive and motivational components of activity. Motivation affects information processing just as well as cognitive processes affect motivation.

In the general model of self-regulation, functional blocks 11 (making a decision about corrections), 10 (formation of a program of a task performance), and 14 (program of performance) are considered separately. In the model of self-regulation of orienting activity, these mechanisms are integrated (compare Figures 3.1 and 3.2). In this regard, we consider blocks 10, 11, and 14 as well as blocks 15–20 that are associated with an evaluative stage of activity self-regulation. Any goal-directed activity includes a program of execution.

Block 10 is responsible for the development of a program of performance, which has the purpose to achieve a formulated or accepted goal of task. We want to stress the fact that the program of performance includes components that are responsible for the execution of not only motor but also cognitive components of activity. The program of activity is formed before the program of performance (block 10 is formed before block 14, see Figure 3.2), which means that the main part of the program of performance program is formed earlier than execution of motor components of activity. If activity is quite complex and variable, then the program of performance can be developed in advance (block 10) only in very general terms. A program of performance is more clearly formed during its implementation. The most important elements of a program of performance are formed before its implementation (see relationship between block 10 and 14). A program of performance includes both conscious and unconscious components. In certain cases, it may be completely unconscious. This is especially true for the regulation of well-established motor actions. Implementation of a motor action program can be represented as a system of nerve impulses. Block 14 is responsible for this stage of activity regulation. Thus, in this model, we distinguish two stages of activity regulation. The first stage is involved in the formation of the program of task performance (block 10) and the second one is involved in the program of performance (block 14).

An evaluative subsystem of activity compares information about external environment and *performance program* information with information about prior performance program. If necessary, a program of performance can be adjusted. It can be modified during performance and evaluation of the ongoing or final result. A program of performance can include hierarchically organized subprograms that are responsible for the performance of some components of an activity. Program formation stage (block 10) can be complex and unfold in time or relatively simple and subjectively proceed instantly. Block 10 (formation of program of performance) depends on other function blocks: it depends on block 12 (past experience) and it can be modified under the influence of block 7 (assessment of sense of task). Indeed, if a task is significant for a subject then she/he can consciously or even intuitively develop a cautious strategy of performance. Block 8 (assessment of task difficulty) also influences block 10.

Block 9 (dynamic model) is of particular importance for the formation of a program of performance (10) because it is shaped and adapted depending on how a subject forms in advance a mental representation of a formulated task. In other words, the formation of a program of performance is determined by the subblocks *situation awareness* and *operative image* that are involved in the conscious and unconscious aspects of the reflection of a situation. Moreover, SA plays a leading role in this process. Block 10 (formation of a program of performance) according to our model includes a decision making process that is involved in sanctioning the program formation process. This is an important stage in the transition from the analysis of a situation or orienting stage of activity to its execution.

There is no independent orienting activity in the general model of activity self-regulation but rather an orienting stage of activity that precedes the executive stage of activity. Executive stage of activity starts when block 10 is activated. Blocks 7 and 8 can be considered as supplementary factors that might influence the program formation process.

It should be noted that the presence of the same dynamic situation or mental task representation (block 9) may lead to different ways of forming a program (block 10) due to various past experiences (block 12) of the subjects, evaluation of difficulty of a task (block 8), and subjective significance of a task (block 7).

The next stage involves block 14 (performance program), which is an important stage in the executive part of activity. It includes cognitive and motor actions. This means that activity can, to a significant degree, be performed in an internal mental plane. In fact, very complex internal cognitive activities can be completed by simple external motor actions. There can be a rigid and dynamic program of performance that is used for stereotyped or automatic kind of activity. A dynamic program is plastic and chainable. Thanks to feedback from block 11 to block 10 a program of performance can be corrected.

Block 11 can also correct block 9, which includes operative image and situation awareness (see interaction between blocks 11 and 9).

The last stage of self-regulation is the evaluative stage that involves function blocks 15 through 19. This stage of self-regulation enables correction of performance by using external or internal feedback, including immediate and mental feedback. A subjective standard of successful results is critically important at this stage of self-regulation (see block 19). It has dynamic relationship with goal (block 2) and past experience (block 12). This standard can significantly deviate from an objective presented by instruction standard. It can be modified during a task performance due to feedback from block 11 and 2. A goal accepted or formulated by a subject does not always determine an exact result of activity due to various factors. For example, a goal often does not have all necessary information about required results of an activity. Moreover, a mental representation of the desired result can be often developed only during performance process. Different subjects can formulate different mental representation of a desired result when they have the same goal. Hence, *subjective standard of successful result* (block 19) is another important mechanism of activity regulation. For example, a precise goal *to react with a particular speed* accepted by a subject can be a source of developing possible drivers and an acceptable desired result. This block is important because the goal of an activity often does not contain sufficient information for the evaluation of result of activity. Subjective standard of successful result can be formed very quickly or its formation can emerge as a complex process of evaluation of interim and final results of activity (Bedny, 1987; Konopkin et al., 1983). Social aspects are also important in the formation of such standards that might depend on a process of social comparison (Bandura, 1982; Bedny, 1981).

A subjective standard of admissible deviation is another evaluative mechanism of activity regulation. It has been discovered that often subjects can develop a range of deviation from a standard. Subjective standard of a successful result is further evaluated by a subject positively or negatively based on an accepted standard of admissible deviation (blocks 17 and 18). This demonstrates that subjects not simply use negative feedback for correction of performance but positive evaluation of an obtained result can lead to further improvement of performance through motivational block 6. Self-regulative mechanisms are not developed simultaneously. Some of the functional mechanisms might be developed first, others later on. Based on the improvement of the self-regulative process, a subject acquires new strategies of task performance and new experience in general (block 20). This block in turn can interact with past experience. The considered model is also the basis for the self-regulative concept of learning. From the standpoint of self-regulation, learning is the transition from well-known strategies to new strategies of performance. The more difficult the task is for a learner, the more intermittent are the strategies utilized by her/him. Learning activity is constricted through a sequence of strategies (Bedny, 1987; Bedny et al., 2012).

Block 15 (information about interim and final result) provides a comparison of ongoing and final result with established standard of performance. We do not discuss all blocks in detail. Additional data on this subject can be obtained in Bedny and Karwowski (2007).

These models of self-regulation are complex. However, the practical application of models does not involve an analysis of all the function blocks and their relationships. It is necessary to consider briefly some recommendations concerning the application of the models in task analysis. In addition in the following chapters, we will look at examples of practical application of these models in task analysis.

Application of models depicted in Figures 3.1 and 3.2 can be understood if we consider each function block as a window that can be opened in order to analyze the same activity during task performance. For example, a researcher can open a window called *Goal* and at this stage begin analysis of the goal's specifications for a task or separate actions and the position of a goal in a hierarchy of goals. Then attention can be paid to such aspects of activity as goal perception, goal interpretation, goal formation, goal acceptance, relationship between verbal and imaginative components of a goal, subject's conscious awareness of a goal, etc. If a goal is imposed by instruction, one can examine the extent to which such imposed goal is accepted and the effect of possible versions of goal interpretation on task performance. A goal can be modified during task performance and therefore it is important how such modifications can influence selected strategies of task performance. Main influences of a goal on other function blocks, specifics of formulation of subgoals, and final goal of task can be studied as well as to what extent a subjectively accepted or developed goal corresponds to objectively presented requirements of a task. The subjective significance of a goal to a subject, correspondence of subjective significance of a goal to an objective value of a goal, and the relationship between goal and motivation can be studied.

If the block 13 (conceptual model) window is opened, at this stage of analysis attention would be paid to various scenarios of possible duties and the most important and most probable tasks that can be performed by an operator. For example, a conceptual model that is created by a pilot before the flight includes basic scenarios of a possible flight mission. It also includes knowledge about possible tasks that should be performed, a final goal of the mission, understanding of possible constrains and difficulties, specificity of team performance, assumptions about changes in flight mission, and environmental changes. Procedural knowledge dominates for a conceptual model of a flight rather than a declarative one. Hence a conceptual model is more specific in comparison with past experience.

In the next step, the analysis window *subjectively relevant task conditions* (function block 9) might be opened. At this stage, one would study aspects or stages of activity that are responsible for the creation of a dynamic mental model of a situation. Relationship between imaginative and verbally logical components in dynamic reflection of a situation should be considered. Interrelationship between elements of the situation, interpretation of situation, prediction of near future events is analyzed at this stage of analysis. A possibility of transformation of unconscious components into conscious ones and vice versa should be analyzed. Feedforward and feedback connections between this block and other blocks are also important. Based on the suggested model we can also consider how such blocks as *goal* influence the formation of *subjectively relevant task conditions* (block 9). If the block 6 (formation level of motivation) window is opened, attention should be paid to how motivation influences the goal formation process (block 2, goal) or how motivation influences block 11, etc. During task analysis, a specialist can consider only those function blocks that are most important for a particular task. Relationships between such blocks should also be considered during task analysis.

In contrast to goal setting theory, expectancy theory, and other theories, SSAT-based models of self-regulation describe psychological data in a systemic way and present some mechanisms such as goal, situation awareness, emotions, motivation, relation between conscious and unconscious, and executive aspects of activity or behavior in a significantly different manner. Such consideration is possible only due to systemic analysis of these mechanisms and analysis of their interactions. Moreover, self-regulation of human activity cannot be reduced to the consideration of such mechanistic terminology as *goal standard, comparator, input functions, output functions,* and *disturbance.*

All models of self-regulation outside of SSAT, the same as human information processing models in cognitive psychology, suggest that feedback is possible only after receiving information, evaluating it and executing motor responses. In other words, a subject can utilize only the final feedback after motor responses. However, such a method of self-regulation is often not acceptable. A subject should anticipate what might happen and use immediate or current feedback during activity performance. Moreover, a subject can perform activity in the mental plane. This means that feedback is not only a result of motor responses. A subject can perform internal mental actions and therefore utilize internal mental feedback. Such feedback can prevent errors rather than correct already performed errors.

Next, one needs to take into account that an external situation is not stable and changes over time, even in the absence of a subject's external motor actions, which means that feedforward and feedback connections are dynamic and their interrelationship should be constantly consistent and coordinated over time. The only block in this model that can influence the external situation is block 14. Therefore, feedback from block 14 to an external situation is not always utilized, meaning that feedback can be mental when a subject evaluates consequences of his/her actions mentally. A subject can use external feedback when he/she performs external motor actions that change the situation. In orienting activity external exploratory actions are aimed at finding out the cause–effect relationships between elements of a situation. However, in most cases, explorative actions are carried out mentally and the externally given situation is not changed. Self-regulation is performed in the internal mental plane. Because such feedback is not used constantly, it is depicted by an arrow from block 14, and from a result by a dashed line. In the model of self-regulation of orienting activity (see Figures 3.1 and 3.2), such feedback is depicted by a dashed arrow that goes from block 10 to the external informational model.

Individuals with different features of personality can perform the same work with equal efficiency by utilizing their individual styles of activity that might be regarded as strategies of activity derived from such individual features. These strategies can be adequate in one situation and not adequate in another situation. Therefore individual feature of personality can play a positive role in one situation and a negative role in another (Hogan et al., 1992). Goal-directed models of activity self-regulation are useful in analyzing the individual style of an activity.

When describing the model, we focused on the use of the proposed self-regulatory models in applied research. Therefore, we did not overload the description of the models with accompanying experimental studies so that it would be easier to understand how to use them in task analysis. In this book, we introduced some additions and clarifications to the considered models.

Analysis of the process of self-regulation of human activity demonstrates that human beings when striving to accomplish a task, consciously or unconsciously seek the maximum probability to achieve the goal of such task with the least time and effort. To achieve this, human beings form a conscious or an unconscious activity strategy. In achieving a goal, a variety of psychological mechanisms of regulation of activity are involved. In cognitive, personal, and social psychology there are attempts to explain the formation of activity strategies by separate mechanisms of activity regulation. For this purpose, such mechanisms as a goal, a motivational factor (Lee et al., 1989), self-efficacy (Bandura, 1982), and situation awareness (Endsley and Jones, 2012) are employed.

However, the formation of human activity strategies cannot be explained by separate mechanisms of activity. To achieve the goal of an activity, human beings invoke various mechanisms of activity regulation that are integrated into a single system with feedforward and feedback connections. Models of activity self-regulation based on systemic approach can explain how people form the most effective activity strategies.

We discuss self-regulation when a subject interacts with a situation, and performs an individual activity. This is *an object-oriented activity*, which is performed by a subject with a material object using tools. There is also *a subject-oriented activity*, which refers to social interaction. Social interaction as well as individual activity is built in accordance with the principles of self-regulation. It is not as simple as a person transmitting information and another person receiving it. In order for the interaction to be successful, one needs a feedback on the results of the interaction. Individuals adjust their activity and specificity of communication based on the feedforward and feedback interconnections in the process of social interaction. Not only verbal interaction, but also object-oriented activity, as well as mutual perception of individuals and assessment of the context in which social interaction takes place, plays an important role in this process. Theory of self-regulation developed in SSAT can be used for the analysis of team activity and social interaction.

We hold that SSAT makes a significant contribution to the study of the SA phenomena. Orienting activity (Section 3.2) or orienting stage of activity provides operative reflection of reality including dynamic orientation in a situation, an opportunity to reflect, not only present, but past and future, as well as not just actual, but also potential feature situations. This dynamic reflection contains logical-conceptual, imaginative, conscious, and unconscious components based on which an individual can develop mental models of external events. A mental model has operative features and is involved in meaningful interpretations of a situation. Such models enable operators to understand a system and predict future states through mental manipulations of model parameters. In contrast to a cognitive approach, a human information processing system is explained not as a linear system of stages but as loop structured system where feedforward and feedback influences are critical. In Section 3.4 we will demonstrate how function blocks and their interactions presented in the model of self-regulation help describe strategies of positioning actions performance.

3.4 Individual Aspects of Activity Self-Regulation

Self-regulation of activity always has individual aspects. People regulate their activity based on their individual characteristics and their strategies are individualized. Thus, we can say that there are strategies that are utilized by a group of people; however, for some people, self-regulation acquires individual coloring. SSAT considers individualized strategies of activity as individual style of performance that should be taken into account during task analysis. Hence, in SSAT, individual style of performance is associated with mechanisms of self-regulation. However, in some situations such adaptation can contradict with standardized requirements of task performance. For example, an operator can utilize her/his individual style of task performance that contradicts with safety or quality of job requirements. Psychologists can identify individual features of personality that are relevant to job performance and those that contradict with the job's requirements. In individual style of activity, people strive to compensate for individual weaknesses with their personal strengths in a given task situation. Such strategies diminish the impact of the negative features of personality. Individual style of performance might be shaped both consciously and unconsciously. Inadequate training ignores individual features of personality when a method of task performance contradicts with the individual features. Individual style of performance is an important concept in training and learning. Students comprehend a situation better and acquire new knowledge and skills more efficiently utilizing their individual style of performance. According to Rubinshtein (1957), human activity is the major determinant of the development of personality. Rubinshtein's basic idea is that activity changes not only the external world but the person as well. In activity, a subject not only changes a situation, but also develops her/his own personality features. Hence, mental development cannot be understood simply as internalization of readymade standards, norms, and rules. External influences always interact with internal personal conditions.

All psychological phenomena can be divided into three groups—psychological processes, psychological states, and psychological features—which are interrelated. For example, memory can influence thinking; cognitive processes are interrelated with emotional-motivational ones. Psychological states have a restricted duration. The persistence of these states is affected by both external contextual and internal psychological factors. Repeated contextual factors may continuously elicit specific psychological states. Such repetitions can transform into stable psychological features. This is an important way of formatting individual features of personality. For example, vigilance that is usually mobilized only in emergency situations may become a personal feature as a result of some traumatic episode or experience. Personality features interact with one another in various ways and are evolved in complex substructures such as abilities and character, social orientation or directedness, and temperament. We will not go into detail about this aspect of personality; it is discussed in the works of Bedny and Seglin (1999a,b). Analysis of activity self-regulation demonstrates that activity is developed in a particular context. Strategies of activity are always situation specific. Cognition and activity in general are not only specific to the situation but also adequate to our individual features. Situated aspects of human performance are also described by Suchman (1987) as the "situated concept of action." However, this concept ignores the fact that situated features of activity or action cannot be studied without considering activity self-regulation and individual features of personality.

Individual style of activity should be distinguished from individual method of task performance. The latter depends on organizational factors and supervisory procedures that can ignore individual features of personality. The following are some methods of formulating an individual style of activity:

- A person involuntarily and unconsciously utilizes methods of work favorable to his individual features.
- A person utilizes blind trails and errors and feedback corrections (unconscious level of self-regulation) attempting to develop strategies of job performance that can help him/her overcome individual weaknesses.
- A person understands that she/he possesses some shortcomings in a particular situation, and consciously attempts to select methods that are more suitable for his/her task.
- A person can consciously or unconsciously utilize her/his positive features for a particular situation to compensate for her/his negative aspects of personality.

A person is often unaware that he/she utilizes an individual style of task performance. For example, some individuals think that everyone else utilizes the same strategies for the performance of a particular task. Psychologists need to discover efficient strategies and disseminate them between those who can use them.

General AT studies individual style of performance utilizing an analytical approach that is based on isolated features of the nervous system and how these features influence individual aspects of performance (Klimov, 1969; Nebilicin, 1976). This approach is based on an idea that each individual feature of personality determines the individual style of performance specific to it. Individual style is associated with artificially selected separate features of the nervous system. Such isolated features of the nervous system are difficult to extract. The systemic-structural approach promotes systemic or macrostructural approach (Bedny, 1976; Bedny and Seglin, 1999a). According to this approach, the same individual style of performance may be based on different features of personality or their combination. We consider an individual style of activity as a number of activity strategies that are formed based on mechanisms of self-regulation and depend on individual features of personality. Thanks to the interaction of various features of personality during the process of self-regulation necessary strategies that have individual coloring are developed.

There are a number of methods for studying individual feature of personality. We only consider those that, according to general AT, are called individual features of the nervous system. For example, there are such concepts as *strength* and *weakness* of the nervous system. The strength of the nervous system refers to the robustness and endurance of the cortical neural cells and their structure, as well as their ability to perform activity in the face of overload and stress. The opposite feature of *strength* is *weakness*, which correlates with limited robustness, poor endurance, and so on. It should be noted that functioning under stress depends not only on *strength* of the nervous system, but also motivation. It was discovered that weakness of the nervous system cannot be considered as a negative factor in many situations. Weakness of the nervous system correlates with the sensitivity of the nervous system. For example, it has been discovered that workers with weak nervous systems tend to develop conditioned responses more quickly in monotonous work conditions (Gurevich, 1970).

In the below presented material, we briefly consider such a feature of personality as the mobility of cognitive processes. This concept is commonly known as mobility of the nervous system (Klimov, 1969; Nebilitsin, 1976). This feature is associated with *speed of reconditioning* when the meanings of conditioning stimuli are altered. The opposite of *mobility* is *inertness*. *Inert* neural systems have reduced mobility. Mobility and inertness are most evident in situations demanding frequent changes of actions or reactions contingent upon changes in the external environment. Mobility is a bipolar dimension; high levels of mobility are described by the term *flexible nervous system*; low levels of mobility are referred to as *inertial nervous* system. The study of mobility helps to forecast how individuals can adapt to a changing environment by altering nervous processes and associated cognitive mechanisms.

The experimental technique that has been developed by Khilchenko (1966) is illustrative of how Soviet psychologists drew on their theories for their application. According to Pavlovian terminology, this technique utilizes two types of stimuli. One type of stimuli is addressed to the *first-signal system* and the other the *second-signal system* with presentation on a screen with three distinct stimuli. Stimuli, such as words, are address to the second-signal system, while graphic symbols such as different kinds of geometric shapes (e.g., square, triangle, and circle) address the first-signal system. In this experiment, subjects held two handles with buttons in both hands. They pressed the buttons with their right and left thumbs. If, for example, a square was presented, the subjects were required to press the left button; if a triangle was presented, they were required to press the right button; if a circle was presented, then they had to press any of the two buttons. The pace of the presentation of stimuli varied from slow to fast. The pace gradually increased until the rate of errors exceeded 5%. The duration of each level of pacing persisted for 1 min. After 2 min of work the subjects rested for about 5 min. The mobility of the nervous system was assessed by the subjects' ability to switch from reacting to one signal to reacting to another signal._

An analysis of this method from the activity self-regulation point of view allows concluding that this method is not physiological but rather psychological because here scientists deal not so much with the nervous system but with the subjects. In this situation, such aspects of activity as goal, motivation, criteria of success, significance of task for a subject, and flexibility of cognitive processes are very important. Of course, mobility of the nervous processes is important for performing the task at hand. This method models real situations that require mobility in information processing. Thus in this method of study, instead of *mobility of the nervous system* we use *functional mobility of psychological processes*. This psychological feature determines complicated integral characteristics of human activity associated with speed of information processing in changing environmental conditions. In our further discussion, we use Khilchenko's terminology. It is obvious that Khilchenko's method is important for application. For example, it has been discovered that this method can predict success in such professions as flight operations. Those who demonstrate better mobility according to Khilchenko's test perform better in this profession. Similar data were obtained from a study on long-distance truck drivers where only a small proportion of the individuals had low mobility. As we have already discussed, the same adaptive behavior might be elicited by different features of personality. In order to prove this hypothesis, we conducted the following experiment.

In Merlinkin's (1977) study two groups of students who took gymnastics classes were selected. One group had an *inertial nervous system*, the other had *flexible nervous system*. The first group had difficulty in adapting to changing environments and the second group did not. Both groups were required to perform three consecutive forward rolls, and complete the exercise at the upright attention position. Merlinkin filmed students' performance.

He discovered that both groups performed the exercise correctly, but each group utilized a distinct individual style of activity. Students with *flexible* nervous systems completed all three forward rolls at the same speed, and stopped immediately. Those with *inertial* nervous systems performed each forward roll at a different speed. They performed the first forward roll quickly. They began slowing down for the second forward roll and slowed down to a large degree on the third one in order to facilitate a crisp finish. Thus they achieved the same result with the same quality by utilizing a different style of performance. The difference in the style of performance was not consciously recognized by the students.

We conducted a similar experiment with some modification. We selected highly experienced gymnasts and tumblers (hereafter referred to as gymnasts), including only students with inertial nervous system in this first group. For the second group we selected sprinters (hereafter referred to as nongymnasts). All selected nongymnasts had flexible nervous systems and had some experience performing forward rolls acquired in physical education class. We defined inertial or flexible nervous systems utilizing a two-step procedure. The first step included observation during physical education classes. As a second step, we used Klimov's (1969) questionnaire. Based on this preliminary assessment, we selected 16 *inertial* gymnasts and 18 *flexible* nongymnasts. All of them were assigned three tasks. The first task was to execute three forward rolls at the subjects' own pace. The second task was to perform three forward rolls with a precise stop. The third task required them to perform forward rolls and a precise stop, but blindfolded. Each subject performed each task three times, and then his or her results were averaged. We also recorded their performance with a 16 fps movie camera. We compared differences in performance time and it was discovered that differences in performance time for all three rolls for the gymnasts was not statistically significant even when they were blindfolded in spite of the fact that they had inertial nervous systems.

At the same time, the nongymnasts who had flexible nervous systems clearly demonstrated slowdown in performance of the third roll, especially when they were blindfolded. The obtained data were statistically significant. The difference between the two groups' individual styles of performance was especially pronounced in the experiment with the blindfold. When we compared our results with Merlinkin's study, we found that the subjects with flexible nervous systems in our study used the same individual style as the inertial subjects in Merlinkin's study. This can be explained by the difference in past experience of both groups. This implies that the same individual style of activity may be unconsciously developed under influence of distinct features of personality.

In Merlinkin's study, the slow roll forward may be explained by the inertial features of the nervous system. In our study, the opposite result may be explained by the past experience interacting with the complexity of task requirements. For nongymnasts, a series of blindfolded forward rolls was a very complex task and they involuntarily slowed down performing the third roll. These results underscore the need to go beyond isolated features of personality to analyze problems of individual style of activity to identify their structural relation. This structural relationship also determines the individual style of activity.

In the Black Sea laboratory, two groups of merchant marine officers of the Black Sea Fleet were observed. One group consisted of navigators and other nonnavigators. The navigators often have to react to constantly changing situations in a flexible and dynamic manner. The nonnavigators have not been facing emergency adaptation to changing situations in their day-to-day job. It was discovered that the flexibility of cognitive processes was much higher in the navigators according to Khilchenko method.

One important aspect in studying the individual style of activity is the analysis of how well the individual adapts to the requirements of activity during the training period. It is important to observe how people with different individual characteristics acquire new knowledge and skills. One difficulty in studying this phenomenon in an experimental setting is ensuring that the participants are not aware of the purpose of the experiment. For this reason, we chose elementary school students, who we felt would be less attuned to the goal of the experiment (Bedny and Seglin, 1999a; Bedny and Voskoboynikov, 1975). The obtained data was used to make projections about how individuals adapt to the objective task requirements in a real work situation. Three groups of students who had completed first grade were selected. The study examined how students fall into distinct groups depending upon the complexity of the task and the individual features of their personality. Frequently, in simple situations, individuals exhibited similar levels of achievement. However, when the task became complicated, individuals began to vary more in their performance.

We selected first-grade students with superior, average, and poor mathematical skills. They were required to perform simple mathematical tasks. This experiment was selected as an illustration because learning and training are crucial to the study of human performance. Further, primary school students are naive enough to perceive the experiment as an extension of their schoolwork. Thus, they are less prone to be affected by the demand characteristics of the experimental situation. The students in the first experiment performed relatively simple tasks. Performance time of similar tasks was measured during the performance of 30 trials. It has been discovered that students with a low skill level spent up to 40 min to complete the task at the beginning of the experiment. Students with a high skill level spent less than 4 min, and those with an average skill level from 6 to 12 min. However, after 19 trials, all students demonstrated approximately the same result. The range of individual differences across groups was from 3 min 10 s to 3 min 30 s. This range was substantially reduced from initial conditions of 37 min 10 s to 3 min 30 s. It is important to note that each group of students utilized different strategies of task performance. Students with the low skill level sharply increased their speed of performance that gradually approached that of the average and highly skilled students.

Highly skilled students exhibited a much flatter learning curve showing only slight improvement suggesting that they were near a ceiling at the outset of the experiment while students with the average and low skill levels gradually continued approaching the task performance times of the highly skilled students. By the 18th and 19th trials (during the 6-day experiment), both groups stabilized their performance time. The point of stability nearly overlapped with those of the highly skilled students, who had achieved stabilization in 13 trials. Between the 19th and 13th trial, there was no observed improvement in performance time for any of the students. The critically important factor is that each group of students utilized different strategies of task performance, which allowed them to demonstrate similar results at the end of the experiment. For example, the student with the low skill level used their memory more efficiently and the highly skilled students used internal mental operations more efficiently.

In another experiment with the same students, we assigned much more complicated tasks that reduced their ability to use memory and required to conduct more complex methods of calculation. Each group utilized different strategies of task performance as in the first experiment. All three groups demonstrated similar diversity in the beginning of task performance. This diversity reduced a lot toward the end of the second experiment. In contrast to the first series of experiments, when all groups converged to a similar level of performance, the second experiment demonstrated significant difference between their performance times at the end of the experiment. In the second experiment, we observed differences in performance times between groups and converging in performance times within groups. Converging in performance times was specific only to each group. When subjects performed complex tasks, individual styles of task performance did not facilitate elimination of differences in the performance times. Thus, in professions that include performance of complex tasks, the selection of workers should be combined with the development of individual style of performance.

In this chapter, we briefly discussed the concept of individual style of activity. Individual style of activity is one the most efficient tools for adapting to a situation. It can be developed consciously or unconsciously. We considered individual style of performance from the perspective of the mechanism of self-regulation. Based on these mechanisms, subjects can develop individual strategies of task performance. These strategies can be flexible, individually specific, and adequate for a particular situation.

3.5 Self-Regulation of Positioning Actions Performance

Positioning actions are important in human task performance. These actions are usually involved in moving small objects from one place to another, such as a robot's arm or a cursor on the computer screen moving to a particular object. Results of positioning actions are widely used during performance of various tasks by operators. Such actions are also broadly utilized in performing production operations. Study of positioning actions also is important in the analysis of perceptual-motor interaction with the computer. Various pointing devices in human-computer interaction (HCI) tasks have led researchers in the HCI field to study Fitts's law (1954) as a predictive tool of the performance time for motor actions. Due to these widespread applications, it is not surprising that many publications are devoted to the study of positioning actions.

In this chapter, the study of positioning actions has been conducted using a functional analysis approach when activity is considered as a self-regulative system. Action precision and error analysis were performed from the activity self-regulation viewpoint in this study. It is interesting in the context of the book not only because the new data in the field of the positioning actions regulation is be presented. This study also gives a clear indication of the potential possibilities of utilizing the activity self-regulation theory in specific studies. The possibility of using functional analysis when activity is considered as a self-regulative system will be discussed in various other studies in the book. Presented in the following is a systemic qualitative analysis of task performance.

In the study of positioning actions, the most important are such characteristics as time, precision, and amplitude. Some researchers have shown that the time of positioning actions performance does not depend significantly on its amplitude. Other experiments could not support these findings (Leplat, 1963). Fitts' studies are well known (Fitts, 1954; Fitts and Posner, 1967) for their attempt to integrate two basic characteristics of positioning actions: amplitude of movement and precision. These characteristics depend on the width of a target w.

The Fitts studies demonstrated that time movement is linearly related to the logarithm of the index of difficulty. This index integrates two characteristics: amplitude and precision.

Temporal parameters are important in predicting performance time of positioning actions during operator work, so some authors suggest using Fitts' law to estimate these performance times (Drury, 1975; Langolf et al., 1976). One specific aspect of Fitts' experiment was that subjects had to move a metal stick between two targets with maximum speed or move a stick from a start position to a particular target position with maximum speed. When transferring this result to a working environment, one can assume each operator's action is performed at a maximum pace, and each performed action does not depend on either previous or subsequent actions. It is difficult to agree with these statements. Multiple actions cannot be considered as independent and isolated from each other. The subject simply cannot move objects or controls with maximum speed during task performance.

Specialists in time study know that one important aspect of time performance is the pace or tempo at which the person is working. Therefore, in every time study, the pace of performance should be taken into consideration (Barnes, 1980). Another important aspect in the study of positioning actions is that they are almost never performed in isolation from other related actions. These other actions are logically organized, which influences time performance of positioning actions. However, in Fitts' experiment, the focus was on how the subject performed the same positioning action multiple times with maximum speed. Fitts' study reflects the behaviorist approach, that is, when human activity is considered a sum of independent reactions to independent stimuli. Thus, these actions and reactions are considered independent from each other.

We demonstrate that theoretical data and methods developed in the framework of self-regulation of activity can be useful in the study of human performance in general, and, study positioning actions in particular. The purpose of this chapter was to study whether there were changes in the performance time for positioning actions in conditions containing two and four targets. Strategies of action self-regulation in various conditions were considered.

There are two main methods of studying activity from the SSAT viewpoint: morphological and functional methods. In the morphological study, action and operations are the major units of analysis. One can distinguish between external behavioral and internal mental actions. External behavioral actions include various motions and transform material or tangible objects. Mental actions transform images, concepts or propositions, and nonverbal signs in the mind.

Functional analysis is based on studying the mechanism of activity self-regulation.

The concept of self-regulation is critical for the understanding of activity as a system.

In AT, self-regulation is not a homeostatic, but rather a goal-directed process where the goal has integrative systemic functions. The concept of self-regulation is meaningful only when the self-regulation model is developed. Such models determine theoretical and derived from it practical methods of task analysis. The self-regulation model is defined in terms of functional mechanisms or function blocks. At any particular time, cognitive processes are integrated to achieve a specific purpose of activity self-regulation.

A subject develops various performance strategies because of the processes of self-regulation that are carried out on conscious and unconscious levels. Strategies of performance change depending on subjective goal significance and the difficulty of task along with other factors. Pace of performance depends on preferable strategies in particular conditions. Study in this chapter demonstrates that homeostatic models of self-regulation utilize terminology such as reference value, comparator, input and output (Vancouver, 2005) are unacceptable in study self-regulation of human activity during task performance. The conscious goals of a human being and the technical system are virtually the same in homeostatic models of self-regulation. This terminology is more applicable for the description of the technical system of the self-regulation process than for understanding the conscious and deliberate process of human self-regulation that includes unconscious components.

So far in this chapter, we have presented in a concise form the basic principles of self-regulation of activity. This will be helpful in understanding the ideology utilized in our analysis of positioning actions. In our study, we mainly used not a general model of self-regulation but the self-regulation model of orienting activity. Very often, in the applied studies it is sufficient to use this model of self-regulation activity (see Figure 3.1). This self-regulative model is very useful in understanding how a worker can develop mental representation of the reality and based on this regulate executive components of an activity.

Thus, with our model, we describe how a person creates a goal and a subjective mental model of the situation, which type of the exploratory actions and operations are utilized, what types of possible mental models are developed, how a subject selects preferable mental models, and so on. In our further discussion, we will demonstrate how function blocks and their interaction presented in model of self-regulation helps us to describe strategies of positioning actions performance. We also discussed how the application of self-regulative models can be understood if a researcher considers each functional block as a window that can be opened to observe the activity during task performance. During task analysis, the researcher usually takes into consideration not all function blocks, but only those that play the most important role in the considered task performance. This simplifies usage of this method of study. The researcher should pay attention to not only separate function blocks, but also their interrelationships. Self-regulation can perform in the internal mental plane. In such situations a subject can use internal or mental feedback during performance of mental actions. The following blocks are especially important in analyzing task performance strategies when subjects perform positioning actions and two or four targets are used. (see Figure 3.1):

- Goal (block 2)
- Subjectively relevant task conditions responsible for the development of stable or dynamic mental models of task or situation (block 9)
- Assessment of task difficulty (block 8)
- Assessment of the sense of task or task significance (block 7)

- Formation of the level of motivation (block 6)
- Criteria of the evaluation block include two subblocks: *subjective standard of successful result* and *subjective standard of admissible deviation* (block 11)

The last two subblocks are explained in more detail in the general model of self-regulation.

We saw that the first functional block is the goal of a task (block 2). The goal should be at least partially conscious. Together, with the motivational mechanism, it creates the vector *motive* → *goal* (more specifically *motives* → *goal*). The goal is cognitive and motives are energetic components of an activity. Overall, the goal of a task performs integral functions and plays an important role in the selection of information. It integrates all the mechanisms of self-regulation into a holistic, self-regulative system of activity. There is no activity without a goal. When the goal is not clearly formulated externally, the subject can formulate it himself/herself. Even though a goal lacks clear definition, it still exists for a subject. During task performance, such a goal can be formulated more specifically.

A mechanism called subjectively relevant task condition (block 9) exists. This mechanism is responsible for the development of stable or dynamic mental models of task.

At this stage, a subject creates a subjective representation of the task. This mental model of the task can vary even if the instructions do not change. The mental model includes imaginative and verbally logical submechanisms or subblocks. A subject develops a mental model in both imaginative and verbally logical forms. He/she can reformulate the task and design his/her own subjective task representation in an imaginative form.

This mechanism often does not provide clear awareness of the mental picture of the task.

The next mechanism is called situation awareness (Endsley, 2000). It includes verbalized components and provides conscious reflection of the task. The interaction of these submechanisms allows creation of a mental model of task or situation. Such a model can significantly deviate from an objectively given task. It is very important to understand that a dynamic mental model can change over time in the process of task performance.

We want to emphasize that such basic concepts as goal, dynamic mental model, motivation, and subjective standard of successful result are considered as mechanisms of self-regulation in our study. Analysis of these mechanisms is carried out by taking into account their interaction with feedforward and feedback interconnections. Further interpretation of concepts such as goals, motivation, and assessment of significance is very different from the traditional approaches in psychology.

From a functional analysis perspective, it is important to study the structure of activity while a subject uses various strategies of task performance. Hence, one should utilize experiments where conditions of tasks performed

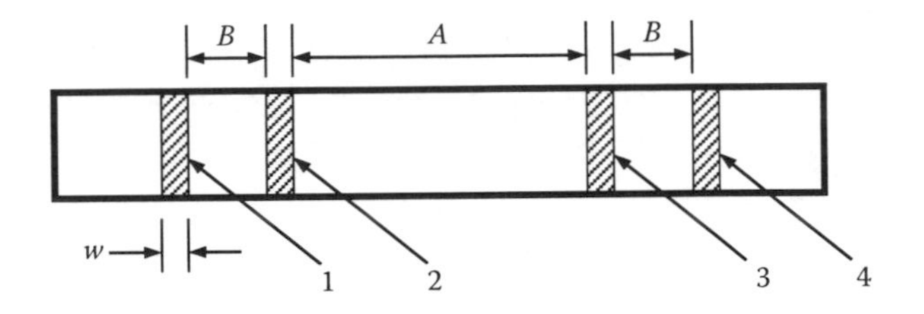

FIGURE 3.5
Layout of four targets: 1–4, targets; *A* and *B*, distance between targets and; *w*, widths of the targets.

vary by including or eliminating elements of activity, change the sequence of their performance, etc., to discover the relationship between them. Objective measurement procedures are combined with such subjective methods as observation, verbal discussion, and error analysis. In our experiment, we introduced two main conditions of task performance: subjects performed positioning actions with (1) two targets and (2) four targets. Distance and width of targets also varied. Figure 3.5 shows the layout of the four targets. In one series of experiments, the subjects were instructed to strike two targets with a metal stick, while in the other series of experiments, four targets were involved. The purpose was to understand the strategies of positioning motor actions regulation, when such actions are performed as isolated actions, and then sequentially. When a subject hits only two targets, this is a repetitive performance of the same action. When a subject hits four targets, there are three positioning actions, which are performed in sequence and repeated multiple times.

To conduct this study, a special device was designed. It contained brass strips (targets) mounted on a base. The number of targets, their width, and the distance between targets were all alterable. There was a meter on the panel of the apparatus, which counted the number of hand movements from one target to another. Another meter counted the number of errors. There were two bulbs on the panel: a red warning bulb used to indicate to the subject that the experiment would begin in 2 s, and another bulb used to signal the beginning of the task. There was a stopwatch on the panel. A green bulb illuminated and the stopwatch started simultaneously. After 10 s, the stopwatch was turned off and a buzzer informed the subject that the task was completed. A video camera was also used to capture the observation, which allowed review of the subjects' performance multiple times after the experiment. The video camera was located approximately 45° to the targets' surface.

As soon as the red bulb was turned on, the subject raised and held a metal stick (connected electrically to the meter) above the target without touching the apparatus (start position). Two seconds later, the green bulb was turned on, the two meters began functioning, and a stopwatch was started. The subject attempted to strike targets at his/her maximum speed. After 10 s, the buzzer activated and the apparatus was turned off. By tracking the entire task performance time and the number of times the subject successfully struck a target, the average action performance time could be determined.

The experimental group consisted of five university students. They were instructed to hit the target with maximum speed and precision. (Fitts gave the same instruction to his subjects.) Two groups of experiments were conducted. In the first group of experiments, only wide targets were used ($w = 50$ mm). In the second group of experiments, only narrow targets were used ($w = 7$ mm).

The experiment was conducted over 4 days. The first 2 days subjects were trained to perform required tasks, and on the third day they were involved in the first group of experiments (series 1–3). On the fourth day, subjects performed the first and second groups of experiments (series 4–6).

The first group of experiments consisted of three series (series 1–3).

Series 1. The subject struck two targets. The distance between targets was 60 mm. The width of each target was 50 mm.

Series 2. The subject struck two targets. The distance between targets was 120 mm. The width of each target was 50 mm.

Series 3. The subject struck four targets. The distance between targets 1 and 2, 3, and 4 was 60 mm. The distance between targets 2 and 3 was 120 mm. The width of each target was 50 mm.

The second group of experiments also consisted of three series (series 4–6).

Series 4. The subject struck two targets. The distance between targets was 60 mm. The width of each target was 7 mm.

Series 5. The subject struck two targets. The distance between targets was 120 mm. The width of each target was 7 mm.

Series 6. The subject struck four targets. The distance between targets 1 and 2, 3 and 4 was 60 mm. The distance between targets 2 and 3 was 120 mm. The width of each target was 7 mm.

Table 3.1 summarizes the parameters for each series in the experiment.

TABLE 3.1

Scheme of the Experiment

Number of the Group of Experiments	Series	Number of Targets	Width of Targets (w)	Distance between Targets (A and B)
1	Series 1	2	50	60
	Series 2	2	50	120
	Series 3	4	50	60, 120, 60
2	Series 4	2	7	60
	Series 5	2	7	120
	Series 6	4	7	60, 120, 60

The main purpose of this study was to discover the characteristics of positioning motor actions conducted in isolation, as well as part of a sequence. Series 1, 2, 4, and 5 reproduced conditions that were observed in Fitts' study. Series 3 and 6 differed from Fitts' due to the introduction of four targets. In series 1, 2, 4, and 5, subjects repeated a single action. In series 3 and 6, the subjects repeatedly performed three different actions in sequence. In the first group of experiments (series 1–3), wide targets were used, while in the second group (series 4–6) narrow targets were used. Subjects performed each series five times. Fitts also used targets with different widths. He utilized the target widths of 7 and 50 mm. The average results were calculated in all series. Subjects were trained to perform the whole task before time was recorded in the experiment. The activity of subjects was observed during the experiment. When the experiment was completed, subjects were debriefed. The goal of the observation and subsequent debrief was to determine the work strategy employed by the subjects and if the subjects realized which work strategy had been used. The questioning was not formulated in detail prior to the experiment and was modified depending on observation results. Such a flexible questioning was selected because the task was not hard to observe and the amount of errors, targets, selected hitting areas, etc., influenced the questions asked to the subjects.

We will begin the functional analysis of activity by discussing the formation of subjective representation or mental model of task. This stage of self-regulation can be understood if we compare experimental data with information obtained from the observations and interviews. After briefly considering some aspects of subjective representation of a task, other aspects of self-regulation of activity can be analyzed in greater detail. The first two function blocks that essentially influenced a strategy of positioning actions performance were the block *goal* (block 2) and *subjectively relevant task conditions* (block 9). Together, these two function blocks are responsible for the creation of a mental model of task or mental representation of task. Findings supported that the instruction "hit the target at your maximum speed and precision" gave the subject an opportunity to vary widely his/her subjective mental representation of the task. Some subjects considered precision to be the main requirement of the task while others considered speed of performance to be more important. Relationships between these two requirements varied substantially among subjects and tasks. As a result, strategies of task performance also varied. In some cases, it was necessary to introduce additional corrective instructions during the training sessions until the subjects adapted to experimental requirements.

The elements of task were presented visually in this experiment. According to Wickens and Hollands (2000), one of the important aspects of subjective representation of the task is perceptual organization of the task elements. As it can be seen in our experiment, the distance between targets 1 and 2 was similar to the distance between targets 3 and 4, but shorter than the distance between targets 2 and 3. This influence on strategies of attention is important for the development of a task's mental representation. Depending on the distance between targets and/or quantity of targets, subjects utilize distributed or switching attention strategies.

In the functional analysis, these cognitive aspects of activity regulation are provided by two function blocks or mechanisms: *goal* and *subjectively relevant task conditions*. This experiment demonstrated that subjects alter their strategies of attention through a series of experiments. When a subject worked with two targets, she/he tried to hold them in her/his field of view (divided attention strategy). However, in the case of four targets, one not only needed to distribute attention, but also switch it appropriately. The subject combined targets into viewing groups based on the distance between the targets. Consequently, the first and third actions with smaller intertarget distance "B" were considered primary, and the second action with the large intertarget distance "A" was viewed as auxiliary, serving only as the means of going from the first to the third action (Figure 3.5).

After contemplating some specifics of the task's subjective representation, the specificity of strategies of task performance in each group of experiments was considered. Function block 8 *assessment of task difficulty* provides an explanation of how subjects achieve required precision of performance when they strike wide targets in the first group of experiments. Tapping 50 mm wide targets is seen by subjects as a relatively easy task, because these targets are subjectively wide enough to comply with precision requirements. These requirements do not contradict with time demands, and the subjects feel they can manage both precision and speed requirements equally. This feeling influences motivational block 6 and emotional-evaluative block 7. The subject is motivated to follow instructions. The significance of speed requirements increases along with precision significance. The subject becomes motivated to avoid errors. This, in turn, influences such mechanisms of self-regulation as subjective standard of successful result and subjective standard of admissible deviation from the standard (the first and the second mechanisms of block 11).

Achieving the required precision of performance, such blocks as *criteria of evaluation* and subblock *subjective standard of success* are particularly important.

In series 1, we have two targets (width = 50 mm) with 60 mm distance between them. It was shown that during the training process and the period of the experiment, the subjective standard of success was formed. This subjective standard does not always match the objective standard of success (objective width of the target). Approximately 80% of the hits were placed in the middle of the target with 35 mm range. This means that the subjective standard of success was much narrower than the objectively given standard. In series 2, the distance between targets was increased to 120 mm. The striking area became a little wider than in the first series of experiments. The variation of positioning actions slightly increased.

In series 3, subjects struck four targets. In this series, the subjects had a tendency to alter their criteria for success. The target zone was expanded but remained narrower than the entire width of the targets. The choice of a subjective standard was determined by the width of targets, distance between targets, individual criteria of success, and personal ability of a subject to achieve the set goal, that is, "to hit targets with maximum speed." The functional mechanism *subjective standard of success* is tightly connected with the concept of *reserve of precision*, which has both objective and subjective meanings (Bedny and Meister, 1997).

Under reserve of precision, we understand the difference between the width of the target and subjectively selected target area that the subject is trying to hit during trials. The greater the width of subjective standard of success (width of hitting area of target), the smaller the subjectively acceptable reserve of precision, and the subjects employ an increasingly risky strategy.

This is manifested by the fact that the hitting area in the target becomes broader and approaches the edges of the targets. Hence, the subjects begin to use a more risky strategy, because they approach the edges or borders of the targets. However, the subjects still avoided approaching the edges of the targets. This strategy allows subjects to maintain the same objective criteria of success ("do not hit areas outside of the targets and sustain required speed"). Such a strategy helps to reach the objectively presented goal—"hit targets with maximum speed." The qualitative analysis of activity strategies is confirmed by error analysis.

The average number of errors per task (trial) is not great and only slightly increases as task complexity increases. For wide targets, on average, participants made 0.68 errors per trial for a distance of 60 mm, 0.96 errors for a distance of 120 mm, and 1.4 errors for four targets. According to a one-way within-subject analysis of variance (ANOVA), the difference is statistically significant ($P < 0.001$).

It is necessary to stress that the objectively formulated and subjectively accepted goals are not the same. Different subjects have a varying understanding of the goals. Some are focusing more on speed and others focus on precision. The goals and criteria of success in our experiments are close but not identical. The goal of the task is something that must be achieved because of task performance (strike targets with maximum speed and minimum errors). The goal is what the subject desires to accomplish during task performance. Criteria of success are what the subject uses to evaluate the task result. Thus, functional blocks 2 (goal) and 11 (criteria of evaluation) allow us to understand that mechanisms such as goal and subjective standard of successful results do not always coincide.

Let us consider function blocks 8 (assessment of task difficulty) and 7 (assessment of task sense or significance). The subject can evaluate the task as very difficult and not significant. In such situations, the subject can be satisfied with a result (subjective standards of successful result) that is much worse than an initial goal. On the other hand, if a task is evaluated as *not very difficult* but as *very significant*, the subjective standard of the successful result will be higher. Hence, the self-regulation model allows a very accurate description of the subjects' strategies during task performance.

It has been observed that externally given instructions do not necessarily determine the strategy of human activity. Despite similar instructions, subjects employ various strategies. They interpret their results using their own subjective criteria of success. Analysis of the strategies of performance demonstrates that some of the subjects try to maintain high speeds of activity at the expense of using a wider part of the target, even though the risk of hitting out of the target's range increases. This is risky strategy. Another subject may use a narrower part of the target and reduce the speed of performance. This strategy protects from errors. Such strategies can vary over a broad range. The traditional method of the speed–accuracy trade-off analysis tells us very little about real strategies of task performance. Mechanisms of self-regulation affect positioning action performance time and precision. In general, results of series 1–3 support that the subjective standard does not expand beyond the size of the target and provides sufficient reserve for precision. Hoppe (1930) showed that especially difficult or especially easy goals fail to become subjective goals, and subjects select other goals instead as requirements. Thus, the goal (block 2) and the notion of a subjective standard of success as an important component of block 11 (criteria of evaluation) have a vital role in understanding how subjects evaluate their own results and plan their strategies of activity.

Not only cognitive but also motivational components are important in the formation of the subjective standard of success. One important motivational variable in the analysis of this mechanism is the level of aspiration. Subjective standards of success lack clearly defined quantitative characteristics. As with precision parameters, the subjective standard of success in this task is the subjectively accepted area of striking distribution. According to temporal parameters, subjective standards of success are a subjectively accepted interval of pace variation. These intervals can vary within a relatively wide range. Precision of performance, including subjective standards of success, are determined not only by limits of the subjects' cognitive capabilities, but also motivational aspects of activity, particularly the aspiration level of the performer. The interaction of the subjects' cognitive capability limits with his or her aspiration level leads to variability in the precision of performance.

In our experiment, it is interesting how the subject corrects his/her errors. In Fitts' experiment, it was suggested that precision can be determined by the width of a target. In other words, the objective criterion of precision is the same as the subjective one, and the latter criterion has not been considered at all. However, we can distinguish between objective and subjective criteria of success. They can be similar or differ significantly. Subjects can choose their own criteria of precision. In series 1–3, when a subject was striking wide targets, he/she immediately corrected errors when approaching the edges of the target. Therefore, in series 1 and 2, subjects avoided risky strategies. Only in series 3, when four wide targets were used, did the strategies become riskier. However, as we show in this case, the pace is affected very little, because the subjective standard of success is sufficiently lower (striking area is wide). The choice of corrective strategies depends also on the relationship between significance of speed and precision. In the case of wide targets, the significance of errors is relatively high. The difficulty of task is subjectively evaluated as low (block 8), and the errors have a greater negative significance. The inability to hit even wide targets lowers the subject's self-esteem. Therefore, the significance of errors for the subject changes depending on the specificity of the task and its significance. The factor of significance is determined by *assessment of sense of task or significance* (function block 7). When the significance of the task increases, this influences a subject's motivation and, in turn, involves another function block called *formation of level of motivation* (block 6). These two blocks influence the level of admissible deviation from an activity standard (mechanism of block 11). *Subjective standard of admissible deviation* as another mechanism of block 11 is responsible for the formation of acceptable deviation from the standard of success.

Positioning actions are widely used in human work. Therefore, the study of time movement of the positioning actions is important. Fitts (1954) investigated the relationship between performance time of these actions and the distance between the targets and precision and described their formal relationship as follows:

$$T = a + b\left[\log_2\left(\frac{2A}{W}\right)\right]$$

where

a and b are constants
A is the distance between targets
W is the width of targets
$\log_2(2A/W)$ is the *index of difficulty*

In Fitts' experiments, subjects had to strike targets with maximum speed. However, in a real situation, a worker almost never performs such actions repeatedly at maximum speed. The *index of difficulty* reflects only a formal concept of difficulty, which depends on the index of complexity being an objective characteristic of a task. There is a probabilistic relationship between objective complexity and subjective difficulty. Subjective perception of difficulty of positioning actions depends also on specificity of task and individual differences of subjects (block 8—*assessment of the task difficulty*).

In real tasks, positioning actions are not performed by a subject in isolation from other cognitive and motor actions, but are affected by these actions. In a real task, a subject can utilize a variety of strategies to perform positioning actions. Fitts formally described only one preferable strategy of positioning actions that is specifically considered in his experiment situation.

Let us analyze performance time of positioning actions in our experimental study when subjects use wide targets. We will designate the action performance time as T_1 for an action with amplitude B and T_2 for an action with amplitude A. Let the performance time of the three actions performed by hitting four targets be T_3.

TABLE 3.2

Action Time (s) When Target Width w = 50 mm

	Two Targets			Four Targets
Subjects	T_1	T_2	$2T_1 + T_2$	T_3
1	0.15	0.15	0.45	0.82
2	0.17	0.18	0.52	0.85
3	0.16	0.16	0.48	0.70
4	0.18	0.19	0.54	0.83
5	0.16	0.15	0.47	0.75
Average time	0.16	0.17	0.49	0.79

For series 1, we obtained performance time T_1 of positioning action, where the distance between targets is B (60 mm) and width of targets is 50 mm.

For series 2, the time for positioning actions is T_2, where the distance between targets is 120 mm and width 50 mm.

If we assume that actions performed in any sequence have no influence on each other, the performance time of three actions can be determined as $T_3 = 2T_1 + T_2$, provided two of them are identical. If $T_3 > 2T_1 + T_2$, then the task with four targets is more difficult and actions are not independent of each other. Table 3.2 shows the results of the experiment where the targets are wide (w = 50 mm).

It is worth noting that the difference in distance between the two broad targets (width of targets 50 mm and distance between them 60 and 120 mm) does not affect positioning actions' performance time. However, it is possible that such variations emerge if differences in amplitude between two actions significantly increase. One can only conclude that there is a zone of low sensitivity to changes in the amplitude of positioning actions with low precision.

Comparing T_3 with $2T_1 + T_2$, we can see that $T_3 > 2T_1 + T_2$. The difference is significant ($p < 0.01$) according to a within-subject t-test.

We determined the differences in performance time for three actions when they are performed sequentially in relation to the performance time for the same actions when they are performed independently:

$$\Delta T = T_3 - (2T_1 + T_2) = 0.79 - 0.49 = 0.3 \text{ (s)}$$

Then the ratio of the change in execution time of three actions performed in sequence to the execution time of these actions when they are performed in isolation can be determined as follows:

$$S = \frac{0.3}{0.49} = 0.61 \text{ or } 61\%$$

This means the pace at which the sequential actions with low precision are performed is significantly slower than the pace of the same actions

TABLE 3.3
Action Time (s) When Target Width $w = 7$ mm

	Two Targets			Four Targets
Subjects	T_1	T_2	$2T_1 + T_2$	T_3
1	0.37	0.45	1.20	1.50
2	0.39	0.47	1.25	1.40
3	0.38	0.50	1.26	1.43
4	0.38	0.45	1.21	1.32
5	0.34	0.37	1.05	1.15
Average time	0.37	0.45	1.19	1.36

performed in isolation. This result can be understood if compared with the information obtained from observations, interviews, and analysis of subjects' strategies. Clearly, the task with two wide targets needs a low level of conscious self-control, which may be described as a predominantly automatic level of self-regulation. When we place four wide targets instead of two, the task becomes significantly more difficult for subjects and needs a higher level of self-control, which would require a conscious level of self-regulation. Activity performed with a conscious level of self-regulation requires more time for its performance. Therefore, transfer from two wide targets to four wide targets leads to a significant slowdown in the pace of performance.

Let us now consider the experiment with narrow targets ($w = 7$). Table 3.3 shows the performance time of positioning actions when subjects hit narrow targets (series 4–6).

Here, as in the previous group of experiments, $T_3 > 2T_1 + T_2$, according to the within-subject t-test difference, is statistically significant ($p < 0.01$). If we change the task from two to four targets, one might expect a greater increase in action time for the narrower targets, where the action is more precise. However, in actuality, the result is the opposite.

The differences in performance time of three actions with higher accuracy when they are performed sequentially in relation to performance time of the same actions when they are performed independently can be determined as follows:

$$\Delta T = T_3 - (2T_1 + T_2) = 1.36 - 1.19 = 0.17 \text{ (s)}$$

S is determined accordingly as

$$S = \frac{0.17}{1.19} = 0.14 \text{ or } 14\%$$

The increase in action time is more significant with the wider targets, that is, for actions with lower precision. In contrast to the previous experiment, the task with two narrow targets needs a high level of self-regulation similar to the situation with four narrow targets. Therefore, the difference in performance time of tasks with two and four narrow targets is less than when subjects are involved in tasks with four wide targets. Tasks with narrow targets require a conscious level of self-regulation in both situations (when we use two or four targets). Of course, difficulty of the task and level of self-control are higher when a subject uses four narrow targets than when two narrow targets are used. When the subject transfers from two narrow targets to four narrow targets, a transition is observed from a simpler level of conscious self-regulation to a more complicated conscious level of self-regulation. When the subject transfers from two broad targets to four broad targets, we can observe transition from an automatic level of self-regulation to the conscious level of self-regulation. In such situations, the difference in the performance time of these two tasks increases more significantly. Thus, assessment of task difficulty becomes particularly important. The greater the task difficulty, the slower is the pace of task performance, and the action performance time increases gradually within one level of self-regulation, but increases sharply, when self-regulation changes from one level to another.

It is also interesting to compare T_1 and T_2, where the distance between targets changes for both tasks with narrow targets and wide targets. If targets are narrow, the time for the larger amplitude action $T_2 > T_1$ and the difference are significant ($p < 0.05$). If targets are wide, the difference is not statistically significant. If the amplitude of a low-precision action increases, it does not significantly change performance time in the work environment. However, if the amplitude of a high-precision action increases, it significantly changes performance time. The precision of positioning actions is associated with the assessment of their difficulty. Thus, the function block *assessment of the task difficulty* plays an important role in the formation of pace of positioning actions. It should be noted that when errors have serious consequences such as in the functional block *assessment of the sense of task or task significance*, these errors can play an important role in the regulation of performance pace.

As mentioned earlier, some studies infer that the time of a movement is constant for various amplitudes (Leplat, 1963). The process of activity self-regulation can explain the contradiction of the results discussed. The performance time of positioning actions depends not only on the distance between the targets, but also on the action's precision. Figure 3.6 illustrates the mode used for moving the hand from one target to another, which can also explain the results.

From Figure 3.6, one can observe three zones of movement. The first zone is the time interval of acceleration of the hand movement. This zone is associated with the programming of motor action. The higher the accuracy of the motor action, the more time is required for programming of action and accelerating stage of action in general. The third zone is the time interval of the deceleration of the hand movement. When the accuracy of motor actions

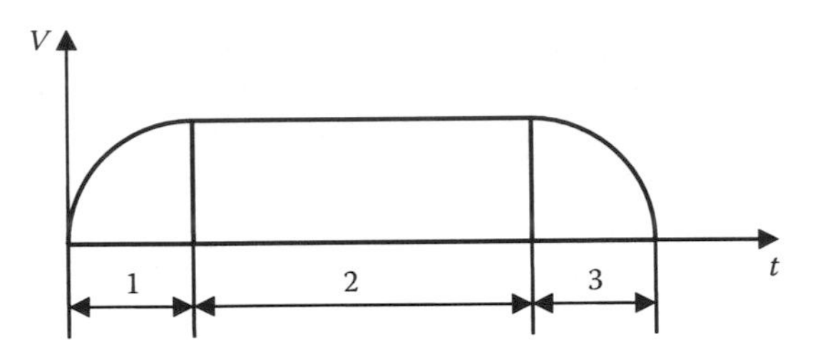

FIGURE 3.6
Three zone of hand movements: (1) acceleration zone; (2) constant speed zone; (3) zone of delay.

increases, it leads to earlier deceleration of action. Hence, zones 1 and 3 take more time to execute. Thus, when increasing the precision of motor action, the time of execution of motor actions in zones 1 and 3 is increased more significantly than during performance of actions with low precision. These findings coincide with the concept of activity self-regulation (Bedny and Karwowski, 2007).

In the second zone, the subject uses an almost constant hand speed at its peak value. Thus, precision of action and width of targets have no noticeable effect on action performance time in this zone. When one makes low-precision movements, zone 2 increases. This interval is characterized by an automatic, unconscious level of self-regulation with a tapering of feedback and feedback corrections. When the distance between wide targets (amplitude of motion) increases, the second zone dominates the action time. This results in smaller time difference due to amplitude variation in the case of low-precision movements.

An investigation of error analysis for the series where subjects hit narrow targets follows. The functional mechanism of block 11, *subjective standard of successful result*, and its relationship with other blocks when subjects use narrow targets becomes particularly important in error analysis. In the case of narrow targets, subjects used the full size of the target when selecting their subjective standard of success. The subjectively acceptable reserve of precision is close to zero. Small deviations from a given mode of action could result in missing the target (risky strategy). Most subjects consider tapping narrow targets a much more difficult task than tapping wide targets. Attempts to use a narrower area inside the target might significantly increase performance time. At the same time, however, an increased number of errors for difficult task performance are considered more acceptable than sharp decreases in the speed of performance. In experiments involving four narrow targets, the errors were considered critical by most of the subjects when they missed several targets during one sequence of movements. When a subject missed just one of the targets with minor deviation, she/he started correcting such errors after only several occurrences. Hence, a mechanism such as *subjective standard of admissible deviation* (block 11) also plays a significant role in the self-regulation of positioning actions.

For increasingly complex tasks, the number and the significance of errors for narrow targets have a tendency to decrease. It is interesting that some subjects, especially during training sessions, expand subjective standards of success even outside of the target area. In other words, subjects accept the situation when they miss the targets and hit areas outside and near the target. This area we consider as a *subjective standard of admissible deviation* and we evaluate it as an acceptable deviation. In this case, the reserve of precision becomes negative; subjects were more tolerant of errors, therefore significance of errors decreased. Time parameters of positioning actions are sustained more than requirements for precision when the significance of errors decreased. The obtained data are confirmed by statistical analysis of errors. For narrow targets, participants made on average (per trial) 5.6 errors when the distance between targets was 60 mm, 5.2 errors when the distance between targets was 120 mm, and 6.32 errors when there were four targets. In a one-way, within-subject ANOVA for narrow targets, there was a significant effect ($p < 0.05$) of target distance/number (i.e., distance 60 vs. 120 mm for two targets and two targets vs. four targets).

We should remember that in the situation with the broad targets, subjects use as *subjective standard of successful result* only an area of the target that is located in the center. This area is narrower than the objectively given target width. Hence, subjects choose as *subjective standard of successful result* not only the width of the target, but the area that is narrower. As a result, the accuracy of actions increases, but the pace of performance is reduced and execution time is increased. We can summarize our data in the following way.

In experiments with wide targets, even a small deviation from subjective standards of success has been evaluated negatively and corrected quickly. In experiments with narrow targets, corrections were made when there were sequences of errors and more deviations that are noticeable. This data also was proved by quantitative and statistical analysis of errors. In a task with wide targets, the general number of errors during performance of 25 trials was as follows: when the distance between the targets was 60 mm, the number of errors was 17, with 120 mm distance, it was 24; when four targets were involved, the number of errors was 35. In the experiments with narrow targets, the number of errors was 140, 131, and 158, respectively. Participants made significantly more errors when they hit narrow targets than when they hit wide targets (within-subject t-tests: two narrow vs. two wide targets, distance 60 mm, $p < 0.0005$; two narrow vs. two wide targets, distance 120 mm, $p < 0.0005$; four narrow vs. four wide targets, $p < 0.0005$).

The qualitative and quantitative analysis of strategies used by the subjects indicated various modes of self-control and corrections of activity, which in combination serve as a self-regulative process. The prior work of Fitts and others cited demonstrated that a person could be compared to an information channel with a limited capacity for process.

According to systemic-structural AT this *"channel"* has specific properties that lead to permanent changes in strategies over a broad domain.

An individual should be considered not as a channel for information processing, but rather as a subject that actively interacts with the objective world and continually develops a variety of strategies based on self-regulation mechanisms.

The presence of rigid, external instructions within a planned environment study often does not uniquely determine the subject's activity. A subject creates her/his own representation of a situation and develops and evaluates strategies based on personal capabilities and the demands of the activity. This often complicates the transfer of laboratory results to field practices, especially in the case of studying isolated actions, movements, and psychological processes.

This research indicates a person is not a reactive system if merely reacting to external factors (stimuli). Subjects actively interact with their environment. They select their own goals, reformulate tasks, and change their subjective standards of success and strategies of activity. In order to be able to transfer experimental results into practical applications, the experiment should be targeted to explore typical strategies of performance in a varied environment.

Human activity is a complex, dynamic, evolving system that is based on self-regulative processes. The presence of self-regulative processes leads to the reformation of the structure of this activity system when we change or introduce a new component of activity. Hence, the analytical research of separate elements of activity should be combined with the systemic research of the entire activity structure. In studying the process of self-regulation, one can outline conscious and unconscious levels of self-regulation that vary depending on task complexity, individual differences, and stage of skill acquisition.

Our studies have shown that positioning actions in the context of the entire activity cannot be seen as independent. This leads to the conclusion that Fitts' law can only be used to measure isolated, discrete motor actions performed with maximum speed. In addition, our studies demonstrated that when one switches from discrete positioning actions (two targets) to sequential positioning actions (four targets), there is a greater increase in action time for low-precision actions (wide targets). Here we found a change from an automatic level of self-regulation (in the case of two wide targets) to a conscious level of self-regulations (when subjects hit four wide targets). Switching from two to four narrow targets causes the change toward a higher level of self-control within the conscious level of self-regulation. This explains why the pace slows down less when subjects transfer from two to four narrow targets than during the transfer from two to four wide targets.

Further, it was noted that distance between targets influences the activity of high-precision movements, but distance is not significant for low-precision movements. These results should be taken into consideration for the design of efficient manufacturing work processes and equipment.

The study of precision in human performance demonstrates that it cannot be reduced to just the analysis of quantitative aspects of error analysis. It is also insufficient to limit analysis to classification of errors. The inadequate strategies of performance causing these errors should also be considered. Discovering these strategies and studying their causes is very important for the improvement of precision of performance and reduction of errors. Emotional-motivational and cognitive aspects of activity are very important for the improvement of precision of performance. Changes in the significance of errors for subjects (emotional-evaluative aspects of error analysis) and motivational state of subjects can dramatically change the precision of their performance.

This aspect of studying precision of task performance cannot be reduced to traditional error analysis of operator performance in stressful conditions. Moreover, in stressful situations, a subject can use a variety of strategies to adapt to emerging conditions.

The analysis of the self-regulation process during positioning actions does not restrict us to just considering the *speed–accuracy trade-off.* This reciprocity between time and precision can happen in a variety of ways. The range of variations between speed and precision has multiple possibilities. We should be able to identify specific strategies when subjects strive to optimize relationships between speed and accuracy. The theory of self-regulation of activity helps us more accurately describe the actual strategies of task performance than existing traditional methods of analysis.

4

Thinking as a Self-Regulative System and Task Analysis

4.1 Meaning and Sense as a Tool of Thinking Process

In the previous chapters, we considered general aspects of activity self-regulation. In this and the two following chapters, we will consider special aspects of activity self-regulation. In this chapter, we will describe thinking as a self-regulative system. Meaning and sense are important mechanisms of activity regulation. We will analyze these mechanisms in the context of the analysis of thinking.

In this chapter, in the analysis of thinking, we try to preserve traditional terminology utilized in general activity theory (AT) when the term *sense* is used. In Chapter 3, the term sense is described as an emotional-evaluative mechanism, which has an impact on the perception and interpretation of human information. Here, sense is considered as a subjective content of meaning. Sense is viewed as a result of subjective reflection of the meaning. We consider this understanding of concept of sense further in this chapter. This ambiguity of interpretation of various terms stems from the fact that the terminology in psychology is not sufficiently standardized and different theories of psychology utilize the same term differently.

Meanings are specific mediated tools in the thinking process. Verbal meaning is particularly important in thinking. However, thinking involves various aspects of human activity such as sensual, volitional, emotional, and other characteristics of activity. Hence, thinking is broader than logical operations with meanings. We will first explore meaning as building blocks of thought. The evolution and development of human culture has depended on the human ability to use sign systems. These sign systems continuously changed, evolved, and became more and more complex. Currently, labor is increasingly dependent upon sign system. With the rapid dissemination of computer systems, the role of sign systems has increased and will continue to increase. Not only is the verbal sign system utilized in human language but also other sign systems. In view of this, much greater emphasis should be placed on the analysis of semantics of work domain (Rasmussen and

Goodstein, 1988). A theoretical foundation for this is the concept of meaning and its role in studying activity self-regulation, and thinking process in particular.

Several scientific directions are connected with the study of signs and their meaningful interpretations. Here, we highlight philosophical-psychological considerations by Frege (1948) and Bühler (1934). In psychology, the study of meaning has its roots in the psycholinguistic and verbal learning. In AT, the founder of psycholinguistics is Vygotsky (1978). Other approaches to this problem are widely available in the English scientific press and are not discussed in this work (Ausubel, 1968; Piaget and Inhelder, 1966; Seel and Winn, 1997; etc.). The integration of various scientific directions that study sign systems led to the science of semiotics, which is considered a theory of signs. This chapter focuses on the psychological aspects of meaning from the perspective of AT.

During the process of mental development, the individual internalizes various sign systems and uses them as internal tools for thought. From this follows the existence of two kinds of signs, one of which exists in the external world and the other in the mind of the subject. The signs in the mind of the subject fulfill the role of an internal psychological tool. According to AT, meaning is the result of conscious mental actions of a person, who uses internal and external tools to carry out these actions. The interrelationship between practical and mental actions forms the interconnections between internal and external sign systems.

One of the most important aspects of the study of signs is the elucidation of the relationship between sign and meaning. Signs do not exist in isolation; they are integrated into language systems. Here, we refer to language in a broad sense as a system of socially fixed signs, gestures, sounds, written images, etc., that allow people to communicate and interpret various real phenomena. From this perspective, language includes not only words but also mathematical symbols, formulas, and geometrical figures.

The three most common semiotic aspects of signs are syntactic, semantic, and pragmatic. The syntactic aspect considers the nature of the relationships between signs, the semantic aspect considers the relationships between signs and referent objects, and the pragmatic aspect describes the relationship between signs and the individual interpreting them (Morris, 1946). The semantic aspect of signs is related to the concept of meaning. While the relationship of the sign and its referent is of extreme importance, it is also important to consider that this relationship is the product of human activity. In order to obtain knowledge about an object, one has to perform certain actions and operations with it. They can include discovering the specificity of that object's interaction with other objects, transformation of that object from one state to another, etc.

Signs can be manipulated in the same way that other objects are manipulated. However, in order to obtain the required knowledge of a given sign system, it is not sufficient to manipulate the material form of its sign.

The most important aspect in subject interaction with sign is meaning. Subject cannot manipulate with sign as with regular object which has such physical features as shape, size, color, etc. A sign can be utilized in human activity according to the laws applied to the meanings of the sign. In order to understand the sign, it is important to consider the sign in relation not only to its referent, but in relation to the activity of which it is part and which grants it meaning and sense (Shchedrovitsky, 1995). A symbol is a sign only because people can interpret the meaning of the symbol. However, this does not mean that the interpretation of a sign is a purely subjective process. The meaning of a sign has an objective character in that it is the result of a sociocultural development. This sociocultural development is what gives a sign a standardized method of interpretation. The fact that people can interpret signs in the same way is proof of the objective existence of meaning, which is independent of the subject interpreting the sign.

The material presented demonstrates that an attempt to determine the meaning of a sign exclusively on the analysis of the interrelationship between the object and the sign is impossible. Objective meaning is formed in the process development of human activity, which has a cultural-historical nature. It is also important to consider that not all signs are related to objects. For example, some signs refer to abstract concepts such as increases in speed or the concept of energy; others may refer to the interrelationship of other signs such as mathematical symbols and verbal syntactic markers. The meaning of some signs is determined by the functional relationship between signs. The meaning of others is uncovered through referents that are real objects.

Currently, the study of sign meaning interpretation focuses on the sign's relation to the object or denotation. This approach derives from the work of Frege (1892). However, from the perspective of AT, sign meaning should be studied not only in relation to the object or other signs, but also in relation to human activity. The relationship between an object and a sign and between different signs exists only in the context of human activity.

One type of meaning is *object meaning*, which can be seen as a network of feelings and experiences associated with a particular object (Rubinshtein, 1957). It has sensual-objective characteristics. Object meaning is derived from an individual's practical experience. Object meaning has situational character. Object meaning, or situational meaning of objects, is determined through the relations of action to a situation, and from this point of view, exist only during the performance of a particular action (Genisaretsky, 1975). For example, during play, children ascribe meanings to objects in a very flexible manner. They can apply the word *car* to the chair because they mentally manipulate with the chair as with a car.

A different kind of meaning is categorical or idealized meaning, which is part of the verbal categories that one masters (Gordeeva and Zinchenko, 1982). Categorical meaning has a stable character and is independent of the situation. In the process of activity with others, object meaning can be transferred into categorical meaning. Categorical meaning has an objective social-historical character. This is the objective property of signs. For example, during human–computer interactions a subject knows the meaning of the various symbols on the screen. The constancy of meaning and its relationship to culture allows us to view culture as a semiotic system, or net of meanings, which is superimposed by the individual on the surrounding natural environment and artifacts (Sokolov, 1974). A variety of fields and researchers have studied the history of meaning development specific to various cultures. These studies have often focused on the material culture of different generations, and through material culture, attempted to penetrate the semantics of their world. This line of research further attempts to understand the history of human thinking and conscious development.

An analysis of material presented demonstrates that subjects act with meaning as with real objects. Actions performed by an individual through the manipulation of sign meanings are necessary for communicating with other individuals, interaction with nature, and creating the products of culture. By extracting the meaning of objects, signs, and natural phenomena, an individual organizes, comprehends, and interprets the world. Such important concept of cognitive psychology as SA includes in its content meaningful interpretation of situation.

There is another interesting aspect of study of the meaning that is specifically important in studying human–computer interaction (HCI). These are the relationships between sign, meaning, and the mythological model of the world. We consider this relationship because it helps to understand the concept of meaning. The interaction of culture and meaning is reflected in the mythological description of the world at different stages of sociocultural development. According to Meletinsky (1976), the mythological model of the world is a global modeling sign system, a system which allows the individual to interpret the self, society, and surrounding reality. This type of representation reflects the deep semantics of culture. Reality is reflected in myth when people incarnate various objects in the surrounding environment and transform them into a quasi-subject, or an object with which one can interact and socialize. Ghosts and spirits are examples of symbolic models, which played an important role in the subjective representation of the world by our ancestors.

Such models anthropomorphize nature and allow an individual to communicate with nature in a language familiar to him/her. This is a manifestation of the symbolic modeling property of the human psyche. This example allows us to distinguish between an objective reality and the symbolic, psychological models associated with it. Such models, which are the products of cognitive activity, are projected onto objects and phenomena in the external world. These external symbols are in turn perceived by the psyche as independent objects with their corresponding meanings. For example, in art, aboriginal people can create different sculptures for designating mythical creatures. This class of cognitive models of reality can be referred to as virtual reality (Maslov and Pronina, 1998). This notion of virtual reality resembles, in some respects, the virtual reality simulated by technology, that is, electronically generated images that approximate reality for training, entertainment, and system control. Mythological virtual reality is an imagined, objectively nonexistent, ideal model of the external world, a model invented by humans to facilitate interaction with the external environment. In the case of virtual reality created through technology, the cognitive model is a result of induced percepts. These two examples of virtual reality are similar in that they comprise a mental model of reality within the individual that has no objective correspondence in the real world. The forms of virtual reality change in accordance with the stages of individual development. These virtual realities have properties and characteristics specific to various cultures. Furthermore, the form of virtual reality is determined not only by the interaction of people with the external world, but also by the character of social interaction between people.

A subject anthropomorphizes nature and enters into a dialogue with it even in the absence of other people. This dialogue facilitates the process of comprehension and interpretation of reality. Bakhtin (1979) noted that this dialogue is like the nature of the comprehension and interpretation process. Dialogue is similar to activity and shares its components such as goals, motive, actions, past experience, context, and roles. This role of sign, meaning, and mental models in the interpretation of reality gives them critical importance in the study of work activity and cognitive activity in general.

In AT, meaning is studied in connection with the concept of sense (Bedny and Karwowski, 2004c). Meaning of an object, sign, or word (a verbal sign) is an objective phenomenon, which can be transformed into the personal, subjective sense. In this transformational process, emotional-motivational components of activity play a significant role. Sense can be considered as a subjective interpretation of objective meaning. It depends not only on emotional-motivational components of activity but also on past experience and current goals of activity. Sense has a more personal character and dependence on general characteristics of activity. It allows for the adjustments of individual to more specific situation and problem.

In Sections 3.2 and 3.3, we discussed the concept of the sense from the perspective of emotional-evaluative aspects of activity regulation or factors of significance. In this chapter, the discussion of the concept of sense has some additional coloring. Sense is considered as a subjective interpretation of an objective meaning or subjective personal meaning. The difference between sense as a subjective personal meaning and objective meaning can be demonstrated by an example where we consider teaching of the history of someone's own country. The interaction between students and a teacher is provided on two levels. The first one is the level of the objective meanings and the second one is the level of senses. If the teacher talks about the United States' involvement in World War II, she/he conveys not only facts about the involvement of the United States (objective meanings), but also seeks to convey to her/his disciples a certain attitude to this event. The objective meaning and the factor of significance appear here in unity, which is a result of such subjective interpretation of objective events. In the model of self-regulation, sense or factor of significance is considered as a mechanism that interacts with objective meaning and brings us to a specific interpretation. Thus the concept of sense in both cases does not contradict itself but is rather complementary.

Meaning and sense often overlap. However, they can diverge in cases where the possibility of variable interpretation of the same facts and data exists. When we consider the notion of meaning, its relationship to the external world becomes central. However, when we consider the notion of sense, we focus on those aspects of meaning that are specific to a given subject. Meaning determines the position or role of an object among other objects. Sense, on the other hand, determines the relationship between objects and the needs of the individual (Gal'perin, 1973). The psychological concept of meaning and sense within the theory of activity has some similarity with the earlier philosophical concept of meaning and sense developed by Frege (1892). He considered denotative meaning as an objective characteristic of the object, which should be distinguished from its idiosyncratic interpretation.

In cognitive psychology, there are logical or objective meaning and psychological or subjective meaning (Ausubel, 1968). However, according to this author, "logical meaning is inherent in certain kinds of symbolic material by virtue of its very nature. Psychological meaning, on the other hand, is a wholly idiosyncratic cognitive experience." In contrast, in AT, objective or logical meaning is a result of the socio-cultural development of human activity. Symbol becomes a sign because people assign meaning to it during the course of their activity and social interaction. Therefore, logical meaning cannot be considered as certain kind of symbolic material that posses inherent meaning. Meaning does not exist without human activity, culture, and historical development. Similarly, psychological or subjective meaning cannot be reduced to a purely cognitive experience.

Meaning and sense not only include cognitive but also emotional-evaluative, motivational, and, closely connected to it, goal-related components of activity. Meaning in various situations goes through certain modifications. The notions of meaning and sense in the theory of activity allow us to analyze why the same meaning, in the context of different situations, acquires different senses for the subject/s. Meaning and sense emerge as different components of activity associated with comprehension and interpretation.

Senses can be viewed as various ways of reflection of the same object in the mind. Each of these reflections represents separate functional characteristics of situational elements, which in turn relate to other elements of the situation. For example, the same verbal expression has several interpretations and meanings.

Sense is a dynamic psychological entity that is developed by involving the same sign or object in different systems of functional interactions. While there is a natural commonality between concept formation and sense formation, the two are distinct.

A concept determines stable nonsituated features and characteristics of an object or situation. Sense, on the other hand, is a dynamic entity that includes the extracted features and attributes of a sign and/or situation that is critically important for one particular time or stage of interpretation. Words, images, and nonverbal symbols can be organized as categorical semantic systems that provide objective interpretations of external phenomena and reality as a whole.

For a long period of time, verbalized and conscious aspects of AT were the major focus of studies in AT. A long line of research in AT has studied the verbal and conscious aspects of meaning and sense. However, since the 1970s, scientists working on AT have started to pay more attention to thinking processes that are performed on the unconscious nonverbal level. We consider these aspects of activity analysis during our analysis of orienting and general models of activity self-regulation. Here, we pay attention to the fact that in AT there are such important concepts as nonverbalized and/or unconscious meaning and sense. Verbalized and conscious meaning is always associated with a goal-directed activity. In contrast, nonverbalized meaning is associated with a goal-directed set. Goal-directed conscious level of thinking is performed by thinking actions, and nonverbalized unconscious level of thinking involves unconscious thinking operations. Thinking operations that are organized by a goal-directed set are not components of conscious actions but rather are independent ones. Such understanding of thinking contradicts with Leont'ev's (1978) concept of activity. According to Leont'ev, mental operations are always components of cognitive actions. These two levels of thinking process interact with each other. Hence, thinking has important components of unconscious reflection of reality. In our further discussion, we consider meaning and sense as important mechanisms of activity regulation.

4.2 Meaning as a Function of Standardized Actions

Essential relationships between the elements of the situation reflect the objective meaning of the situation. However, this objective meaning should always be considered in relation to the potential subjects who can interpret it. Specific situations may include the potential meaning for the subject. In cognitive psychology, meaning that is inherent in certain situation is called logical meaning. Psychological meaning is an idiosyncratic cognitive experience. Psychological meaning is the result of a particular kind of activity, the goal of which is not the transformation of a situation, but rather its understanding and interpretation.

Based on performed actions, workers discover functional relationships between perceived objects and elements of the situation and therefore interpret its meaning. The leading role in this process belongs to thinking actions. Based on an analysis of a situation, the subject interprets its meaning in terms of the goals and actions that he/she can perform. In this activity of comprehension, symbolic systems have a special role. Human activity in general and comprehension in particular, involves the substitution of real objects by sign systems, each of which requires a specific system of actions and operations. Here, the sign begins to serve the function of the real object of activity. The subject performs different actions not only with the material form of sign but also with the meaning of sign. During the activity, sign systems allow the subject to extract and fix relevant aspects of objects and phenomena. Having expressed certain aspects of the objects in sign form, we simultaneously determine a system of actions and operations that correspond to that particular form of sign.

An example of this reciprocal relationship between activity and sign is activity with numbers. In counting objects through the use of signs, one extracts those aspects of the object's meaning that pertain to the objects' quantity. The number as a sign makes quantity an entity independent from the objects to which it refers. The number sign system determines the system of operations or actions, which allows us to reflect the quantitative aspects of the objects at hand. In a task with numbers, the subject fixates particular content in sign form, and then performs actions with these signs, actions which in turn are determined by the sign system being used. Subjects can manipulate not only with real or material objects, but also with signs that exist in the human mind or can be externally presented. This requires various types of material or mental actions and transformation from one type of action to another. In AT, one method of actions categorization can be presented in the following way: (1) object-practical actions with real objects; (2) object-mental actions, which are performed with images of actions; (3) sign-practical actions, performed with real sign, for example, receiving symbolic information from various devices as well as transformation of these signs; and (4) sign-mental actions, which are performed mentally by manipulating signs.

The interconnection and transformation of one action type into another demonstrates that material aspects of activity and the semiotic aspects of activity cannot be considered separately. The separation of material aspects of activity from the semiotic aspects leads to the reduction of human activity to that of animals. In the same way, semiotic aspects of activity are impossible without interaction and interconnection with material aspects of activity. Material aspects of activity cannot be separated from semiotic aspects.

Vygotsky (1962) focused on the semiotic aspects of activity. On the other hand in the work of Rubinshtein (1958) and Leont'ev (1977), the focus is on material activity.

For a long period of time, these two approaches were viewed as contradictory. For example, Brushlinsky (1979) criticized Vygotsky for viewing culture as represented by sign systems as autonomous from object-oriented, material-practical activity. In contrast, Kozulin (1986) criticized AT because according to him it neglected social interaction and mediation of activity by signs. He wrote that the work of Leont'ev suppressed Vygotsky's idea of semiotic mediation.

However, the interconnection of practical or material actions with mental, including thinking actions, demonstrates that the development of human activity and its major component—consciousness—become possible only when the individual acquires both material actions and actions with various sign systems.

Meaning and sense is the result of the interaction of these two kinds of actions and operations. Thus, meaning and sign systems should be considered in unity with object-practical activity. Furthermore, the meaning and function of the objects under consideration are tightly interconnected. For example, the shape of a chair determines how we can use it or what kind of actions we can perform with it. The relationship between shape and function has logical and associative interconnections. Changing shape, therefore, can result in a change of function. However, very often minute changes of an object's image do not influence our understanding of that object's functions. On the other hand, when an object's shape is significantly changed, a reconsideration of the object's function must often follow.

While the interpretation of the object's meaning by the subject is often influenced by the object's shape, very often in operator or user activity, the image and function of objects do not match each other. Not only does this make the interpretation of information difficult, it also hinders the regulation of executive actions performed by the operator or user. Their interpretation of the object's meaning in turn partially determines their activity.

In any particular period of time, a work situation can be more or less potentially meaningful for a worker. Interpretation of meaning depends on past experience and utilized strategies of performance. Depending on these strategies, a worker can interpret the same situation totally differently. Therefore, self-regulation of activity plays a critical role in the meaningful interpretation of work situations. In conclusion, meaning provides not only orientation in a situation, but it also regulates the executive actions of an operator. The importance of the notion of action to meaning becomes obvious in noting the relationship between the meaning and the function of an object.

To further elucidate the relationship of meaning and action, let us consider the meaningful interpretation of situation in comparison to Gibson's (1979) concept of direct perception. In the same way that perceptual features can lead to direct perception, semantic features of the situation can lead to direct interpretation. However, in cases when the semantic identifying features of situations are hidden, deliberate thinking operations and actions are required for the interpretation of the situation. The more hidden these essential indicative features are from the perspective of an activity's goal, the more complicated the gnostic actions and operations involved in the process of interpretation and comprehension. In general, meaning and sense are major operative units of thinking activity. Meaning and sense can also be considered as mechanisms of activity self-regulation. When considering sense as a mechanism of activity self-regulation, we mostly focus on the emotional-evaluative aspects of sense, which is a different aspect of the concept of sense in comparison to studying it in thinking. We consider thinking in the following section.

4.3 Study of Thinking in the Framework of Task Analysis

Thinking is directed to the discovery of new properties, relationships between phenomena, and objects of reality that are not directly given in the perceived situation or are unknown to the subject. Thinking is aimed at the transformation of data, including ideal objects in order to discover their properties and relationships. It plays an important role in the study of human work. Thinking is necessary for learning, understanding, and interpreting various situations; forming hypotheses; formulating new problem-solving tasks; and finding ways for their solutions, among others.

An analysis of the literature in ergonomics and psychology shows that now often much attention is given to the mechanisms of memory while thinking is not considered enough when studying human work. Usually, consideration of thinking is reduced to studying decision making processes. However, decision making in thinking is possible only when the problem is understood by an operator. Problem solving cannot be reduced to decision making.

Thinking is a particularly important cognitive process for various problem-solving tasks (task-problems). In AT, scientists distinguish between a problem solving situation and a problem solving task. Problem solving is required when a subject encounters something new and unknown during her/his work activity. For example, a pilot notices an unusual noise coming from the engine. In order to understand the situation, the pilot tries to formulate or create a task-problem. The problem-solving task or task-problem arises from a problem situation but is different from it. A problem-solving situation is something vague and not entirely conscious to the pilot. The pilot begins to realize that something is wrong with the engine, but the specific cause of the problem is unknown to the pilot and, therefore, he/she does not know what type of actions should be taken. At the first stage, the pilot has task conditions (givens) and requirements for what should be achieved. At the second stage the pilot formulates the goal of the task based on givens and requirements. Thus, the task-problem very often is not given to a subject in readymade form and should be formulated by him/her independently. Such tasks usually have their origin in a problem situation. In science, for example, the formulation of scientific problems in a number of cases is more difficult than solving them.

The concept of goal plays a specifically important role in problem-solving tasks. When the task-problem is formulated, conditions of the task or givens and what should be achieved or requirements can be identified. Based on givens and requirements, a subject formulates the goal of a task. In a situation when a task is presented to a subject the goal of the task is formulated for him/her in advance (goal of task problem is presented to a subject in a readymade form). In this situation, the goal should be interpreted and accepted by the subject. A subjectively accepted goal can deviate from an objectively given goal. The process of interpretation and acceptance of a goal is more complex in problem-solving tasks than in skill-based tasks.

Sometimes, task formation is a multistep process. Suppose, a pilot faces not only the issue of noise from the engine but also failure of the displays that should present information about engine malfunctioning. The pilot can hear the strange noise from the engine but does not know its cause. So, she/he has to formulate an orientation hypothesis about failures, select the most subjectively acceptable one, and needs to check this hypothesis. In order to do that, the pilot formulates a goal for the diagnostic task and performs the required actions. If the solution to the diagnostic task did not confirm the orientation hypothesis, the pilot would put forward a new hypothesis, and the cycle would repeat. Only after finding the solution to the diagnostic task it is possible to formulate the task-problem related to the ways of dealing with engine malfunctioning. In this example, we can see the importance of not only instrumental but also noninstrumental information. Information presented by displays is instrumental while information that is not presented by displays is noninstrumental. So, in our example, noise from the engine is an example of noninstrumental information. Very often, tasks that include noninstrumental signals are ill-defined problems.

According to cognitive psychology, any problem includes at least three components (Ormrod, 1990, p. 340): *givens*—pieces of information that are provided when the problem is presented; *goal*—desired end state that the solution to a problem should accomplish; *operation*—actions that can be performed to approach or reach the goal.

However, the last two terms have no precise meaning in cognitive psychology. As discussed before, goal cannot be considered as end state of solutions because this definition ignores such stages as interpretation, acceptance, or formulation of a goal. In problem-solving tasks, a goal often cannot be precisely formulated at the beginning of the problem-solving process but is rather formed in an approximate manner, and as the solution of the problem gradually advances, the goal becomes clearer and more specific.

This definition of goal does not tell anything about consciousness of the goal, its relation to motives, or the goals of actions or tasks. Cognitive operations or actions are frequently described in terms of productions. They are basic units of procedural knowledge (Anderson, 1985). However, in AT, instead of the term *production*, the term *cognitive and behavioral actions* is used. The term production has a mentalist orientation. Thoughts are considered to be in the head of a person without its relevant environmental context. For example, Anderson (1993b) considers our thinking as a system that actively manipulates internal rules in the mind. These rules are triggered automatically in working memory. They define the character of practical actions performed by a human. Utilized rules and practical actions provide new information in working memory. As a result, the sequence of logical rules, and interconnected with them, behavioral actions are developed. However, human activity is *object oriented*. Object-oriented activity is performed by a subject using tools on a material or ideal objects. A tool can also be material or mental. Object-oriented activity is possible only by utilizing cognitive and behavioral actions.

The other approach assumes that one reasoning by association activates another reasoning (Eisenstadt and Simon, 1997). Then, thinking is based not so much on logical operations, but on the associations between related cases from past experience (precedents). These approaches present human thinking as internal rules or associations. The proposed approach describes thinking as automated mental operations that are performed in human memory. These mental operations are considered as automatically triggered mental rules or associations.

However, mental associations in the thinking process are formed gradually. A subject reveals properties of objects and their relationships, which are essential for certain types of problems independently or under the supervision of an instructor. He/she acquires mental actions and operations needed to solve a problem. Mental associations are a result of learning when various associations are formed between elements of a problem and thinking actions involved in solving these problems while working on numerous solutions to specific types of problems. As a result of multiple solutions of specific problems and automation of mental actions, they can be transformed into mental associations, which should be considered as thinking operations. Such associations or thinking operations are not conscious for the subject.

Hence, Eisenstadt and Simon (1997) do not take into account the fact that mental associations are the result of the automation of mental action with repeated solution of typical problems. Thinking, according to AT, involves active manipulation of internal representation of external world according to the goal of an activity. Our intuition is always in play in conscious goal-directed activity. Thinking is a combination of intuitive mental operations that are combined with internal mental and external behavioral goal-directed actions. However, some thinking operations may be transformed into conscious mental actions again. The ability of thinking to operate not only material but also ideal objects with the aid of mental actions and operations contributes to the formation of the internal structure of thinking, which is unique and individual specific.

Cognitive theory does not sufficiently consider motivational aspects of thinking. A problem-solving situation that is not interesting for a person will not activate the intuitive thought processes. Our thinking is an object-oriented process, where objects can be internal ideal or mental and external or material. In Bedny and Karwowski (2007), we describe object-practical, object-mental, sign-practical and sign-mental actions. In our further discussions, we present standardize descriptions of cognitive and motor actions. An analysis of these actions demonstrates that thinking involves manipulation of internal ideal objects or external material objects.

Thinking facilitates transfer from actions with external material objects to actions with their internal representations and vice versa. For example, when a subject manipulates with real objects or external models of these objects, it is an object-practical action, which is basis of concrete-practical thinking. However, when he/she manipulates internal symbols or words, about it becomes verbal-logical thinking. Symbols and words have specific meanings and therefore a subject operates with objects designated by these symbols.

Thinking depends on our past experience or knowledge. Ander knowledge, we understand information encoded in the long-term memory in various ways such as verbally, through images, propositions, and productions (Klatsky, 1975). Verbal knowledge can represent an object as a set of its characteristic features. Therefore, whether an object belongs to a particular category or concept can be determined by using an identification algorithm (Landa, 1984). For example, a general algorithm of categorization, when an object's attributes are connected via the conjunction *and*, can be described in the following manner. The subject has to check in sequence all required attributes. Even if one attribute is absent, then this object does not belong to a particular category. Therefore, strategies of categorization and concept formation process can be described by utilizing various types of identification algorithms.

It should be noted that some scientists do not give a precise definition of the term *concept* and the term *operation of categorization* associated with it. For example, Baron (1992) defined concept as follows: "Mental categories for an object or an event that are similar to one another in a certain respect." However, mental categories are similar to one another not in a certain respect, but according to some essential features that are important for solving specific problems. For example, some objects can be of the same color, which is not important to a concept. During the categorization process, a subject identifies the most essential features of a functionally equivalent object. When he/she encounters these features in a new unknown object, it becomes possible to conclude if this object belongs to the same category. This is an example of how one can go beyond given information thanks to the thinking process (Bruner, 1957).

Therefore, the discovery of the most essential features of an object from the point of view of the considered concept or category should always precede the development of the identification algorithm. Otherwise, unrelated objects can fall into one category. Availability of adequate actions and operations of thought provides effectiveness of thinking when a person is operating with knowledge. Hence, thinking is not knowledge about facts, data, and phenomena but what one does with that knowledge (Bedny and Meister, 1997).

In AT, similar to cognitive psychology, we distinguish declarative and procedural knowledge. Declarative knowledge includes images, concepts and propositions. Propositions are units of knowledge that reflect the relationship between different objects and considered as separate statements or assertions. They can be judged to be either true or false. Concepts and propositions are interrelated but not the same. A student can acquire the correct concepts of an object, for example, he/she can list its characteristic features, but may not be able to give a correct definition of it. Individual items of declarative knowledge, without being organized hierarchically as a complex structure, cannot be efficiently employed in practice.

Procedural knowledge is knowledge about cognitive and motor actions and operations.

A subject can have a large repertoire of knowledge about concepts, images, and propositions but only a small repertoire of mental actions and operations and that would limit her/his ability to apply this knowledge. Such person is knowledgeable but cannot apply that knowledge in practice. A subject needs knowledge not only of the facts but on how to act on them to form his/her thinking process. This is the operating system of human thought. The operating structure of thought is formed by integrating various systems of mental actions. Algorithms of mental activity are an effective tool of forming thoughts. Procedural knowledge, which is required to achieve a goal, is not sufficient for a trainee to perform certain procedures. Knowledge should be transformed into skills during extensive practice. In cognitive psychology, knowledge of the facts is known as declarative knowledge and knowledge of operating with them is considered as procedural knowledge (Anderson, 1993a; 1993b). It should be remembered that knowledge and actions are not the same. A subject can have knowledge on how to perform, but cannot perform.

In cognitive psychology, there is an attempt to use Anderson's cognitive architecture for virtual product design (Carruth and Duffy, 2008). For this purpose, the authors recommend integrating cognitive and digital human models. However, the material presented earlier demonstrates that cognition cannot be considered as a system that manipulates by internal rules. Human mind actively interacts with the external environment (Gibson, 1979; Rubinshtein, 1959). Hence, systemic-structural activity theory (SSAT), which utilizes the concept of cognitive and motor actions, provides an ecological approach to ergonomics. The ecological framework has already some impact on the field of ergonomics (Dainoff, 2008; Hancock et al., 1995).

According to Rubinshtein (1958) and his follower Brushlinsky (1979), thinking is a process that unfolds in time. During this process, a subject continuously reveals new conditions and requirements needed to complete a thinking task. However, activity is not only a process but also a structure, which also unfolds in time (Bedny and Meister, 1997). This structure can be described as a logically organized system of cognitive and practical actions and operations. When a worker performs a task that includes thinking components, the sequence of actions to be performed is not known in advance. The sequence of mental and practical actions to be performed depends on information perceived by the worker. Moreover, determining what type of information should be selected for problem solving is an important stage in the thinking process. In each case, the worker should evaluate the presented combination of signals, select an appropriate decision, and perform appropriate executive actions. The worker has to evaluate each stage of performance and correct his/her actions if necessary. Such task problems can be described using deterministic or probabilistic algorithms. It is necessary to uncover algorithmic mental processes, which should be utilized during solving specific class of problems and then they have to be explicitly described in terms of mental actions and operations. Human algorithm describes a logical sequence of human cognitive and behavioral actions. If such algorithms do not guarantee success, they can be seen as heuristic prescriptions of an algorithmic type. Algorithms are presented to trainees as prescriptions. If such prescriptions cannot guarantee correct solution, they can be considered as heuristic prescriptions. Hence, many thinking strategies can be discovered and then described as algorithmic processes. Knowledge of such algorithms helps trainees to solve all problems of a certain class.

Without explaining the concept of action as it was done in SSAT, it is very difficult to describe thinking by utilizing the concept of human algorithm. A subject should know the meaning of individual actions and operations and how they have to be performed. Actions include mental operations. Hence the content of mental actions should be discovered for describing deterministic or probabilistic algorithms when analyzing work performance. Fully creative problem-solving tasks are rare during a production process or in a controlled automatic system. For example, if astronauts encounter an unknown problem in an emergency situation, the Command Center would be involved in resolving the situation.

AT scientists distinguish different types of thinking processes depending on the work that is under study. Practical thinking is usually divided into perceptually manipulative thinking, perceptually imaginative thinking, theoretically logical thinking, and operative thinking (Bedny and Meister, 1997). This division of thinking is relative because such types of thinking can be encountered in various combinations. However, distinction between these types of thinking on the basis of their dominance allows for a more efficient analysis of work-related problem-solving tasks.

Perceptually manipulative thinking utilizes a *trail-and-error* method. Usually it is not a blind trail-and-error method as in behavioral theory of learning. Such a method includes some predictions and even promotion of hypothesis. Only in very stressful and time-limited conditions this method approaches chaotic manipulation. The trail-and-error method is a self-regulative process in its exteriorized form. A subject evaluates the consequence of his/her actions and corrects them. Thinking is based on the perception and evaluation of changed objects and alternates with the behavioral action aimed at changing them. Also, mental actions aimed at evaluating the identified changes are involved in this process. A worker normally uses such a thinking method when there is uncertainty and lack of information about a problem-solving task.

Perceptually imaginative thinking is important for an operator when he/she interacts with various devices and controls and manipulates with perceptual images of objects and secondary images in his/her memory. The key to solving a problem is to transform images of objects, regroup them and imaginatively anticipate results, and so on. Therefore, imaginative actions play a leading role in such thinking. Verbal and real actions are of secondary importance, which is specific to various types of controllers. With this type of thinking, an operator uses cues from instruments and manipulates controls.

Verbally logical thinking is based on the use of abstract concepts and determining relationships between them. This type of thinking is used by an operator for preparatory, intermediate, or final calculation or problem solving according to certain rules. Such thinking is directed at establishing links between abstract concepts.

Pushkin (1978) introduced the concepts of operative task and operative thinking. Usually operative tasks are performed based on visual information. Task conditions (givens) represent a stable structure with some embedded dynamic elements. The goal of a task is for the subject to transfer initial position of dynamic elements into the required pattern. Such tasks are encountered in games where information is presented visually. For example, in a chess game, the subject formulates problems specific to a particular stage of the game based on visual information. The main aspect of such tasks is the identification of the relationship between the dynamic elements of the situation. Such tasks are called operative tasks and the thinking involved in their solution is called operative thinking. These tasks require mental restructuring of the situation. They can very often be encountered in the study of human work. One example of such a task is that of a railroad dispatcher (Pushkin, 1978; Pushkin and Nersesyan, 1972). The tasks that are solved by the dispatcher have a problem-oriented character and are performed based primarily on visually presented information supplemented by verbal information. The instrumental panel has both static and dynamic components. Dynamic components of the panel include visual indicator that depicts movement of trains. The dispatcher creates a dynamic mental model of the situation based on visually received information from the panel and performs mental actions that enable him/her to coordinate the movement of trains at the railroad. In this task, trains are the dynamic units, which move upon a fixed structure—in this case, the railroad. Hence successful performance of the task depends on the gnostic dynamics of the operator. Under gnostic dynamics Pushkin understood the ability of a person to constantly change his/her mental picture of a situation in spite of its external constancy. Gnostic dynamic is a result of the system of conscious actions and unconscious mental operations of manipulation with externally given unchanged information. In our example, presented on the panel data is referred to as informational model of the situation which is an external tool of the thinking process. Based on this information and past experience, a dispatcher can develop his/her internal or mental model of the situation. Another distinctive feature of this task is that a train moves almost continuously except for occasional predetermined stops. However, information about the continual movement is discrete in nature. Here we can observe contradiction between continual movement of the controlled object and the discrete information about this movement. Such a description of a task allows to develop a practical recommendation for the control panel design. The study of operative thinking was developed further in SSAT where it is considered to be a complex self-regulation process responsible for the creation of a dynamic mental model of a situation. For more details, see Bedny and Karwowski (2004c, 2007, 2008b).

4.4 Self-Regulation Model of Thinking Process

From an SSAT perspective, we can describe the thinking process as a self-regulative system that is comprised of various stages presented as interdependent functional blocks (see Figure 4.1).

The self-regulation model of thinking process takes its roots in multiple studies of thinking process in AT and cognitive psychology. The source of thinking in a production environment is information that is received by a worker about ongoing work processes. When a worker perceives information from a computer screen or a display panel, the essential data about a task can be considered as an informational model. A worker selects information that is subjectively most important in a presented situation. He/she also can select required information directly from the environment. The selection of information also depends on emotional-motivational aspects of the worker's activity. Therefore not only externally presented data but also the internal mental state of the worker, his emotionally motivational state, is a key factor in selecting the required information. When studying the thinking process, externally presented data or data extracted by the subject analyzing the situation are called conditions (givens) and requirements.

In the first stage, at the core of the receiving process are perception and sensation. This stage of receiving information is closely related to thinking that is involved in the understanding of the nature of perceived phenomena. The stage of obtaining information about an ongoing work process is an active one, when a worker continuously shifts her/his attention from one object or element of situation to another. Thus the process of receiving information

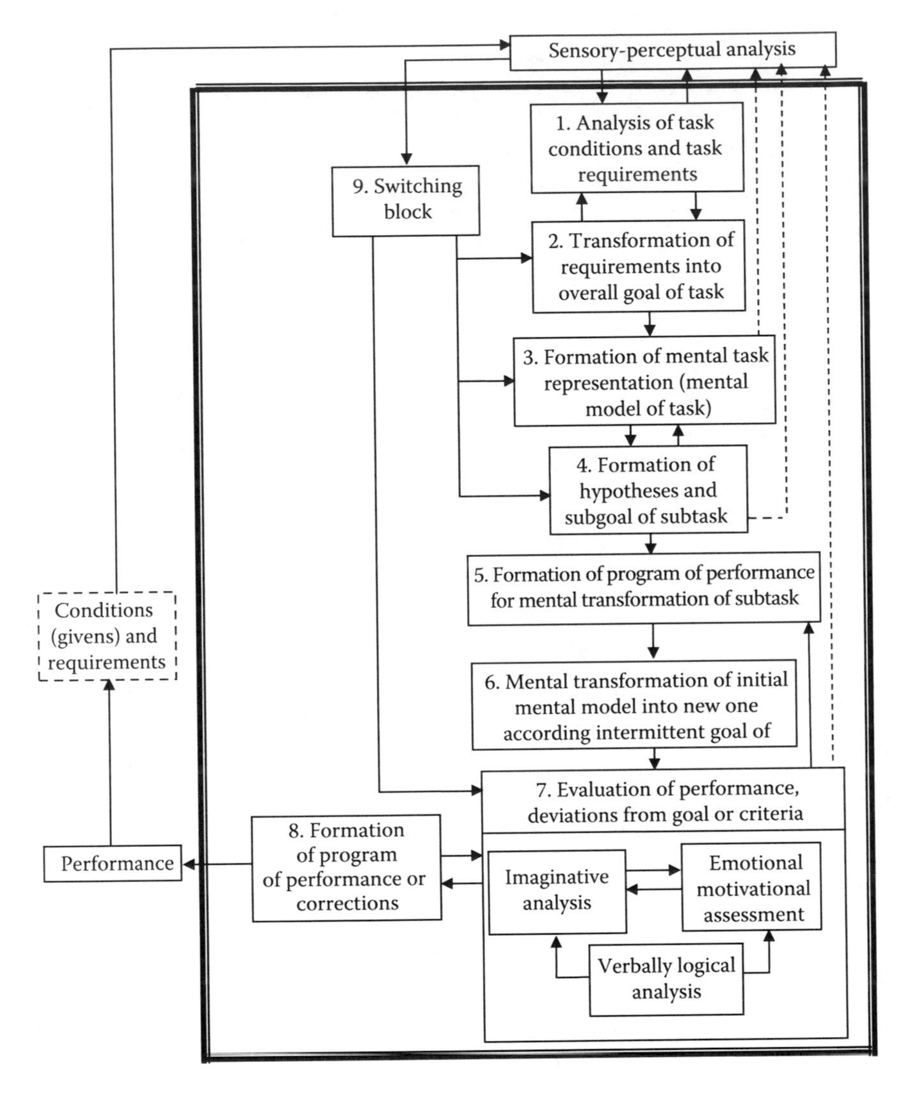

FIGURE 4.1
Self-regulative model of the thinking process.

involves not only sensation and perception but it is also closely linked to past experience or memory, attention, and thought. The content of perception also depends on the goals and motives of an activity (see feedback from block 2 to sensory-perceptual analysis block, Figure 4.1). However, at this stage of task analysis, the main focus of study is sensation and perception. At the stage of the sensory-perceptual analysis of a problem-solving task, verbal components of thinking are presented in a reduced form. Verbal processes are mainly connected with the mental categorization of elements of the situation, that is, assign them to a class of related objects. At this stage of task performance, a subject does not evaluate the relationship between elements of a situation. In some cases, the success of subsequent stages of thinking depends on how information is organized and perceived. The considered stage is not directly involved in thinking. This is a stage of activity that precedes the thinking process. Therefore, we present the stage of receiving information as functional block that has been left outside of the boundaries of the thought process.

When function blocks that are directly involved in the thinking process are activated, the role of speech significantly increases. The perceptual process begins to be monitored by functional blocks that are directly involved in the thinking process. This provides reorientation of attention to externally presented elements of information.

The first functional block that is directly involved in the thought process is block 1 (analysis of task conditions and requirements). At this stage of task performance, a subject defines task conditions and task requirements that can be most relevant to solving a problem. A subject restructures a task and assesses the obtained results according to the subjective significance of the elements of a problem-solving task. At all considered stages of restructuring, such thinking operations as analysis and synthesis are most clearly revealed. Restructuring of a problem can be performed not just practically but also mentally.

Mental restructuring of the situation is performed based on a combination of intuitive and logical forms of thinking. The inner speech—not spoken out loud but internally—plays an important role in these stages of thinking regulation. The opportunity to manipulate elements of the situation that cannot be directly perceived but only mentally constructed emerges. Major operative units of thinking actions are concepts, propositions, and mental images. Thinking actions and operations are involved in the manipulation of these operative units of thinking. Such actions have specific meaning and reflect the external material world. Often, practical thinking also involves manipulating real objects. Therefore, thinking in this model is described as an object-oriented process that cannot be reduced to manipulating symbols according to internal rules of the brain. Rules are not yet actions. A subject might know the rules but not be able to utilize them. The subject should learn how to act according to the rules.

The formation of mental task representation has a preliminary character because the goal of a task is not yet formed or accepted by the subject. This becomes possible only after block 2 (transformation of requirements into overall goal of task) is activated and the final goal of the task is formed and accepted. Blocks 1 and 2 have forward and backward connections and therefore these blocks are mutually adjusted (see Figure 4.1). A goal is of particular importance in the regulation of the thinking process due to the complexity of goal formation, its interpretation, and acceptance by a subject. Problem-solving tasks often are not given in a ready form. A subject has to formulate a task independently. The success of solving such tasks depends largely on adequacy of the goal of the task formed by subject. In cases where the goal is given to the subject, there can be difficulties associated with the interpretation and acceptance of an objectively defined goal. As can be seen from the presented model, a goal can be revised at the later stages of the thought process regulation in cases when it is discovered that the thought process was incorrect or inaccurate. Very often, the goal of a problem-solving task can be formulated in a very general manner at the beginning of its performance. Only during such task performance the goal gradually becomes more specific and clear.

Let us consider as an example of the well-known and popular game Sudoku. This game is a logical, number-placement puzzle. A puzzle setter provides a partially completed grid. The goal of the subject is to fill a 9 × 9 grid according to the specified rules. However, the exact positions of the numbers are unknown to the subject until the end of the game. Hence, the goal in the beginning of the game has a very general form. Thanks to this generalized goal the subject knows which direction he/she has to take. Gradually the goal becomes more specific. From this example, one can see that a goal is not a readymade standard to which human activity is directed, as stated by some scientists. The concept of the goal and subgoal formation process has fundamental meaning in the analysis of thinking and problem solving (Anderson, 1993a; Bedny and Meister, 1997; Newell and Simon, 1972; Pushkin, 1978; Tikhomirov, 1984, etc.). However, understanding of a goal and the process of its formation in AT is different from its understanding in cognitive psychology.

Based on the analysis of task conditions and accepted goal, a subject develops his/her mental model of the task. A mental model, similar to a goal, includes verbally logical and imaginative components. The model can be corrected if necessary. This stage of the thinking process is associated with block 3 (formation of mental task representation or mental model of task). This block has some similarity with block 9 (subjectively relevant task conditions) in models of self-regulation of activity (see Figures 3.1 and 3.2).

When a mental model of task is created, the solution stage begins. The initial mental model of a task often cannot facilitate the achievement of a desired result. Hence, after understanding a task at hand, a subject divides a task into subtasks. He/she begins the formation of subtasks by formulating various hypotheses. Each hypothesis has its own potential goal. Based on comparison and evaluation of such hypotheses, a subject selects one and formulates the first sub-goal associated with the selected hypothesis (see functional block 4 *formation of hypothesis and subgoal of task*).

Comparing a new subgoal with an existing mental model of a task-problem allows transforming an original mental model into a new one that is adequate for a new subtask. Therefore, this stage of thinking process is associated with block 5 (formation of program of performance for mental transformation of subtask) and block 6 (mental transformation of initial mental model into a new one based on intermittent goals of task). The program of performance in block 5 is used only for the mental transformation of a situation. Block 6 is responsible for the mental transformation of the initial mental model of a subtask into a new one according to a new hypothesis and a task subgoal. The obtained result is evaluated in block 7 (evaluation of performance, deviations from goal or criteria) based on its correspondence to the goal or sub-goal of a task and a subjective criteria of success. The last one can deviate from the objective goal, and therefore a subject can evaluate her/his own result as a successful one even if it is lower than required. For example, a blue-collar worker can lower quality of performance in order to increase quantity.

We have to distinguish goal from the subjective criteria of success. A goal is formulated in advance, but subjective criteria of success can change during task performance. For example, there are two methods of problem solving. One method is easier and less precise. The other one is more difficult but more precise. The second method fits the requirements but a subject can select the first method of problem solving in order to reduce fatigue and evaluate obtained result as subjectively acceptable. The development of subjective criteria of success can be based on verbally logical analysis of deviation from a goal or imaginative and emotional-motivational analysis of such deviation. An emotional-evaluative analysis is not always conscious and precise while a verbally logical evaluative process is more precise. At the same time, subjective criteria of success during problem solving can coincide with a goal. The more significant a problem-solving task is for a subject, the more rigorous are the subjective criteria of success. If a task is specifically significant for a subject, he/she can introduce subjective criteria of success that is even higher than the existing requirements or goal. Hence, function block 7 has three subblocks. The first one is involved in verbally logical analysis and evaluation of the thinking process. This block depicts a predominantly conscious and precise evaluative process that utilizes conscious verbalized criteria. The subblock on the left-hand side (imaginative analysis) is involved in not sufficiently conscious and therefore sufficiently precise evaluative process. The most imprecise is the third subblock on the left-hand side *emotional-motivational assessment*. This process of evaluation is often not verbalized and sometimes even unconscious. Relationship between these three subblocks determines specifics of evaluation of thinking at this stage of regulation process. Function block 7 has a feedback connection with block 5 because performance program for a subtask can be corrected and mental transformation of initial mental model into new one can be repeated based on such evaluation. This cycle can be repeated several times until evaluative stage (block 7) is perceived by the subject as positive. Therefore, function blocks 5–7 form a closed loop subsystem, which, allows to carry out multiple mental transformations of the situation and its evaluation. However, the real situation during such transformations remains unchanged.

Only after completion of the evaluative stage, a worker can develop her/his program of performance that can be utilized not for mental but for real transformation of the problem-solving situation (see block 8, formation of program of performance or corrections). Block 8 can be corrected based on feedback from block 7. At this stage, the evaluative process is based mainly on verbally logical analysis, which is a predominantly conscious process.

After this stage of thinking regulation is complete, the performance block is activated. This block is outside of the thinking stages of activity regulation because it depicts an executive stage where a subject modifies a task and adds newly obtained data to it. At this stage, changes are made not just mentally but also in reality. Such changes can be perceived and evaluated in a sensory-perceptual block.

Results of an evaluative stage can be used in two ways. One involves transmitting information into block 1 and then the whole cycle of thinking process can be repeated. The other one involves switching block 9 where the result of sensory-perceptual analysis can be used by blocks 2 or 3 or 4. For example, if data is transmitted to block 4, then this information bypasses blocks 1 through 3. Switching block 9 can also transmit information directly to block 7 (evaluation of performance, deviation from goal or criteria). The actual performance result of a problem-solving task is evaluated in block 7 and the obtained data can be used for corrections of performance in block 8.

If received data is evaluated positively, the thinking process cycle is complete. If the result is evaluated negatively, information can go from block 7 to block 8 or block 5 or all the way back to the sensory-perceptual block. If information from block 7 goes to block 8, the program of performance can be corrected. When information from block 7 enters block 5, a subtask can be mentally corrected. If information from block 7 goes to the sensory-perceptual block, the whole task is reevaluated. Hence, switching block 9 and block 7 can selectively influence various blocks. Connections between the blocks demonstrate that the model can very flexibly describe strategies of the thinking process. Therefore, the circle of thinking regulation can be shorter or longer, depending on the specificity of a problem-solving task and the thinking process strategies selected by the subject.

The sensory-perceptual block can periodically interact with blocks 3 and 4 through the switching block 9 and their feedback to sensory-perceptual analysis (see dashed lines). Similarly, the sensory-perceptual block interacts with block 7. This happens when internal mental activity needs support for the thought process from externally presented data. Due to the interconnection of the considered functional blocks or stages of thinking regulation, a problem-solving task is continuously modified mentally or in reality until the required solution is obtained.

Analysis of the self-regulation model of the thinking process shows that there are two types of feedback. One of them involves the presence of external feedback used after the actual practical implementation of actions that change the external situation. This type of feedback is associated with the completion of some stage of the thinking task or the final evaluation of its results. Another type of feedback is mental, where a subject manipulates elements of the situation in her/his mind and evaluates the result of such manipulation with ideal objects. This internal mental feedback arising during mental transformation and assessment of the situation allows prevention of real errors. It is important to understand the difference in the interpretation of feedback between cognitive psychology and SSAT. In cognitive psychology, feedback is possible only when a subject receives information about the result of external behavioral actions execution.

Description of functional blocks in the model discussed demonstrates the range of issues that need to be considered at various stages of analysis of the thought process.

Each block in the self-regulation model of the thinking process requires its own specific method of analysis. Thinking activity is considered as a multidimensional system. Therefore, analysis of thinking as a self-regulative process also requires the development of interdependent and supplemental methods of analysis that should be organized into stages in accordance with the selection of analysis functional blocks. Forward and backward connections between such blocks show relationships between stages of thinking and allow conducting a more effective analysis of strategies of the thought process. Depending on the nature of a task-problem, a practitioner can choose the most relevant blocks for her/his analysis. Here, as well as for the analysis of the self-regulation process of the entire activity, each functional block can be compared to a window. We open each window one at a time and consider the same object from different perspectives. Thanks to such analysis various aspects of the thinking process can be considered. All aspects of such an analysis can be compared, and therefore this method of study can be qualified as a qualitative systemic analysis. Analysis of the presented material demonstrates that the described concept of thinking is important for the study of human work.

The presented concept of thinking can be used for the analysis of computer-based tasks. Computer-based tasks should often be considered as problem-solving tasks. Some of these tasks have well-defined attributes: the overall goal of the task, the method of task performance, etc. Some computer-based tasks can be ill-defined. Studying mechanisms of human thought is critical for an understanding of the nature of computer-based tasks and strategies of their performance.

An informational model presented on a computer screen can be considered as a structurally organized system of tools and objects that can be modified and compared with a user's mental models. Such externalized tools and objects can be viewed as an external support of the thinking process. In general, computer is a powerful means for a user's thinking process. A computer-based task has one final or overall goal and multiple intermediate or subgoals associated with corresponding subtasks and human actions. The number of subgoals of a task depends on the strategies of performance and an overall task complexity. The more complex a task is, the more intermediate subtasks users utilize in order to achieve the final goal.

Therefore, a computer-based task has a lot of similarities with the tasks that are performed based on dynamic visual information and involve operative thinking. Information is presented on the screen in visual form. Data on the screen consists of a number of discrete elements and can be considered as a structure with static and dynamic elements. A user continuously changes this structure. Based on the obtained result the data are immediately corrected. Performance of a HCI task can be treated as a sequence of cyclical processes, the purpose of which is the reduction of differences between a current state and an end-goal state. A problem solver works on the subgoals as a means of reaching the ultimate goal. This is similar to what Newell and Simon (1972) described as a means-ends analysis. However, there are also some differences. First of all, the concept of goal in AT differs from the one in cognitive psychology. In SSAT, thinking is considered to be a self-regulative process, which includes various functional mechanisms. Final and intermediate goals always interact with a corresponding dynamic mental model of a situation. Not only verbally logical but also imaginative components are important in this analysis. During various stages of task performance, a subject can formulate subjective criteria of success that is adequate for each stage of problem solution (Bedny and Meister, 1997). These criteria facilitate feedback influences that are necessary for the thinking process. A subjective standard of success usually is formulated on a mental plane. Hence feedback is not only a result of material but also of thinking mental activity. Self-regulation can be performed on the internal mental plane. Thanks to such feedback each result of thinking process is evaluated based on the subjectively developed standards of success. These subjective criteria can be incorrect and subjects can discover this later on. If such correction does not take place, a performance result can be inadequate or a subject could fail in solving the problem at hand. Hence, thinking evolves as a goal-directed self-regulative process with dynamic and constantly corrected subjective criteria of success, which can be either material or mental. Stages of thinking during performance of computer-based tasks can be described in a general form as follows:

1. The first stage includes formulation or acceptance of the goal of task and development of mental representation of the task.
2. In the next stage, the subject promotes a hypothesis in the frame of the formulated subtask and performs actions that provide solution to the considered subtask.
3. In the final stage, the subject formulates a final subtask and evaluates actions that lead to finding the solution to the problem in general.

General logic of task performance has a sequence of stages. These stages have feedforward and feedback connections, which can be developed on the internal plane of the thinking process. Each preliminary stage is the source for the next thinking stage. The basis for problem solving is the continuous reformulation of a task and the development of its corresponding mental models. A program of task performance for each particular subtask and a task performance in general depend on these models. The overall goal of the task and the subgoals of the subtasks are considered as conscious cognitive mechanisms that perform predictive and regulative functions at each specific stage of performance. After performing a qualitative analysis of strategies of a thought process based on this data, the considered task should be described algorithmically.

4.5 Integration of Cognitive and Activity Approaches in the Study of Thinking

The presented material demonstrates that there are some differences between activity and cognitive approaches in studying thinking. At the same time, they have some similarities and can be integrated into a unitary theory of thinking. In this section, we give a short analysis of the cognitive approach to thinking and compare it with the activity approach. This allows outlining the ways of integrating data both approaches.

The most influential theories in cognitive psychology were suggested by Newell and Simon (1972), where problem-solving thinking process is described as a means-ends analysis.

Such analysis involves dividing a problem into a series of smaller subproblems each of which is then solved. As a result, the distance between the original state of a problem and the final goal is reduced. The main concepts in this approach are the problem space, problem-solving state, the goal, the operator, etc. The concepts of state and operator define the concept of a problem space. Problem-solving states are understood as transition from an initial state of a problem to intermediate states and finally arriving at a state that satisfies the final goal. The major mechanisms of thinking are difference reduction and sub-goaling. Problem-solver tries to select operators that produce state more similar to the goal state. A means-ends analysis is organized as a cyclical process. However, the concept of self-regulation of thinking or cognition is not considered by this approach. A means-ends analysis uses such terminology as *match current state to goal state to find difference, eliminate difference, search for operator relevant to reducing the difference*, which present useful but still very general ideas about thinking.

These authors extensively use such concepts as goal and operations when describing a solution process. However, these terms do not have a clear meaning and differ in their meaning from AT. Interrelationship of goals with motives and consciousness and differences between the goals of individual actions and that of the whole task is not discussed. Moreover, AT distinguishes conscious thinking actions and unconscious thinking operations (Bedny and Karwowski, 2007). In cognitive psychology, these terms are used as synonymously. In AT, these terms are used to describe subconscious and unconscious thinking processes. When thinking operations are not integrated into conscious thinking actions, such automatic thinking processes do not require executive control from the prefrontal cortex of the brain. However some mental operations can be integrated by the goal of an action into conscious thinking actions and these components of the thinking process become conscious (Bedny and Karwowski, 2007). Usually, the conscious components of thinking can be verbalized. Unconscious thinking components more often involve manipulation with images. These components of thinking overlap and can be transformed into each other. Manipulation with images based on the goal of actions should be considered as imaginative actions that play a critical role in the thinking process.

Goal formation and goal acceptance processes are also not identified clearly in human information processing when studying thinking. Even in cases when the goal of a problem-solving task is presented to a subject in a ready-made form, it still has to be accepted and interpreted by the subject. Being one of the leading experts in the field of thinking, Anderson (1993b) also did not pay enough attention to these aspects of the study of goal. A goal, according to him, is simply something that a subject seeks to achieve. Goal formation and goal acceptance are also not discussed in spite of the fact that they are important steps of the thinking process. Anderson, like Newell and Simon, does not discuss the relationship between goal and motives, goal and consciousness, and so on. All these aspects of goal should be clarified during the analysis of the thinking process. It is not incidental that some practitioners working in applied fields suggest eliminating the concept of human goal (see, e.g., Diaper, 2004, p. 16). Diaper wrote that goal should be assigned not to people but rather to a work system. Therefore, human goal and system goal are not distinguished by some practitioners. Such scientists as Austin and Vancouver (1996) and Carver and Scheier (2005) expressed similar views on the concept of goal. The attempts at eliminating the basic concept of human goal is easily explained by the fact that there is no clear definition of this concept in psychology outside of AT and SSAT. Data presented by Anderson (1993) and Newell and Simon (1972) contradict data presented by Austin and Vancouver and Carver, Scheier, and others. Such disagreements make it difficult to study the thinking process in applied studies. The interpretation of the concept of goal in SSAT can significantly reduce contradictions in the understanding of thinking in applied, SSAT, and cognitive psychology.

In cognitive psychology, thinking is described as a process of manipulating symbols. According to this approach, thinking provides transformation of signs and symbols.

However, when a subject manipulates symbols he/she manipulates their meaning and through them manipulates objects of the real world (Brushlinsky, 1979). Moreover, such factors as culture, beliefs, significance of information, emotions and motivation, experience, and individual differences are critical for human thinking. This aspect of study of thinking is not clearly analyzed in the human information processing approach.

A disadvantage of this approach is that the interpretation of the thinking process is largely limited to formal manipulation of symbols based on logic. Such basic terms as goal, subgoal, operator, rules, etc., do not have a clear psychological meaning. Problem solving can be considered as finding some sequence of problem-solving operators that allow traversal from an initial state to a final goal state. According to Newell and Simon, operator is an action that transforms one state into another. However, the concept of action is not clear in their theory. Not only goal but also action has totally different meanings in various psychological approaches. AT considers cognitive and behavioral actions as the building blocks of activity. Activity can be described as a logically organized system of cognitive and behavioral actions that are classified according to certain criteria.

Anderson (1993b) has elaborated on Newell's and Simon's ideas introducing such new important concepts are declarative and procedural knowledge. According to this author, thinking takes place within a means-ends problem-solving process. Problem-solving states can be understood as transition from an initial state of a problem to intermediate states and finally arriving to a state that satisfies the final goal.

In AT, this is the transformation of a situation from an initial state to an intermittent state and finally to a state that corresponds to the goal of the activity. If the intermediate state does not satisfies the intermediate goal, this step of solution should be reconsidered. Formation of adequate subgoal, mental model and hypotheses, decision making, and subjective criteria of success are involved in each intermediate state.

Thinking should be considered as a complex self-regulative system with a variety of strategies for solving the same task-problem. Such a self-regulative process includes various stages or function blocks. Each stage is evaluated based on feedforward and feedback influences between functional blocks of the self-regulation process. In SSAT, thinking is described as a self-regulative process. Function blocks or functional mechanisms can be considered as important units of analysis of thinking. Self-regulative models of thinking depict functioning of the thinking process. The more precisely we can describe functions of the thinking process at various steps of its regulation the more precise we can describe strategies of thinking. Due to self-regulation it becomes possible to describe the transformation of the preliminary state of thinking into a new state much more precisely.

Anderson (1993a) described origins and nature of problem-solving operators that are utilized in the means-ends analysis. The origin of operators includes three stages: the interpretive stage, the knowledge compilation stage, and the tuning stage. Let us briefly consider the first two stages. The first stage is known in AT as the orienting stage of activity. It includes the interpretation of the meaning of information, the creation or acceptance of goal, the formation of dynamic models of the situation, and the evaluation of its difficulty, significance, etc. These mechanisms are interconnected through feedforward and feedback influences. Hence, Anderson's interpretive stage of the origins of the problem-solving operator can be more precisely described as an orienting stage of activity. The knowledge compilation stage is the process of transiting from an interpretive stage to a procedural stage. The procedural stage, according to Anderson, is a system of production rules that are condition–action pairs. At this stage, logical rules are activated in memory and based on them various actions are performed. The results of such actions add new information to the working memory. Thanks to this a chain of production rules is created. The production rules system is a computer-like deterministic algorithm. However, human thinking cannot be reduced to an automatic actualization in memory If-Then rules. A subject might know rules but is not able to perform the actions. Conscious and unconscious thinking processes are interrelated. Intuitive processes can be activated when a subject preliminary consciously attempts to solve a problem or at least thinks about it. Motivational processes are also involved in such interrelationships. The theories proposed by Newell and Simon (1972) and Anderson (1993b) do not pay sufficient attention to the emotional-motivational aspects of thinking.

Activity approach considers thinking as manipulation not only with symbols but also with mental objects such as images, meanings that correspond to certain objects or events, etc. Operating with material and ideal objects can be achieved by practical and/or cognitive actions and operations. From this process, a subject does not only manipulate by a logical structure but also creates mental models of a situation in which imagery and sensual experience is critically important. Mental model is a critically important concept in the self-regulation of activity and contemporary task analysis (Bedny and Karwowski, 2007; Norman, 1988). It has also become an important construct in ergonomics. Gnostic dynamic plays an important role in the creation of a mental model of a problem-solving task. This process often cannot be verbalized, and, to a significant degree, can be a subconscious or even unconscious process. Gnostic dynamic does not involve actual changes in a problem-solving situation. This mechanism of thinking is not discussed in cognitive psychology. Human algorithm in AT cannot be reduced to the rules that guarantee a solution of specific types of problems. It is a logically organized system of cognitive and behavioral actions. In SSAT, there are clearly developed procedures for the development of such algorithms that does not exist in cognitive psychology.

A subject does not simply manipulate with signs and interpret their meaning in the thinking process. Meaningful interpretation of a sign system depends also on the subject's motivational state. If the motivational state changes within the same situation, it alters the subjective meaning of that situation. Objective meaning is transformed into subjective meaning depending on the sense of actions and task (Bedny and Karwowski, 2004). The same meaning in the context of a different situation and motivational state leads to a new interpretation. Goal and its relationship to emotional-motivational components of activity determine the way objective meaning transforms into subjective meaning. Hence the same objective meaning can be interpreted by a subject in a number of different ways (Bedny and Karwowski, 2007).

Theories of thinking developed in cognitive psychology are very interesting. These theories provided general direction for further studies of thinking. However, these theories lack a well-developed conceptual apparatus adequate to psychological study during task analysis. Everyone can understand goal, action, operator, strategies, etc., in their own way. In thinking process production rule system remained computer program. The study of thinking concentrates on the study of memory processes. All these differences in the understanding of thinking can be eliminated by the integration of cognitive psychology and AT. There is a possibility to establish a unitary conceptual approach to the study of thinking based on such integration.

In this chapter, we consider thinking from a functional analysis perspective when thinking is presented as self-regulative system. The major units of analysis are function block. The concept of a function block is a productive theoretical construct, which emphasizes the importance of the integration of mental processes in the analysis of thinking process regulation. Description of the functions of each block in the presented model specifies the content of analysis of the thought process at different stages of its regulation. The significance of each block in the regulation of the thinking process is determined by the specifics of the problem-solving task being analyzed.

However, the thinking process can also be studied from a morphological analysis perspective where the main units of analysis are cognitive and behavioral actions. At this stage of analysis, thinking process can be described by utilizing the concept of human algorithm.

Analyzing how people solve various task-problems demonstrates that subjects utilize strategies of thinking that can be described as human algorithm. Such algorithms describe logical organization of cognitive and behavioral actions during problem solving.

Therefore, thinking strategies can be described by using the human algorithmic description method. For example, human algorithmic analysis can describe the structure of human activity during a subject's interaction with the computer. The structure of computer interface can, in a probabilistic manner, influence the strategies of the thinking process and task performance in general. In this framework, the main principle of design is the comparison of the structure of the computer interface with the structure of the user's activity, which can be described algorithmically. The analysis of the thinking process in combination with the algorithmic analysis of the user's performance gives us an opportunity to describe the explorative activity of the user and her/his incorrect actions that cause errors.

In cognitive psychology, there is no such concept as human algorithm. As a result, such concepts as computer algorithm and human algorithm are not distinguished clearly.

Analysis presented material demonstrates that self-regulative model of thinking process can be useful tool for analysis strategies of performance computer based tasks and development more efficient presentation of information on the screen develop better instructions and training programs, etc.

5

Attention as a Self-Regulative System

5.1 Mechanisms of Attention and Strategies of Information Processing

Relationship to the study of attention in psychology changed depending on the dominant theoretical approach in psychology. For example, behaviorism, which for long period of time dominated in psychology, rejected the concept of attention as mentalist category in psychology. However, with the emergence of cognitive psychology, and especially after the work of Broadbent (1958) *perception and communication* interest to the problem of attention sharply increased. It should be noted that there are some difficulties associated with the study of attention. This is explained by the fact that attention encompasses a broad range of phenomena. This field of study has no clear boundaries and it often is difficult to separate attention from the consideration of other mental processes. The term *attention* encompasses such phenomena important for the study human work as selectiveness, concentration, switching, dividing, sustaining, etc. Due to the variety of phenomena of attention and its properties, it becomes important the selection of it's the most important aspects in the study of human work. In this chapter, we want to demonstrate the usefulness of integrating ideas of cognitive psychology to the ideas of activity theory (AT) in the study of attention. This integration is especially evident in the study of human work. More specifically, we pay attention to application SSAT and some ideas of cognitive psychology in the study of attention. Here, cognitive approach will be combined with functional analysis when attention is described as a goal-directed self-regulative system.

The study of attention is important from both theoretical and applied perspectives. It can be useful to study such factors as automation, prediction of multi-task performance, cognitive efforts, and complexity of task. Task complexity is a major factor in creating a challenge in the operator's performance. A complex task demands greater cognitive efforts (Bedny and Meister, 1997). The more complex the task is the more is the subject's concentration during its performance. Depending on individual features of the subject, the same complex task will be evaluated by subject as relatively more or less difficult. An increase in the complexity of task will increase the performer's mobilization of mental efforts and the concentration of attention. The subject can perform time sharing tasks differently depending on the complexity of each task.

For this, two different types of attention models are used. One discusses attention as a mental effort (Kahneman, 1973). The other treats attention as an information-processing system. The mental effort models can be related to either *single resource theory* or *multiple-resource theory* (Wickens and McGarley, 2008). Kahneman's model can be related to single-resource theory. Kahneman suggests that there is a single, undifferentiated pool of resources available to all tasks and mental activities. If the task's difficulty is increased or the person is performing two tasks simultaneously, it requires more resources. It also requires proper allocation of these resources. The more difficult a particular task is, the fewer the resources that are available for a second task. One limitation of *single-resource theory* is, for example, in interpreting the well-established empirical finding that when concurrent tasks are in different modalities or use different codes (spatial, verbal), the allocation of resources becomes much easier. Single-resource theory is only able to predict variation in task difficulty.

Multiple-resource theory argues that instead of one single, undifferentiated resource, people have several differentiated capacities with distinct resource properties. For example, it is easier to perform two different tasks that require different modalities than two tasks that require the same modality. In this situation, time-sharing tasks will be more efficient (Wickens and McGarley, 2008). With regard to the cognitive information aspects, Norman's model is of interest (Norman, 1976). In his model, two mechanisms are important: data-driven and conceptually driven processing. Data-driven processing depends on input information. This is an automatic process. When this process predominates, we use the notion *involuntary attention* as per AT (Dobrinin, 1958). Expectation, generation of hypotheses about the nature of sensory signals, conceptualization, and past experience are also important. The information from memory is combined with the information from sensory data. This is conceptually driven processing that we will call *voluntary attention* as per AT. Attention includes not just cognitive but also energetic components. The model suggested by Norman ignores energetic components.

The next theoretical problem that is also important in the study of attention is the relationship between peripheral and internal processes. One of the deficiencies of *multiple-resource theory* is its inability to determine whether the advantages of cross-modality tasks over intra-modality tasks are attributable to central or peripheral processes. For example, time-sharing may not in fact be the result of central resources, but rather the result of peripheral factors that constitute the two intra-modal tasks. Two visual tasks may pose confusion and masking, just as two auditory messages may mask one another (i.e., this exhibits peripheral over central affects). Wickens and Hollands (2000) wrote that the degree to which peripheral rather than central factors are responsible for cross-modality interference, or better cross-modal time sharing, remains uncertain. In one part of this study, it has been shown that when visual scanning is carefully controlled, cross-modal displays do not always produce superior performance. This can be explained by the fact that attention can be considered as a self-regulative, adaptive, and adjustable system. The study also demonstrates that attention can be characterized not only by attention limitations, but also by the ability of the subject to use attention features efficiently in any particular task or in a specified period of time. An ability to adapt and tune different features of attention to certain task requirements is provided by mechanisms of self-regulation of activity. Self-regulation mechanisms are in charge of not only tuning, sustaining, and regulating attention, but are also in charge of all other cognitive processes. The self-regulative process we discuss here is not homeostatic, but rather goal-directed (Bedny and Karwowski, 2004a,b). Self-regulated mechanisms of attention as a goal-directed process are not sufficiently studied.

Many single-channel theories of attention are based on research of the psychological refractory period (Meyer and Kieras, 1997; Pashler and Johnston, 1998, etc.). However, the psychological refractory period is only one mechanism of attention. Attention also depends on consciously regulated strategies. For example in our study (Bedny, 1987), it was discover that a subject can program a second motor action while performing the first motor action. Such a strategy depends on the complexity of the two motor actions. For example, if the first action is very complex, or associated with danger, the subject can utilize sequential strategy. As we will attempt to demonstrate in this chapter, the ability to perform different elements of an activity in parallel depends not only on mechanisms of the psychological refractory period but also on strategies of self-regulation of activity in general. Attention is not *a performer* of two tasks. The subject with his past experience, motivation, conscious goal and strategies, etc., is the performer of these tasks.

Thus, at present, no one theory of attention completely explains the phenomenon of attention. The goal of this study is to present some new data that can be useful in understanding attention mechanisms. In this work, we consider the attention processes of the subject when he/she performs sequential tasks of various complexities (Bedny and Karwowski, 2011). As a result of this study, models of attention are developed. In order to understand the principles involved in our study of attention, it is necessary to briefly review the existing experimental methods of the study of attention.

One of the most important procedures for studying attention is the use of two simultaneous tasks, or time-sharing tasks. The time-sharing method refers to situations when an individual simultaneously receives two messages (Novan and Gopher, 1979). For example, each ear is used as a separate input channel. Typically, the instructions introduce the goal of tracking either one or both of the channels. In the first situation, we talk about switching attention, in the other situation about allocation of attention (Norman and Bobrow, 1975). The other widely accepted method of studying attention requires performance of two tasks in sequence. Experimental studies usually assume fulfillment of the choice–reaction tasks. The interval between two tasks (stimulus-onset asynchrony, or SOA) can be changed during the experiment. Performance time for the second task is increased for the short SOA. However, complexity of choice–reaction tasks on which the difficulty of task performance depends is also a critical factor in dual-task performance. Usually, this characteristic of task does not change during the experiment. A more complex choice–reaction task has been described by Pashler and Johnston (1998) in their experimental study. The first task included two acoustic stimuli and the second one three visual stimuli. The interval between these two tasks was changed from 50 to 450 ms. It was discovered that the lesser the time interval between the two tasks the more was the response time for the second task.

We approached this from a different perspective (Bedny, 1979; Bedny and Karwowski, 2011). Subjects were asked to perform two tasks in sequence. The second task was presented immediately after the first task was completed. The complexity of the tasks varied. We analyzed how the complexity of the first task influences the performance of the second task. Moreover, we asked how this influence changes when the complexity of the second task also changes. The choice–reaction tasks employed in this study include a number of alternatives for the first and the second tasks in different sets of experiment. The left hand reaction is a response to an auditory stimulus. The right hand reaction is a response to a visual stimulus. The number of auditory stimuli varies from one to four. The number of visual stimuli varies from one to eight. Therefore, the complexity of both tasks can be changed by changing the number of presented stimuli. The right hand reaction is performed immediately after the left hand reaction, and the number of possible choices for the right hand varies from one series of experiments to another. The interaction of complex choice reactions, performed sequentially, is analyzed from the point of view of the concept of activity self-regulation. In conclusion, we note that the complexity of the reactions was determined not only by the feature of the presented stimuli, but by the nature of the required responses as well.

For the purpose of studying the interaction of complex choice reactions, we developed an experimental bench that had a subject panel on one side and a researcher panel on the other. The subject panel had a digital gauge with numbers that light up and a sound device to create clear tone sound signals. There were two start positions on the panel in front of the subject: one for the left hand fingers and another for the right hand fingers. Push switches were located at different radial directions from the start position. Four switches were used for the left hand and eight for the right hand. Two meters were located on the researcher's panel. The left hand reaction time (RT) was registered by the left meter and the right hand RT by the right meter.

The researcher was creating the program for the subjects using the keyboard on his panel. The task consisted of the following sequence of events:

1. Subjects kept their eyes on the digital indicator. When one of four sound signals was given through a headphone on the left ear, they had to react by pushing the corresponding switch with the left hand. During the execution of this action, subjects had to keep looking at the digital indicator.
2. Immediately after the sound signal and subject's response, one of the eight numbers lit up and the subject had to react with the right hand by pushing the corresponding switch. The time of each reaction was measured separately. The number of alternatives in each set of experiments was the same.
3. A warning signal was turned on to signify the start of the next sequence of signals.

Four male university students were selected for this experiment. Prior to testing, they were trained for 2 days (a 1 h session per day) to work with the panel. The experiment was conducted over 3 days.

Let us look at the design of the experiment. In the first set of experiments, only visual stimuli were presented to the subject and the subject had to react only with the right hand. The number of stimuli was increased from one to eight. This set was marked 0–8 with 0 meaning there was no previous reaction of the left hand on the sound stimulus and 8 signifying a reaction with the right hand on the visual stimulus with eight alternatives. The average result of all measures was calculated on the basis of 40 reactions of each subject on the corresponding signals (10 preliminary reactions were not considered for the calculation of average RT). Erroneous reactions were not considered. The next set of experiments consisted of a measure of simple RTs when reaction was executed with the right hand following left hand reactions. The number of sound stimuli was been gradually increased from one to four. (The given sequence was 1–1, 2–1, 3–1, and 4–1, where 1–1 means one sound stimulus for the left hand reaction and one visual stimulus for the right hand reaction, 2–1 means two sound stimuli for the left hand reaction and one visual stimulus for the right hand reaction, etc.) Subjects were instructed that after receiving the sound signal they had to react using their left hand. Immediately after that, a "1" would appear on the visual indicator. The subjects would have to react to sound using the left hand and to the visual stimuli using the right hand as soon as they could. The same instructions were given for each set of experiments.

In the third set, after the simple reaction to the sound signals, the visual signal was given and the subject pushed the corresponding switch. In this set of experiments, the following programs or signals were used: 1–2, 1–4, 1–6, 1–8. For all trials, only one sound signal was used and the number of visual stimuli was varied from 2 to 8.

On the second and the third day, three sets of experiments were conducted (sets 4–6). In the fourth set, the following programs or signals were used: 2–2, 2–4, 2–6, 2–8. Subjects had to execute left hand choice reactions with two alternatives and the number of stimuli for the right hand was increased from two to eight. In the fifth and sixth sets, the following programs were used: 3–2, 3–4, 3–6, 3–8, and 4–2, 4–4, 4–6, 4–8, respectively. We uses partial counterbalance schema of experiment. Two subjects started to perform the simple combination of tasks and finished with the complex ones, as described earlier. The other two subjects started with the complex combination of tasks and finished with the simple ones (started with 4–2, 4–4, 4–6, 4–8 and finished with 2–2, 2–4, 2–6, 2–8).

An additional set of experiments was performed on the fourth day. We offered two signals with tones so close that subjective discrimination was evaluated as difficult. The following programs of signals were used: 2′–2, 2′–4, 2′–6, 2′–8. The result of this set was compared with the results of the previous sets for the same subjects. Table 5.1 represents the general plan of the experiments.

Subjects received information about time reaction for visual and acoustic stimuli after each trail. The significance of the experimental results was checked by a two-way factorial analysis of variance.

Let us consider the results obtained on the first day. In the preliminary experiment, we measured simple right hand reaction for visual stimulus. The average time for simple right hand reaction when only one visual stimulus was presented was equivalent to 0.23 s. After that we conducted the main sets of experiments. In the first set of experiments, the number of visual signals was changed from two to eight (sequence 0–8). The average times for right hand reactions to the visual signals (without reaction to the sound stimulus by the left hand) for each subject when the number of stimuli were changed from two to eight are shown in Table 5.2.

If the number of stimuli increased, RT also increased. The interdependence of the RT and the number of stimuli can be depicted by a logarithmic curve (see Figure 5.1, curve with symbol ▲—no acoustical signals). This function is called Hick's law (Hick, 1952).

TABLE 5.1

General Plan of the Experiment

Day of Experiment	The Number of the Set of Experiment	Program (Relationship between Sound and Visual Signals)
Day 1	1	0–1; 0–2; 0–4; 0–6; 0–8;
	2	1–1; 2–1; 3–1; 4–1;
	3	1–2; 1–4; 1–6; 1–8;
Days 2 and 3	4	2–2; 2–4; 2–6; 2–8;
	5	3–2; 3–4; 3–6; 3–8;
	6	4–2; 4–4; 4–6; 4–8;
Day 4	7	2′–2; 2′–4; 2′–6; 2′–8;

TABLE 5.2

Average Time for Left- and Right-Hand Reactions to the Acoustical and Visual Signals

Left Hand		Right Hand Reaction Time Number of Visual Stimuli			
Number of Acoustical Signals	**Reaction Time**	**2**	**4**	**6**	**8**
0	—	0.37	0.52	0.6	0.62
1	0.33	0.34	0.5	0.6	0.61
2	0.62	0.41	0.59	0.69	0.71
3	0.75	0.43	0.63	0.75	0.77
4	0.78	0.46	0.66	0.82	0.82
2′	0.74	0.43	0.65	0.77	0.8

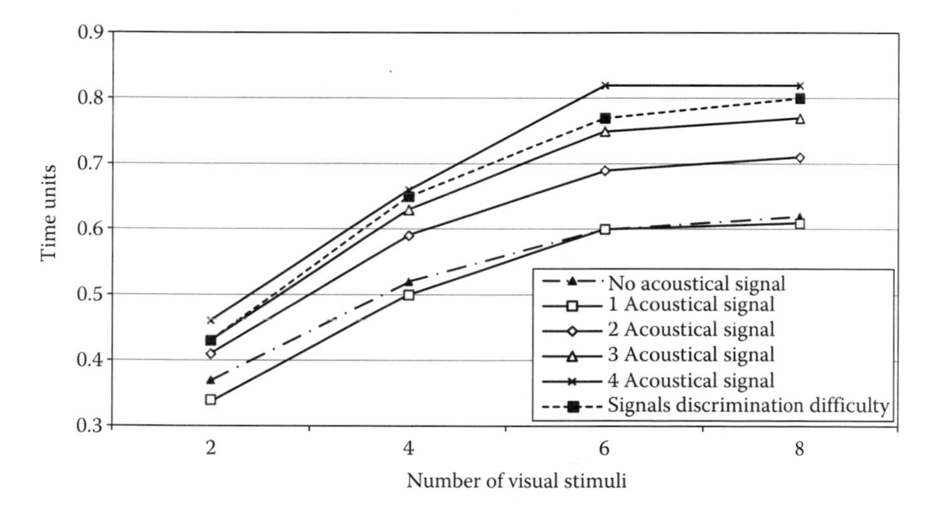

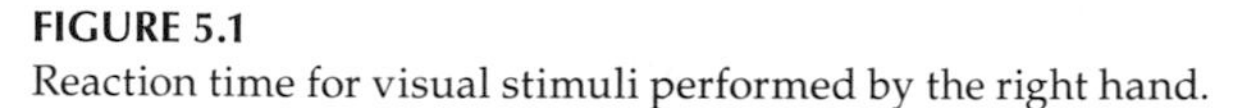

FIGURE 5.1
Reaction time for visual stimuli performed by the right hand.

The result of the second set of experiments, when the number of sound signals for left hand reaction was changed from one to four and only one visual stimulus was given, shows that the average right hand RT for a visual stimulus was 0.12 s. It is sufficiently less compared to the right hand RT for simple digit signals (compare 0.23 and 0.12 s). The difference between the means according to the *t* criterion was statistically significant ($p < 0.05$).

The result of the third set of experiments, when only one sound signal was used and a number of visual signals for the right hand reaction has being changed from two to eight is shown in Table 5.2. The curve with the symbol □ in Figure 5.1 was drawn on the basis of these data. It was discovered that the RT for right hand after the simple left hand reaction to the sound stimulus has a tendency to decrease in comparison with the RT of the right hand without previous reactions of the left hand. This difference is not statistically significant. Nevertheless, this result is interesting because curves ▲ and □ do not intersect.

On the second day, the fourth, fifth, and sixth sets of experiments were conducted. The results of the fourth set of experiments, when the number of sound stimuli was two and the number of visual signals were changed from two to eight is shown in Table 5.2. The results of the fifth and sixth sets of experiments, when the number of sound signals was three and four, respectively, and the number of visual stimuli was changed from two to eight are also shown in Table 5.2. The curves with symbols ◇, △, and × were drawn on the basis of data from Table 5.2 (see Figure 5.1).

The results of the fourth day of testing (when subjective discrimination of two sound signals were evaluated as difficult) are shown in Table 5.2 (last line where number 2′). The curve ■ illustrates the right hand RT to visual stimuli when subjects preliminarily react to two poorly distinguishable sound signals. In this case, we can see that the discriminative features of an acoustical stimulus have approximately the same effect as the number of acoustical stimuli. One can see that the curve ■ is positioned between the curves △ and ×. The position of this curve is much higher than the position of the curve ◇, when subjects preliminarily react to two well-distinguished sound stimuli.

The data demonstrate that the more complicated the previous reaction to the sound signal is, the higher is the curve position that describes the RT of the visual stimulus. We compared the significance of the differences between RTs for when the reaction was performed only by the right hand (curve ▲), and by the right hand with previous left hand reactions (curves ◇, △, ■, and ×).

The within-subjects two-way analysis of variance (ANOVA) was calculated to identify the effects of the number of visual stimuli, acoustical stimuli, and their interaction with the RT to the visual stimulus.

The within-subjects ANOVA revealed that as the number of visual stimuli increased, the RT increased: $F(3, 45) = 54.51$, $p < 0.0001$. Furthermore, as the number of preceding acoustical stimuli increased, RT to the visual stimulus increased as well: $F(3, 45) = 6.11$, $p < 0.01$. The effect of the interaction between acoustical and visual stimulus difficulty was also significant: $F(9, 45) = 3.02$, $p < 0.01$. These results show that as the number of the preceding (acoustic) stimuli increases, the acoustic stimuli have progressively more interference on the RT to the following visual stimuli. Furthermore, there is an interaction between the difficulty of the acoustical stimulus and that of the visual stimulus. This suggests that as the number of the visual stimuli increases, the preceding stimuli have a progressively greater effect on the performance of the second task.

Let us analyze the results. Curve ▲ illustrates how the right hand RT changes when only visual stimuli (from 2 to 8) are given. This curve is a logarithmic function. If we scale the *x*-axis as log *N*, Hick's law applies. It describes the speed of human information processing when the choice reaction is accomplished. Such research is the foundation of the implementation of the information theory in psychology. All of the following results will be compared with curve ▲ in Figure 5.1.

In the second set of experiments, two sequential reactions were performed. One reaction was performed by the left hand as a response to a sound stimulus when the number of stimuli varied from 1 to 4. The second simple reaction was performed with the right hand as a response to a visual stimulus. A prior reaction to a sound signal reduces the second simple RT. This was proved statistically and does not depend on whether the left hand reaction is a complex or a simple one. This is caused by the partial time overlap between the first and second responses. The observation shows that it is very difficult for subjects to act quickly with the right and left hands without partial overlapping of reactions. They take place despite the subjects being instructed to use the visual stimulus as their start signal for the right hand reaction. In other words, the start signal for the right hand is not actually an external stimulus. Rather, it is the start of the left hand movement. Subjects, however, did not realize the overlapping factor.

For the third set of experiments, the overlap between the first and second responses is eliminated (program 1–2, 1–4, 1–6, 1–8). The reason was that decision making (what switch to press by right hand) could be made only after the digital gauge is lighted up. We assumed that when a subject switched his attention from the first reaction to the second one, it should increase the time for a selected right hand reaction. We assumed that when the left hand reaction to a sound signal is simple the increase of the right hand RT to visual signal would be insignificant. However, the results of the experiment showed the opposite. The right hand response time with previous simple left hand reaction tends to decrease comparing with the single right hand reaction (see curves ▲ and □ in Figure 5.1). In order to understand this result, we analyzed the subject's behavior. The observation of the subjects' strategies and their debriefing shows that all subjects considered a single right hand reaction to be more difficult than a right hand reaction with a previous simple left hand movement.

The following two factors that influence the time of the second reaction have been noticed: the elimination of time uncertainty of presenting the second (visual) signal after the previous left hand reaction and the shift of attention from the first reaction to the second one. The first factor increases the speed of the response to the second signal, but the second factor decreases the speed of this response. In this set of experiments, a left hand movement is a simple reaction to a sound signal and performing this reaction requires a low level of attention. Due to this fact, the second factor has a weak influence and practically does not increase the second reaction response time. At the same time, the elimination of time uncertainty of appearance of the second signal reduces the second reaction response time. It means that in this case, the factor of elimination of time uncertainty of presentation of visual signal overrides the factor of the shift of attention from the first reaction to the second one. As a result, we can see the insignificant reduction of the response time in the case of a visual signal (compare curves ▲ and ◻). However, these differences are not statistically significant.

On the second and third days of the experiment, the program 2–2, 2–4, 2–6, 2–8 was used in the fourth set, 3–2, 3–4, 3–6, 3–8 in the fifth set, and 4–2, 4–4, 4–6, 4–8 in the sixth set. In fact, the shape of the curves did not change (see Figure 5.1). However, as a result of the increasing complexity of a previous reaction, the location of curves is higher in this case. Therefore, increasing the complexity of the previous reaction influences the performance of the following one. The level of the second reaction complexity is also very important. The more complex two successive reactions are, the more they influence each other. This outcome can be interpreted on the basis of attention theory. During the time-sharing tasks performance, we have two streams of information. One source of information was presented to the right ear and the other one was presented to the left ear. In our experiment, two streams of information have been presented sequentially, so that with each time period a subject dealt with one source of information only. The more complicated two successive pieces of information and response selection processes in both tasks were, the more difficult it was to shift active attention from one source of information to another. The results of the conducted experiments prove this conclusion.

The increase in difficulty by shifting attention from one reaction to the other overrides the elimination of the time uncertainty of the onset of the second reaction. As a result, the time of the second reaction increases. During the debriefing of the subjects, an interesting fact was discovered. During the performance of the most complicated task where four acoustical signals were presented, there was a subjectively noticeable break between identifying the digit and making a decision about the performance of the second reaction. (Subject said, "I see the digit, but can't make a decision and move my hand. My hand sticks to the switch.")

The obtained data offered an explanation of how mechanisms of attention influence the strategies of information processing. Recognition of stimulus is made by a passive automatic process using a low level of attention, but making a decision is linked with active processes using a high level of attention. Active processes reorganize automatic processes that are slower than passive. The reorganization of attention mechanisms, when sequentially disconnected portions of information are presented, is the same as when the subject shifts attention from one portion to another portion of simultaneously presented information. In our experiment, a subject cannot keep all of the information about two reactions in the short-term memory. It becomes necessary to use information from the long-term memory. The search of information in the long-term memory using a scanning device can start at any node point in a structure of the information base of the long-term memory (Norman, 1976).

Hence, in the experiments, during the process of extracting the information from the long-term memory, the alphabet used by a subject constantly changed. This alphabet is dynamic. As a result, the speed of the information processing changed as well.

Let us analyze the result of the experiments conducted on the fourth day when two acoustical signals subjectively difficult for differentiation were presented. The similarity of curves △ (three acoustical signals are presented) and curve ■ (acoustical signals that are difficult to differentiate were used) in Figure 5.1 allows us to conclude that deterioration in the differentiation of the sound signals influences the RT on the visual signal the same way the increase in the number of alternative sound signals does. We discovered that the increase in the RT on the visual stimulus performed by the right hand is a result of the deterioration in differentiation of the sound signal for the first reaction performed by the left hand. This situation makes it more difficult to decide which sound signal was presented. The first task (reaction by the left hand to a sound stimulus) becomes more complex and therefore more difficult for the subjects. As a result, the process of adjusting the mechanisms of attention from the first to the second task becomes more complicated. Hence, the second RT depends not only on the amount of information presented for the first reaction, but also on the differences between signals. This demonstrates that the complexity of the two considered subtasks determines the strategies of shifting attention from one task to another.

The outcome is that two tasks performed sequentially cannot be considered as independent. When subjects repeated the experiment with two reactions, they combined these reactions into a holistic structure, and complex activity strategies were developed. These strategies can be conscious or unconscious. For instance, when the program 1–1, 2–1, 3–1, 4–1 was performed, premature right hand reaction was performed unconsciously. It contradicted the instructions. Subjects developed their own strategies to optimize their activity. Therefore, none of the given instructions can strictly predetermine the possible strategies of the actual activity (Bedny and Seglin, 1999a,b). In our experiment, a significant increase in the speed of the reaction to a sound signal led to the delay in the reaction to a visual signal. Subjects tried to choose the optimum speed of reactions, which allowed them to respond quickly to both signals. The strategies of activity were optimized through the coordination of external conditions and internal capabilities of the subject. As may be seen, the subjects did not just react to various independent stimuli, but rather developed distinct strategies to reach the specific goal of the unitary task: *react with maximum speed with left and right hands.* The conscious goal influences the strategies of activity and such strategies can be conscious or unconscious.

Comparing the results of the observation, debriefing of the subjects, and analysis of experimental data shows that individuals do not react to various stimuli but actively select and interact with the information. Depending on these results, individuals reformulate goals and strategies of activity.

This results in the transformation of conscious contents of activity into unconscious contents and vice versa. Voluntary attention by an individual is the mechanism through which consciousness is attained. Our observation showed that during the experiment subjects shifted their attention between two tasks, allocating their attention and efforts in an attempt to perform one task quicker, slowing down other tasks, correcting errors, and attempting to enhance the strategy of activity. As a result, individual actions are integrated into a holistic structure based on self-regulative mechanisms. Thus, describing an individual behavior in terms of stimulus-response is crude and inaccurate. Individual behavior cannot be explained as the sum of independent reactions to a series of independent stimuli. Experimental data demonstrates that a model of attention should incorporate the self-regulation process and take into consideration various voluntary and involuntary attention mechanisms.

In cognitive psychology, the delay in reaction in the second task is called psychological refractory period (PRP). It is assumed that processing of each task can be divided into three stages: *early stimulus processing stage central processing stage*, and *late processing stage* (Pashler and Johnson, 1998). The first and the third processing stages can be performed simultaneously for both tasks. However, central processing stages of two tasks cannot be combined or performed in parallel.

In systemic-structural activity theory (SSAT), these two tasks are considered as a single task, which integrates interdependent and to a significant degree voluntary, regulated actions. Each task (more precisely sub-task) has its own subgoal. A high-order goal of a general task is to perform each subtask as quickly as possible. During the experiment, subjects realized that the first and the second tasks were not independent. Subjects developed complicated subjectively suitable strategies for holistic task performance. Limitation in information processing was only one of the factors that influenced PRP. This factor always interacts with the conscious and unconscious mechanisms of activity regulation. In SSAT, these stages are associated with different cognitive and motor actions. *Early stimulus processing stage* can be described as a simultaneous perceptual action. The *central processing stage* can be related to a group of actions called decision-making actions at a sensory-perceptual level or explorative thinking actions that are performed based on sensory-perceptual information. The *late processing stage* could be considered as a motor action that usually requires a low or average level of concentration.

Simple perceptual actions and average complexity motor actions can be performed in combination with other similar actions. At the same time, decision-making actions at a sensory-perceptual level, decision-making actions at the verbal-logical level, explorative-thinking actions, etc., cannot be performed simultaneously (Bedny, 1987; Bedny and Meister, 1997). There are also complicated perceptual actions that consist of a chain of subsequent perceptual actions that are involved in the recognition of unfamiliar stimuli or in complicated motor actions that require high precision or performing under stress.

Sometimes, motor actions also can require a high level of concentration. Such perceptual or motor actions cannot be performed simultaneously. Therefore, as per SSAT we do not discuss the possibility of one or several bottlenecks in the attention mechanism. Rather we distinguish between automatic and voluntary conscious processing mechanisms of attention. Automatic processing mechanisms of attention facilitate performance of different elements of task (actions or operations) simultaneously. In contrast, mechanisms of attention that are involved in voluntary and conscious processing of information very often provide the possibility of sequential processing only. As can be seen the concept of *stages processing analysis* introduced by Sternberg (1969a), and data that describe an opportunity to perform these stages in sequence or simultaneously are in agreement with data presented by SSAT. At the same time, in contrast to data obtained in cognitive psychology, formal rules that describe the possibility of combining mental and motor components of activity (actions or operations) during task performance were developed in SSAT (Bedny and Karwowski, 2007). The data obtained in the framework of SSAT provides a link between theoretical data and applied research.

5.2 Self-Regulative Model of Attention

Studies show that attention functions as a self-regulatory system. It integrates and organizes the strategy of cognitive components of activity. In the self-regulation of attention not only cognitive but also emotional-motivational mechanisms are important. Therefore the model of attention described in this chapter includes both these mechanisms. The basic units of analysis of self-regulative systems are functional mechanisms or function blocks. Function blocks can be considered as specific stages information of processing, which have feedforwarded and feedback connections with other stages. Each stage has a particular purpose in attention regulation. Depending on task specificity, cognitive processes can interact differently at different stages of the attention process. Hence the content of each stage or block depends on the specifics of the task. During the development of the cognitive model of attention, we pursued both theoretical and practical goals. A model should not be overloaded with insignificant details. At the same time, it should be able to explain and predict behavior or activity of a person performing the time-sharing tasks and tasks of various complexities. The model must also account for the possibility of combining elements of activity, depending on their complexity.

There are two factors in our experiment that can reduce mutual interference of the considered tasks. The first factor includes the following. We have used two information-processing channels. One of them was auditory and the other was visual. This means that we used two modality-defined resources.

According to multiple-resources theory, this should alleviate the process of simultaneously performing two considered tasks (Wickens and McGarley, 2008). The second factor that can reduce interference of the tasks is that the subjects performed two tasks in sequence. They should perform the second subtask only after completing the first one. However, our research showed that an increase in the complexity of each task increased the effect of their interference even when they were performed in sequence and subjects utilized two modality information-processing channels.

This contradicts with multiple-resource theory, wherein people have several different capacities in terms of resource properties. Under this theory, the two tasks in consideration are independent and should not influence one another, but our research data show that it is not the case. The interference of two tasks, especially when their complexity has been increased, allows us to conclude that resources are shared. Complicated tasks make it difficult to access resources and reduce the ability to allocate resources. It means there are undifferentiated, limited resources and there is a mechanism that regulates allocation of these resources. We named the mechanism that is responsible for the investment of required resource of attention in performance as *available level of arousal*. The second mechanism or block is designated as *regulator integrator*. The latter is responsible for voluntary allocation of resources and their coordinated usage. Physiological studies consider specific and nonspecific arousal. When we evaluate task difficulty, the nonspecific arousal is especially important (Aladjanova et al., 1979). The more difficult the task is for a subject, the more mental effort it takes. These efforts demand nonspecific arousal. Therefore the difficulty of the task to be performed is a critical element of time sharing. The more complex the task is the higher is the probability that this task will be difficult for the subject. Hence the higher the task complexity, the higher the degree of limited energy resources it requires.

Each information-processing task should have an appropriate level of energy support not only at the physiological but also at the psychological level. Hence there is another block called *evaluative and inducing level of motivation*, which is responsible for energy supply at the psychological level (see Figure 5.2). The first one is considered to be a physiological mechanism and the second one a psychological mechanism. The last one regulates emotional and motivational states of a person during task performance. This block is involved in the creation of the vector *motives* → *goal*, which gives activity its goal-directed feature (Bedny and Meister, 1997).

The evaluative and inducing level of motivation block consists of two subblocks, which mutually influence each other. One subblock is called *sense* (significance) and the other one *motivation*. Sense is responsible for the evaluation of personal significance of goal and various components of activity. Motivation refers to the inducing components of activity (Bedny and Karwowski, 2004). The factor of significance influences the method of information interpretation and the creation of motivational forces associated with

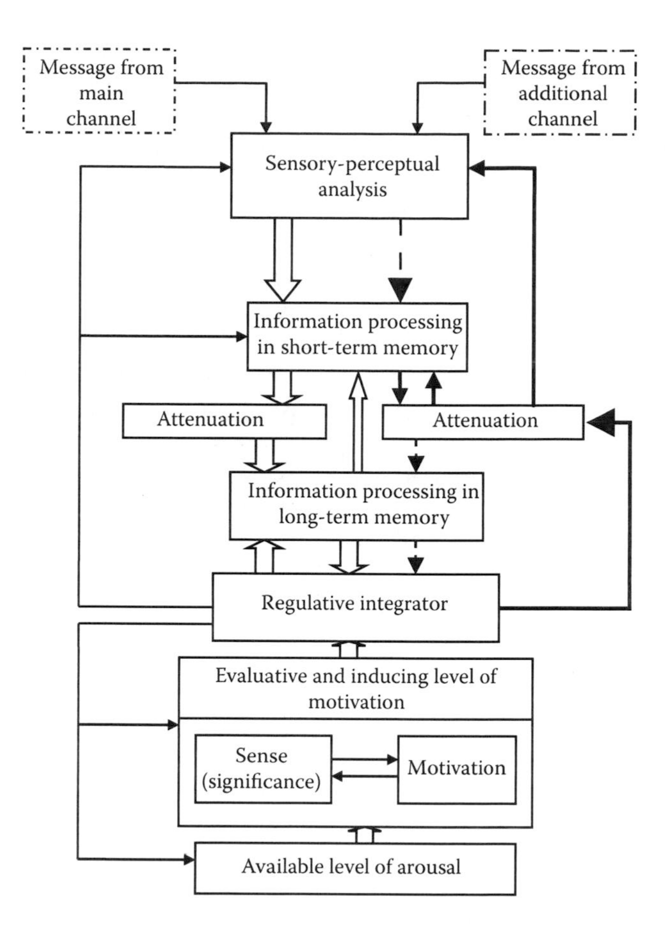

FIGURE 5.2
Self-regulative model of attention.

the main channel of processing information. The function block *regulative integrator* includes mechanisms that are responsible not only for unconscious but also conscious processing of information and first of all for the development of the conscious goal of an activity. Hence, the interaction of *regulative integrator* block with *evaluative and inducing level of motivation* block provides voluntary regulation of attention, selects the more significant goal of activity, and switches attention from one task to another. This block includes the mechanism of thought.

Through the feedback, *regulative integrator* activates and regulates functional blocks *available level of arousal* and *motivational level*. Figure 5.2 demonstrates these three interrelated blocks. The material presented demonstrates that we can create a model of attention utilizing single-resource theory without taking into consideration specific arousal. According to functional analysis of activity, any self-regulative model includes not only energetic but also cognitive mechanisms that we describe in the following text.

We start our discussion with the analysis of relationship between peripheral and central processes of attention and their effect on our model of attention. Sometimes cross-modal time sharing is better than intramodal (Wickens and Hollands, 2000). It can be due to central or peripheral processes, where central processes are associated with separate perceptual resources and peripheral processes with visual scanning, imposing confusion, masking, etc. In our experiment, the subject was given only one task for a certain period of time, which involved visual or auditory processing. In this case, interference of the peripheral processes was eliminated. The process of switching attention became complicated only because central processing is a critical factor in performing two tasks. However, our model of attention does not ignore peripheral processing. Peripheral and central processes are interacting with each other. This relationship is shown in our model by feedbacks from *regulative integrator* to information-processing blocks—to the *sensory-perceptual analysis* block in particular. The model of attention presented in Figure 5.2 has two channels of information processing. One of them is the main channel and the other one is an additional channel. Each channel of updating information has a limited capacity. The complexity of updating information by each channel depends on quality and quantity of information and on the speed of its update. There has to be a mechanism that is responsible for the coordination of information processing between these two channels and switching of attention from one channel to another. The more complex the listed characteristics of information and the loading of channels are, the more difficult it is for the *regulative integrator* to allocate informational resources. Hence, the *regulative integrator* is responsible not only for the allocation of energy resources but also for the coordination of information processing between two described channels. It includes conscious and unconscious processes of thinking. The overload of channels updating information leads to a decreased efficiency of this coordinating block. Our model of attention allows to utilize data from both single-resource and multiple-resource theories. The notion of *resources* is tied to the energy aspects of activity. The notion of *regulative integrator* is tied to the regulation of informative and energy processes.

It shows the unity of the informational and energetic aspects of attention. There are situations when the same demands are presented for energy resources that have different demands for their allocation and coordination with information processes. For instance, it is easier to distribute attention between visual and auditory processing resources than between two visual resources. It can be explained not just by different modality-defined resources, but also more likely by the easier coordination of demands of various resources of information processing. Such coordinating function is tied to the *regulative integrator* function block. Therefore, the allocation of attention depends not only on energy resources, but also on the complexity of coordination of information processes and their coordination with energy processes when there is an undifferentiated pool of such resources.

The suggested model presents attention as a self-regulative system and includes various function blocks that are responsible for information processing.

In our model, the first stage of information processing is the *sensory-perceptual analysis* block; automatic processes dominate this block. However, this block also includes elements of conscious, controlled information processing. Function blocks *regulative integrator* and *attenuation processes* influence the function block *sensory-perceptual analysis* through feedback. This fact is supported by the studies that demonstrate the effect of facilitation of processing of attended information and suppression of unattended information at the sensory-perceptual stage (Wickens and Hollands, 2000). Therefore, selectivity, to some extent, takes place at the sensory-perceptual level.

In contrast to Broadbent's (1958) filter model, Treisman (1969) introduced the concept of attenuator as an important mechanism of attention. She has proposed that the attenuator as a filter weakens, rather than entirely rejecting, unattended information. There is an inductive relationship between attenuators (Norman, 1976). If one of them becomes more active, the other one becomes less active and vice versa. In our model, instead of the *attenuator* block we have an *attenuation processes* block. This means that different psychological and physiological processes are involved in this mechanism. The data obtained in neuroscience also supports the fact of the existence of the attenuation mechanisms. For example, it has been discovered that attention operations are associated with the activation of some neural attention network of the brain and inhibition of the other neural structures of attention (Fafrowicz and Marek, 2008; Sokolov, 1969). Hence, the attenuation block is not a separate mechanism of the brain. This block simply demonstrates that there are complex relationships between activation and inhibition processes in the neural structure of the brain.

The next stage of attention involves a combination of various cognitive processes in short-term memory. This combination is provided by feedback influences of the *regulative integrator* block. Conscious thinking operations are particularly important at this stage.

The next function block is called *information processing in long-term memory*. It presents a combination of various cognitive processes in long-term memory. This block interacts with *regulative integrator*. The message from the main channel is not attenuated. On the contrary, the information flowing along the additional channel is partly attenuated in short-term memory and mostly during the transformation to long-term memory. Information flowing along an additional channel is processed automatically. Hence, meaningful interpretation of complicated information in the short-term memory is possible only in the main channel. Memorization and efficient manipulation of information in the long-term memory can be performed with attended information. The analysis of the relationship between these functional blocks demonstrates that *regulative integrator* activates and regulates mechanisms involved in the processing of information from the main channel and turns off inhibition processes for this channel. At the same time, *regulative integrator* turns on the inhibition processes for the additional channel and is not involved in active control of these functional blocks along this channel. Information processing along the additional channel involves nonvoluntary attention according to AT terminology. For this type of attention, the most important features are those of external stimulation instead of internal willing processes linked with conscious goal-directed activity.

The functional block *regulative integrator* enables switching from one channel of information to the other using attenuators. This functional block continually constructs and revises expectations and controls and corrects sensory messages. According to Norman (1976), a conceptually driven analysis should be distinguished from a data-driven analysis. A conceptually driven analysis has a limitation on the number of units of information that can be processed at any given period of time. Regulative integrator is the most important function block in the conceptually driven analysis. This block creates the conscious goal to which our attention is directed and compares the system of expectations with input information. Based on this comparison, this function block generates feedback influences (depicted by the thin line) and is connected with consciousness, language, and speech. The integration between central and peripheral processes is provided by feedback connections. As has been shown, regulative integrator also coordinates energetic and informational processes. Hence, attention is directed toward attaining an established goal in the given period of time through the main channel. At the same time, the additional channel is involved in the attainment of information based on the existing set of activity. The set can be transformed into conscious goal and vice versa. There is also a system of expectation and anticipation. The systems of expectations and anticipation are connected with the set (Uznadze, 1967) and goal of activity. In the functioning of this system, feedforward and feedback influences are critically important. When the set and goal of activity are altered, the system of expectations is altered as well. Specificity of the goal and set determines the content of a subject's expectations. It is well known in perceptual psychology that when people perceive ambiguous pictures, altering the goal of the perceptual process results in a modification of their expectations and the outcome of the perceptual process is different. In other words, once again we see how goals determine the specificity of the selection of information (Bedny and Karwowski, 2007). The process of comparison of sensory data with the system of expectations depends on the goal of observation. The feedback that derives from this comparison can influence every function block involved in the attention process.

As may be seen in our model, expectations are integrated with the constitution of the goal or set and feedback processes. In general, regulative integrator has complicated functions. It includes formation of a conscious goal or an unconscious set that plays an important role in tuning all other function blocks involved in the attention process. The information that goes through additional channels is associated with an unconscious set and can only be partly updated. This information is connected with automatic processing requiring little conscious attention (involuntary attention).

The feedback influences for the main channel shown on the left side of Figure 5.2 depicted by the thin line should be switched to the additional channel on the right. This happens when the information obtained from an additional channel becomes more significant and an unconscious set is transformed to a conscious goal by the regulative integrator. At the same time, the regulative integrator activates attenuation processes by means of feedback influences in the preceding main channel. As a result, the main and additional channels switch places. Switching of channels is carried out by the regulative integrator, attenuation processes, and feedback influences. Therefore our model of attention functions as a self-regulative system.

The *regulative integrator* has an impact not only on different blocks of a higher level, but also on blocks of a lower level. The blocks at the lower level are responsible for motivation and activation. The regulative integrator governs information processing and matches it with resources of energy. Due to self-regulation, the coordination between energetic and informational processes is accomplished. This is realized most effectively through a main channel. The more coupled and complicated the informational processes are, the more energy resources are required and the less is the ability to allocate resources necessary for an additional channel.

The self-regulative model of attention demonstrates that allocation of resources depends not only on constrains imposed by resources' limitation but also on the subject's ability to consciously regulate strategies of attentions. The notion of *the strategies in allocation of attention* (Navan and Gopher, 1979) is important for the support of the self-regulative model of attention. Additional data that prove that attention should be considered as a self-regulative system have derived from the work of Young and Stanton (2002). They introduced the concept of *malleable attention resources pool*. The main idea of this concept is that attention capacity can, to some degree, change in response to changes in task demands. This can be possible only if the attention system can regulate its functioning, and therefore should be considered as a self-regulative system. According to the self-regulative process, people invest only as much effort as they deem appropriate.

Using cognitive psychology data provided by other authors as well as principles of functional analysis of activity developed in SSAT, we created a model of attention that allowed us to describe the process of performing cognitive tasks that can interfere with each other. This model allowed us to combine data from single-resource and multiple-resource theories. It also shows the relationship between central and peripheral processes and between voluntary and involuntary attention. Our model of attention explains why the combination of channels with different modality (visual and auditory) and spatial and verbal processing make it easier to perform time-sharing tasks even when a subject has only one undifferentiated pool of attention resources. In such situations, it is easier for the regulative integrator block to coordinate the information processes and to make these processes agree with the energy processes. The existence of the interference of the informational processes, increasing in complexity and energy support restrictions, further complicate the work of the coordination block.

The goal of this research was to analyze an interaction of complex reactions performed sequentially and to create a model of attention. We also wanted to analyze how independent reactions integrate into the whole system of organized activity. In this study, we utilized two approaches of SSAT. The cognitive approach was utilized through parametrical methods of study and functional approach was utilized as a systemic method of analysis. The conclusion is that the more complicated the previous and/or following reactions are (subtasks), the more they interfere with each other. This interference can be explained by utilizing the model of attention presented in this work. The described experiment is different from prior experiments conducted by other authors. In our experiment, two information channels have different modalities and two messages are sent sequentially. The time interval between two tasks was the same and was equal to zero. The complexity of both tasks was varied. In other words, during a given period of time, a person has to process only one portion of information (perform one task). The more complex the previous and current portions of information are (first and second tasks), the more complex is the process of adjustment of the attention mechanisms and the more they influence each other. This experimental data demonstrate that attention can be better explained by single-resource theory.

The obtained result allows us to conclude that the development of a model of attention that can be successfully used in a real work situation should be based on single-resource theory. These data are in agreement with Kahneman theory. Allocation of resources is taking place during performance of interdependent tasks. Coordination of these resources with informational processes can be executed by a specific mechanism we call *regulative integrator*. This mechanism is, to a great extent, tied to our conscious and verbal-thinking processes. It also includes unconscious components associated with a set. The existence of this functional block can explain the data revised in single- and multiple-resource theories. Intensification of information processing and increase in the amount of used energy further complicate the functioning of the coordinated block (regulative integrator). Difficulty in the allocation of attention between two tasks is caused not only by energy restriction but also by the ability to coordinate different informational currents, and to match them with energy supply. For example, perceptual modalities can influence the strategy of attention. It is easier to divide attention between eye and ear than between two visual channels. It can be seen from this model that coordination may be complicated due to peripheral and central processes. The complexity of the task is the main element that determines the possibility of coordinating and matching energetic and informational processes.

Features of attention put some limitation on the ability to perform time-sharing tasks. In accordance with these limitations, people create various strategies of activity in order to achieve a set of goals. Some subjects paid more attention to the first task, others to the second one, and still others tried to distribute their attention evenly between the two tasks. Subjects varied allocation of their efforts during the task performed from trial to trial. All of the strategies got feedback based on the instructions given, their subjective understanding, individual characteristics, and an estimation of achieved results. The core is the self-regulation process, due to which various reactions are combined into a single activity that has a systemic structure. This means that activity cannot be represented simply as a set of independent responses to a set of independent stimuli. Due to self-regulation, the subject develops various strategies, using resources of attention, in trying to coordinate cognitive, executive, evaluative, and motivational components of activity. Therefore, the study of the mechanisms of attention and the process of their self-regulation during the performance of time-sharing tasks is very important. The self-regulative model of attention demonstrates that it is possible to voluntarily regulate attention strategies. Resource limitation can influence choosing preferable strategies to allocate or switch attention. However, the subjects can be unaware about their resource limitations. The complexity of task and derived from it the difficulty of performance are critical elements of concurrent time-sharing task performance. Increasing task difficulty causes unspecified activation of the nervous system.

In cognitive psychology, the ability to perform different elements of cognitive processes simultaneously is considered from the cognitive processing stages perspective. It is assumed that processing of each task can be divided into several stages. Some of them can be performed simultaneously. Only the central processing stages of two tasks cannot be combined or performed in parallel. In SSAT, this problem is considered as a subject's ability to combine cognitive and behavioral actions and operations. Strategies of activity, which derive from the mechanisms of self-regulation, determine the specificity of their combination during task performance. From an AT's perspective, a subject cannot be considered only as a device for information processing. Characteristics of attention depend on the goals of the activity, the significance of the task and the motivational state of the subject, and the strategies of self-regulation. In turn, the limitation of processing resources influences these previously listed components of activity. In this model of attention, function blocks are considered as relatively independent stages of processing of information. Our model of attention also rejects the concept of bottleneck as a possible mechanism of the attention process. We prefer to explain the interference between various elements of activity as a result of activation and inhibition of various structures in the brain.

In conclusion, the proposed model differs from others because it includes three separate subsystems: informational, energetic substructures, and coordination mechanisms. The existence of forward and backward interconnections between functional blocks allows depicting the formation of a strategy of attention directed to achieve the conscious goals of activity. The central mechanisms influence the process of time sharing. Attention is considered as a complex self-regulative system. The suggested model of attention allows explaining the data obtained in single- and multiple-resource theories from a unified perspective and takes into consideration a person's ability to voluntarily regulate her/his own activity. This work demonstrates that SSAT and cognitive psychology are considered as interconnected approaches. The combination of SSAT and cognitive psychology data help us to develop more comprehensive models of attention.

The data obtained from the experiments discussed are used to further develop the time structure of activity when it is necessary to determine which elements of activity can be performed sequentially or simultaneously. The material in this chapter is also utilized for the evaluation of complexity of work activity elements, depending on the level of attention concentration.

As discussed in Chapters 3 through 5, there are models of self-regulation that describe strategies of activity in general and the ones that describe self-regulation of individual cognitive processes. When there is a need to study strategies of activity as a whole, the general self-regulation models should be utilized, and when the attention should be focused on strategies of separate cognitive process, the latter type of models should be applied.

Section II

Design

6

Cognitive and Behavioral Actions as Basic Units of Activity Analysis

6.1 Description and Classification of Cognitive Actions

In this book, we focus on operator's interaction with computerized systems. Computer-based tasks should be well designed in order to use computers efficiently in various work conditions. Therefore, the main purpose of studying HCI system is design of computer-based tasks. The second volume of this book concentrates on discussing operator's performance on various tasks in highly automated technological systems and interaction with various displays and controls and performance of manual components of work (Bedny, 2014).

The term *design* has emerged from the field of industrial engineering. This term has been used in reference not only to material systems and technological processes but also to work methods' design. These aspects of design are interdependent. For example, equipment design can be evaluated based on the analysis of human interaction with equipment. Therefore, the structure of human activity during interaction with equipment is an important indicator of design solution. Moreover, we can have the same equipment and use different methods of task performance. The description of the best method of task performance is known as work method design. However, in ergonomics, *design* is often reduced to experimental procedures. It should be noted that design can be performed based purely on analytical procedures. Experimentally obtained data should be described analytically. There is no design without an analytical model, it is simply an experimentation. The description of models of human performance can be compared to the development of drawings in engineering. The history of equipment design or a method of task performance can be easily traced based on the analysis of models of human activity.

Examples of such models are drawings of the human body and equipment. However, when one needs to design cognitive aspects of activity, anthropometrical methods do not suffice. This requires units of analysis and a language of description of human activity during task performance. Human activity is a process that makes such design more complex.

Systemic-structural activity theory (SSAT) creates a hierarchical system of units of analysis allowing to use the analytical stages of design during system modification or when designing a new system. Analytical aspects of psychological design in ergonomics are relevant to product safety and to safe methods of task performance (Bedny and Harris, 2013). Analytical methods are critical when studying computer-based tasks. Each task in the work process is regarded as a situation-bounded activity that is directed to achieving the goal of a task under given conditions. Only when objectively given or subjectively formulated requirements of a task are accepted by a subject as a desired future result do they become the goal of a task. There are also task conditions that include means of work, raw materials, input information, instructions, and past experience of a performer. Task conditions determine possible constrains of task performance. The vector *motive* → *goal* determines directness of activity during task performance. A goal is a mental representation of a future desired result. It is a cognitive component of and a motive is an energetic component of activity. The more intensive the motive is the more efforts a performer is putting in reaching her/his goal. Every task has one final goal and intermittent goals of human actions. Thus, a task can be described as a logical system of cognitive and behavioral actions. Morphological analysis of task describes logical organization of actions during task performance. This is a critically important method of activity structure description and analysis. Cognitive and behavioral actions are the main units of analysis for the creation of such models of activity. The design utilized in the AT concept of action is significantly different from understanding the concept of action outside of AT because AT considers cognitive and behavioral actions as basic elements of activity during task performance.

Hence, standardized description of cognitive and behavioral actions is the basis for designing activity during task performance. As we have already discussed, there are two basic terms in activity theory (AT). The first one is the word *deyatel'nost'*, which is translated into English as *activity*. The second one is *dejstvie*, which is translated as *action*. In the West, the concept of activity and action are used interchangeably. The term *action* in AT is understood as an element of activity and its main building block. An action can be defined as a discrete element of activity that is directed to achieve a conscious goal. Actions can be further divided into unconscious operations, the actual nature of which is determined by concrete conditions under which activity takes place. The achievement of the goal of an action and the assessment of its result is the end point of an action that separates an action from a following action. Actions can be cognitive and behavioral. Therefore, cognition is not just a system of cognitive processes; it also is a system of cognitive actions and operations. A standard description of cognitive and behavioral actions is necessary for describing activity structure and particularly for design purposes. Actions consist of operations (psychological operations). Motor and cognitive actions should be considered as complex self-regulative systems. Even cognitive actions that have a short duration emerge as complex self-regulative structures that adapt to a particular situation. Motor and cognitive actions are not just tightly interconnected but include cognitive components.

The additive factor method developed by Sternberg (1969a,b) contributed to a number of studies that discovered some cognitive stages of motor actions regulation (Meulenbroek and van Galen, 1988; Sanders, 1980). The principle of self-regulation of motor actions in AT also implies existence of cognitive mechanisms of regulation of motor actions. Gordeeva and Zinchenko (1982) extracted three basic stages of motor movement regulation: program formation stage (latent stage), executive stage (motor stage), and evaluative stage (evaluation of the result of movement and correction of movement). The similarities between visual perception and sense by touch demonstrate a strong relationship between cognitive and behavioral actions. Cognitive actions are developed during interaction with behavioral actions. Turvey (1996) demonstrated that *dynamic touch* provide perception of external information. The continuous streams of feedforward and feedback influences provide development and construction of cognitive and behavioral actions (Anokhin, 1962; Bernshtein, 1966). Therefore, motor and cognitive actions are interdependent. Motor actions provide not only transformation of the material world, but also its reflection.

Study of cognitive and behavioral actions demonstrates the principle of unity of cognition and behavior that is critically important in AT (Bedny et al., 2001). Cognitive actions sometimes have very short duration and it is often not easy to extract mental operations out of the content of cognitive actions. Therefore, in our further discussion, we offer a standardized description of holistic cognitive actions. There are direct connection actions and transformational cognitive actions (Zarakovsky and Pavlov, 1987). Direct connection mental actions unfold without distinctly differentiated steps and require less attention than transformational cognitive actions. These actions are less consciously directed and experienced subjectively as instantaneous. Recognition of a familiar object is an example of such actions. Because they have a short duration, direct connection mental actions are often called mental operations. In comparison, transformational mental actions involve more deliberate examination and analysis of stimulus as, for example, perception of an unfamiliar object in a dimly lit environment. Cognitive and behavioral actions have some duration. Therefore, the classification of cognitive actions should always be complemented by analyzing their duration. Duration of cognitive actions can be obtained from psychology or ergonomics handbooks or from special experimental studies. In our further discussions, we show how to determine the duration of a cognitive action by utilizing eye movement data.

Mental actions can be classified based on dominant cognitive processes and the ultimate purpose as follows (Bedny and Karwowski, 2007; Bedny and Meister, 1997; Zarakovsky, 2004).

6.1.1 Direct Connection Actions

Direct connection actions unfold without distinctly differentiated steps and require a low level of attention. They can be further distinguished as follows:

Sensory actions—detection of noise or decision about a signal at a threshold level; obtaining information about distinct features of objects such as color, shape, sound, etc.

Simultaneous perceptual actions—identification of clearly distinguished stimuli well known to an operator that only requires immediate recognition, perception of qualities of objects or events (recognition of a familiar picture).

Mnemonic (memory) actions—memorization of units of information (UOI), recollection of names and events, etc. Direct connection mnemonic actions include involuntary memorization without significant mental efforts.

Imaginative actions—manipulation of images based on perceptual processes and simple memory operations (mentally rotating a visual image of an object from one position to another according to a specific goal).

Decision-making actions at a sensory-perceptual level—operating with sensory-perceptual data like decision making that requires selecting from at least two alternatives (detecting of a signal and deciding to which category it belongs out of several possible categories).

6.1.2 Mental Transformational Actions

Mental transformational actions deliberate examination and analysis of stimulus (perception of an unfamiliar object in a dimly lit environment), exploration of situation based on thinking mechanisms, etc. They can be further distinguished as:

Successive perceptual actions—recognition of unfamiliar stimuli and creation of perceptual image of an object that require deliberate examination and analysis of stimuli.

Explorative-thinking actions are based on sensory-perceptual information, involved in deliberate examination of various elements of tasks, discovering specificity of their interaction, extraction of subjectively significant elements of situation, interpretation of obtained information and creation of mental pictures of a situation.

Thinking actions of categorization are based on analyses of features of signals or situations with further logical analyses of their relationship followed grouping into two or more categories or classes (binary or multi-alternative categorization); can be performed based on various strategies of categorization that might change during training or self-learning.

Logical thinking actions are based on manipulation with concepts, major and minor premises (deductive actions, syllogisms, reasoning, etc.); use various strategies that can change during the skill acquisition process.

Decision-making actions at a verbal thinking level are based on an algorithmic level of regulation and heuristic level of regulation of the thinking processes (after receiving information, an operator has to determine which steps out of several possible to take based on a logical analysis of the situation.

Decision-making actions that involve emotional and volitional components are performed in combination with verbal thinking components; include conflict of motives and volitional process (in a dangerous situation, a subject has to decide between "I have to act" and "I don't have to act").

Recoding actions—transformation of one kind of information into another (e.g., translation of meaningful verbal expressions from one language to another).

6.1.3 Higher-Order Transformational Actions

Higher-order transformational actions include a complex combination of thinking and mnemonic actions or creative actions. They can be further distinguished as follows:

Creative-imaginative actions are an empowering combination of logical and intuitive operation with images.

Combined explorative-thinking and mnemonic actions are complex manipulation of information in working memory based on mechanisms of thinking, extracting information from the long-term memory, storing requisite information, and maintaining information in working memory.

Creative actions are operations that generate new knowledge either logically or intuitively; involve divergent thinking vs. reproductive actions that involve convergent thinking.

When using eye movement data, it is important to find out the difference between successive perceptual and explorative-thinking actions. The main difference between them is that the purpose of successive perceptual actions is developing a perceptual image of an object or percept (e.g., categorization of objects based on their shape, color, and size), while the purpose of explorative-thinking actions is to discover a functional relationship between elements of a situation based on available sensory-perceptual data. Frequently, functional property of objects can be discovered only through the analysis of the relationship between various elements of a situation. Sometimes perceptual properties directly demonstrate functions of an object. At the same time, shape of an object and its function can deviate from each other. In the first case, the thinking process is almost entirely eliminated and we classify such actions as successive perceptual, while in the second case, the thinking process dominates and we classify the actions involved as explorative thinking actions that are based on sensory-perceptual data. Cognitive psychologists obtained some interesting data that also demonstrate the importance of the concept of cognitive actions although they do not use this concept in their study. For example, Kosslyn (1973) and Cooper and Shepard (1973) measured the time of manipulation with mental images. They found that time of mental rotation of objects was similar to the time for actual external rotation. According to SSAT, if an individual intentionally turns a mental image of an object to the position according to a required goal, it is an imaginative action or an object-mental action.

Cognitive actions have a certain analogy with motor action in terms of a number of features. They are goal directed, have a beginning and an end, function according to the principle of self-regulation, and so on. Motor actions presuppose the existence of material objects with which a subject interacts. Cognitive actions transform not material objects but information. More precisely, cognitive actions manipulate not with material objects but with operative OUI or operative units of activity. These UOI perform functions that are similar to those of material objects' for motor actions. Such internalized operational units of cognitive actions should be regarded as internal mental tools of activity. Operational units of activity are semantically holistic entities that are formed during the acquisition of a specific activity. A person can mentally manipulate images while thinking, extracting UOI from memory, even without an external representation of data. For example, when receiving information, a person can almost simultaneously perceive and structurally organize some features of an object and manipulate them as unitary UOI in a process of constructing a perceptual image. Such units are formed during the acquisition of specific types of activity in a person's past. During the thinking process, such units are connected with the ability to interpret them as separate meaningful UOI. Working memory is also an important mechanism in the formation of operative units of cognitive activity.

It is necessary to distinguish between two ways of describing cognitive and behavioral actions. One way involves utilizing technological terms or terms that describe some task elements associated with a considered action. Taking a reading from a pointer or a digital display are examples of perceptual actions that are described based on technological principles (technological units of analysis). Depending on the distance of observation, illumination, and constructive features of a display, the content of mental operations and the time of action performance can vary. Based on such descriptions as "taking a reading from a pointer on a display" we do not know exactly what action is performed by the subject because conditions of reading can vary. If in addition we use such descriptions as *simultaneous perceptual action* with duration 0.30 s, we can really understand what action is performed by the subject. This is an example of perceptual actions that is described based on psychological principles (psychological units of analysis). Another method is by utilizing MTM-1 (methods-time-measurement) system. This system describes the action as *eye focus* or *fixation* (EF). According to SSAT, this is a simultaneous perceptual action, or a simple decision-making action at the sensory-perceptual level, with duration 0.27–0.30 s, which is an example of a psychological unit of analysis. Anybody who is familiar with MTM-1 or with the standardized description of cognitive actions would understand what was done by the subject if we use EF to describe this specific component of activity. Actions such as *detection of signal* checking conjunctive (AND) or disjunctive (OR) logical conditions are examples of mental actions that are related to psychological units of analysis. The more standardized the conditions of actions that are described according to technological principles, the more often they become similar to standardized actions that are described according to psychological principles. It can be explained by the fact that the content of mental operations of these actions also becomes similar. Initially actions are described in terms of technological units of analysis. Then they are transferred to psychological units of activity description. The introduction of such concepts as technological and psychological units of analysis have important meaning in the development of models of activity for design purposes.

6.2 Principles of Cognitive Actions Extraction in Task Analysis

Currently, there has been significant progress in the development of technical means for recording eye movements. However, the interpretation of eye movements and related issues in the use of these data to studying performance have not been satisfactorily. So, there is still a great need for the development of appropriate interpretation of such data. Cognitive psychology examines a possibility of using eye movement data in the analysis of complex cognitive components and of thinking in particular. However, there are real theoretical problems associated with the interpretation of eye movement data. Vertegaal (1999) wrote that eye fixations provide some of the best measures of visual interest; they do not provide a measure of cognitive interest. Eye movements according to this author simply determine whether a user is observing certain visual information. As a result, eye movements have been commonly assumed to simply predict attention (Rayner, 1998). Attention shifts in most cases is associated with shifting of gaze into a corresponding area. Psychologists who study eye movements and fixations generally avoid considering such cognitive processes as memory, thinking, and decision making. Thus, in spite of the tremendous amount of data on the eye movement registration, there are still methodological problems in the interpretation of eye movement data in cognitive psychology. Some scientists suggest that usability researchers do not always have a strong theory to perform eye movement analysis (Goldberg and Kotval, 1999).

In AT, knowledge of the principles of visual system functioning has important theoretical and practical meaning. This information is critical not only for understanding the nature of perceptual processes, but also for analyzing various cognitive processes, and for the optimization of interfaces between human and artificial systems. In AT, the analysis of eye movement during performance of visual tasks is an important source of information for studying human cognition during task performance. Eye fixations are tightly linked to step-by-step, goal-directed cognitive actions. A subject actively selects the needed information for a momentary goal of his/her actions. The sequence of fixations is determined by goals of actions that often cannot be predicted. Sequence of eye fixations should be considered as flexible cognitive strategies of task performance that can also be combined with behavioral actions. Eye movement data is very important for the analysis of problem-solving strategies during an operator's interaction with the equipment or a user's interaction with the computer.

Yarbus (1965, 1969) is the founder of the modern methods of direct analysis of eye movements. Analyses of his work in the West concentrated on the study of perceptual processes leaving out other important aspects of Yarbus' studies, which were carried out in the framework of the concept of activity. The objective of his study was not only to analyze principles of functioning of the visual system while receiving visual information but also included exploring the evidence that cognitive processes should be considered as mental activity of the subject. This suggests existence of mental actions that are involved in manipulating perceptual data, mental images, and verbally logical material. Special attention was given to the dependency of eye movement strategies on activity goal and motivation. Further, his studies were aimed at proving the principle of unity of cognition and behavior. Such studies are useful for the application of eye movement data to various practical domains and particularly to task analysis. From an AT perspective, eye movement registration should be performed in the context of a specific task. Depending on the goal of a task, motivation, and subjective significance of information, strategies of eye movement can vary. For example, inducing different goals in the observation process can change the pattern of eye movement. Many eye movement and usability studies have discovered eye tracking differences between novice and more experienced subjects (Aaltonen et al., 1998; Yarbus, 1965). Therefore, eye movement analysis can be useful in the study of skill acquisition processes.

An operator may have a task the performance of which requires a combination of complex cognitive and behavioral actions. The other type of task is when an operator monitors, observes, and evaluates the situation. Sustaining vigilance is the main function of such operators. Anticipation is also one of the main components of activity when an operator has to predict events. Usually in such situations, an operator formulates her/his task independently. Eye movement analysis helps predicting the strategies of activity for tasks where anticipation is required. In general, eye movement analysis is important in any task where visual information is needed for task performance. AT research demonstrated that eye movement is an indicator not only of perceptual but also of higher cognitive functions. Mental activity involves transformation of images, searching information in memory, logic operations, and so on. Based on the analysis of eye movement it is then necessary to determine the content of perceptual, cognitive, mnemonic, thinking, and other actions and operations.

Yarbus (1965, 1969) invented an ingenious method for direct eye movement registration during receipt of information from a specific scene. This method allowed observing successive fixations of the gaze, which helped to discover eye movement and eye fixation patterns specific to the scene presented to the subject. Such fixations and movements are not specific to a presented scene but rather to the task that is presented to a subject or formulated by a subject independently. For example, Yarbus presented to a subject a painting of the famous Russian painter Ilya Repin (see Figure 6.1).

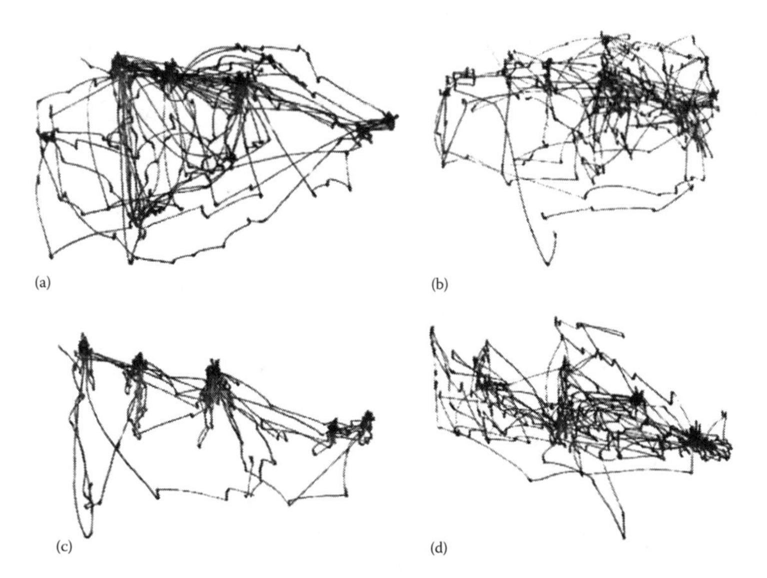

FIGURE 6.1
Eye movement during perception of the picture *the unexpected visitor.* Subjects were asked to: (a) examine the picture; (b) estimate the material circumstances of the family in the picture; (c) determine ages of the family members; and (d) describe what the family members were doing before the unexpected visitor entered to the room. (From Yarbus, A.L., *The Role of Eye Movements in the Visual Process*, Science Publishers, Moscow, Russia, 1965.)

This picture depicts unexpected return of a family member who was imprisoned for putting forward the ideas of socialism in the nineteenth-century Russian Empire. The first task for the observer included free examination of the picture. It was discovered that eye movement and fixations were not random. The subjects tended to follow the contour of the picture.

In a subsequent recording experiment, subjects were asked to: (a) examine the picture; (b) estimate the material circumstances of the family in the picture; (c) determine ages of the family members; and (d) describe what the family members were doing before the unexpected visitor entered to the room. Several other tasks were also presented. It was discovered that different questions and, therefore different tasks, produce different patterns of eye movements and fixations (see Figure 6.1). According to Yarbus, different questions create different task goals and a subject selects perceptual features of the scene that are relevant to the specifics of a task, his/her goal, and motivation. It was also discovered that the significance of perceptual features of the scene and the goal of activity can change the strategies of eye movements.

It has been discovered that without eye movements in relation to the scene, the scene disappears. Yarbus developed a devise that has a projector mounted on a contact lens that should be attached to the eye's cornea. A slide projector is connected to the contact lens. When a stimulus is presented on the screen, the eye with the contact lens looks at the image. Since, the lens and projector move with the eye, the image presented to the retina is stabilized in relation to the moving eye. Therefore, retinal image impinges on the same retinal receptors regardless of the eye movement. It has been discovered that when the slide projector is turned on, the subject can see a stimulus for a while, but within a few seconds the image begins to fade and then totally disappears because the light from the stimulus strikes the same retinal receptors. This produces an inhibitory effect and receptors stop firing. These studies demonstrate that for visual perception it is necessary to ensure relative movement of the object along perceiving elements of the retina. From general AT point of view, this proves that motor processes are directly involved in perception and also serve as a proof of the unity of cognition and behavior.

However, in further studies, the logical/reasonable relationship between eye movement and perception of objects could not be established. Moreover, studies of Zinchenko and Vergiles (1969) demonstrate that in certain conditions it is possible to maintain vision of an object when its position is stabile in relation to the subject's retina. This study suggests that there are special mechanisms of the visual system that facilitate movement of attention along an object in the absence of eye movement. Zinchenko and Vergilis called it an internal or mental visual scanning. In other words, their experiment demonstrates the possibility of moving visual sight along a stabilized, in relation to retina, visual stimulus, which is achieved by mental movement of attention. A slight shift in attention created the impression of movement of the subject's gaze on a stabilized object. Such eye movements helped to change the sensitivity of the eye's receptor area. Changing the sensitivity of the elements of the retina facilitated receiving information about an available stimulus on the retina. Receiving required information about an object was considered as a perceptual action. The system of perceptual actions that has as its source trace on the retina is called vicarious, that is, replacing external perceptual actions. Vicarious actions provide a subject with the ability to manipulate images in the internal mental plane. Such actions are involved in transforming the image of a situation and are used in imaginative thinking.

Other scientists have suggested that the activity of the visual system may be partly connected with changes of sensitivity of receptors of the eyes (Granovskaya, 1974). These changes in sensitivity can partially replace external motor eye movements involved in perception. Changes in sensitivity of perceiving elements of an eye act as a beam that is involved in palpating a perceived object. In humans, this beam is within the receptor surface of the eye. However, in some animals, such as dolphins, a perceiving system can send signals outside of the receptors surface. Such a beam *touches the perceived object* by analogy with the sense of touch. Thus, an active nature of perception is preserved even in those cases where there are no external motor actions of the eye. There is a constant readjustment of an internal state of the visual system in accordance with the changes of external stimulation due to the feedforward and feedback connections between a visual sense organ and an external environment. Visual system is an active, self-regulatory system. This type of functioning is another example of the principles of activity regulation (Bedny and Meister, 1997).

The thinking process is involved in problem solving and is associated with two types of eye movements (Zinchenko and Vergilis, 1969). The first type is external—eye movements with relatively high amplitude. These motor eye movements, and the sensory components associated with them, are integrated into perceptual visual actions. Formation of perceptual images of the situation is facilitated by these visual perceptual actions. In the second stage, at the time of fixation, vicarious actions are accompanied by an amplitude of eye movements. Sometimes, these movements are so mild that they can be considered as micromovements of the eyes. Such eye movements are involved not in the perception of information, but in the mental transformation of the situation needed to solve the task at hand. Vicarious actions that are involved in transforming the image of a situation are components of the thought process. These data explain that during *blind fixations,* a subject performs mental operations that can be highly automated and unconscious. A subject looks at the stimulus and does not see it. Understanding the nature of these micromovements or vicarious actions of the eyes is important for correct interpretation of eye movement. Kamishov (1968) discovered similar data in his study of the eye movements of a pilot during a real flight.

Zaporozhets (1969) and Turvey (1996) demonstrate that touch perception is accompanied by complicated hand and finger macro- and micromovements. Comparison of hand and finger movements with eye movements shows that both kinds of movements perform similar functions in the perceptual process (Zinchenko and Ruzkaya, 1962). The study of the visual perception process and the sense of touch are examples that demonstrate the relationship between external and internal activity. Bedny et al.'s (2008) study also demonstrates that eye movements are involved not only in perception but also in the thinking process. Kochurova et al. (1981) performed a microstructural analysis of motor actions and motions. They found cognitive components in motor activity. These studies demonstrate that eye movements involve cognitive functions. Eye movements can be used as reliable indicators of an operator's cognitive activity.

Strategies of eye movements depend on individual features of observers. This means that similar tasks can produce task-specific strategies depending on individual features of a subject. Eye movement pattern also demonstrates considerable similarity between subjects, when they do not have significant individual differences. Recently, some scientists obtained data that demonstrated the task-specific nature of the strategies of eye movements (Ballard, 1991). In AT, a task has a conscious goal, which is given in specific conditions. This goal can also be formulated by a subject independently. The direction of activity during task performance is determined by the vector *motive → goal*. Personal significance of various features of a task influences eye movement strategies. These factors explain individual strategies of eye movements that depend not only on cognitive features of tasks, but also on emotional-evaluative and emotional-motivational factors of activity.

Specialists who study eye movements and fixations usually use such data as cumulative fixation time, number of fixations, distance of eye movement, time spent in areas of interest, transitions between areas of interest, scan path analysis, and total fixation time. These data are useful but not sufficient, because such methods ignore the cognitive aspects of task performance.

The basic characteristics of eye movement analysis are saccades (i.e., rapid intermittent eye movements occurring when the eyes are fixed on one point in the visual field after another) and fixation. They usually include scan paths, frequency fixation, fixation duration, and transition between areas of interest. There are no standard methods for identifying the fixations and saccades. At least three processes are assumed to take place within a typical fixation with a duration of 250–300 ms (Viviani, 1990). These processes include the analysis of the visual stimulus in the fovea field, the sampling of peripheral field, and the planning of the next saccade. Observation of the participants' field behavior has shown highly task-specific eye fixation strategies and considerable regularity in fixation patterns between subjects. This data demonstrates dependence of eye movement strategies on the features of the interface. Observations of natural behavior have demonstrated the highly task-specific nature of eye fixation patterns (Henderson, 1993).

Further, we will provide a detailed description of using eye movements for task analysis. Here, we want to present only general principles using eye movements for analyzing tasks where visual information is essential.

We will use these data further to prove the validity of our method of interpretation of eye movements during performance of various types of tasks, including computer-based tasks.

Let us consider a study conducted by Zinchenko et al. (1973). Subjects were presented with various tasks associated with the perception of visual information, selection and evaluation and comparison of elements of the situation, and decision making. Tasks varied in complexity depending on the specifics of operating with visual elements of the situation, the nature of information retrieval from memory, and the specifics of decision making. The study used the following performance indicators: (1) measured fixation duration and amplitude of eye movements to determine the specifics of the external scan and the character of investigation of the external situation; (2) reviewed electroencephalogram (EEG) of the occipital cortex in terms of the total energy value of the alpha rhythm, which was considered as an indicator of internal complexity of manipulating with internal or mental images of the situation—the level of electroencephalogram depression of the occipital brain region (total energy value of the alpha rhythm) serves as an indicator of internal logical operations and operations of comparison of information from external situation with templates kept in memory; and (3) registered electromyogram (EMG) of the lower lip as a measure of internal verbal activity. When analyzing the data, the authors considered quantitative values of these parameters, relationship between these activities in time, transitions from one type of physiological electro-activity to another. For example, the relationship between the activity of the visual system and EEG has been studied. The results showed that depending on the complexity of the task, the relationship between analyzed indices changed. Considering such indicators as duration of eye fixation for the simplest, mostly perceptual tasks, such duration can change in the range from 0.2 to 3 s with a dominating duration of up to 0.5 s. Fixations with longer durations were observed relatively seldom. For the tasks of moderate complexity, fixations with duration of up to 1 s prevailed. Fraction of fixation with period of time up to 5 s also increased. When subjects performed more complex tasks involving thinking components, short duration fixations significantly decreased to 30% of the time. Long fixations from 3 to 8 s dominated in such tasks.

Comparison of the EEG data with external eye movements showed that if external scanning is reduced, intrinsic activation increases because the intrinsic activity of the brain, such as mental manipulation of images, comparing them with the standards of memory, decision making, etc., increases during this time. An analysis of the articulation apparatus showed that the EMG of the lower lip was rarely longer than 1 s. This activation usually precedes subjects' verbal responses. This means that the articulation apparatus played a role in the final stages of task performance.

The second series of studies involved performing a task in the control room simulator for a power system (Zinchenko et al., 1973). This task closely imitated real operators' tasks problems. The subjects had to evaluate the status of the individual blocks and the general state of the power system. In this experiment, activations of the visual system and EEG activity of brain during mental manipulation of presented data were determined. In these studies, the authors found that when the mental plane for solving the problem is increased, the activity of the visual system (eye movements) decreases and the duration of visual fixations increase, which is important from a practical point of view. The internal mental activity associated with the semantic processing of information leads to an increase in the duration of eye fixation and is accompanied by a depression of the alpha-rhythm on the EEG.

These data confirm the validity of our method of eye movement interpretation. Eye movements can be used as an indicator not only of perceptual but also of more complex cognitive activity, for example, when a subject performs various types of tasks, including computer-based tasks. In this section, we discuss only the general principles of eye movement interpretation that are used in our further detailed description of the proposed approach. In our proposed approach it is necessary to determine the duration of perceptual actions. If the duration of fixation exceeds duration of perception, additional time for fixation is attributed to the more complex cognitive processes. Therefore, it is important to properly determine duration of visual perceptual action. We can identify simultaneous perceptual actions and complex successive perceptual actions. Simultaneous perceptual actions are usually associated with the identification of a familiar object. The duration of such actions is usually well known, and is presented in a variety of sources. For example, according to some authors (Jacob, 1991; Just and Carpenter, 1976), gaze fixations associated with perception typically vary between 200 and 600 ms. Saccades or ballistic eye movements last for about 30–120 ms. Not much information is available on the duration of perception of complex and unfamiliar objects. Strategies of complex perceptual actions include extraction of various features of an object, analysis of perceptual properties of the object, structuring of obtained data, and constructing a perceptual image of an object. Perceptual image of the object is developed based on successive eye movements. As a result of the consistency of eye movements and fixations on elements of an object or the situation, a subject integrates obtained data and performs categorization. In more complex cases, the perception process also includes decision making to facilitate categorization. In such situations, decision making process at the sensory-perceptual level is also involved in the final stage of the perceptual process (Zabrodin, 1985). Only after such analysis, and based on the duration of fixation, more complex mental actions can be determined. A complex perceptual action involves no more than three to four simple perceptual actions. This limitation is due to the capacity of working memory. The goal of a simple or complex perceptual action is constructing a perceptual image of an object, but no further manipulation with the obtained perceptual data. It is important to know that we are not usually conscious about all our eye movements. Hence, eye movement registration helps us to discover some cognitive components of work that are not conscious.

Analysis of data shows that if the duration of fixations does not exceed duration of the perceptual process, or is roughly the same then, the duration of fixations should be attributed to perception. However if the duration of fixation exceeds the duration of perception of an object, the extra time of fixation should be referred to the mental components of work related to manipulation with mental images, logical operations, and extraction of information from memory. The duration of mental actions that begin after the completion of a perceptual action is determined by the following formula:

$$T_{ment} = T_{fix} - T_{per},$$

where

T_{ment} is the duration of higher mental actions (mnemonic, thinking, decision-making actions)
T_{fix} is the duration of fixation
T_{per} is the duration of perceptual action

T_{ment} is determined based on qualitative analysis of activity during this time period. We identify what information was known to a subject at the time, if she/he was aware of the course of events that preceded the fixation period; what type of cognitive and behavioral actions were performed before eye fixation took place; what actions should be performed after receiving information; and what type of cognitive and behavioral actions were really performed by the subject after fixation was completed. The ability of the subject to forecast future events in a given time period is also important for determining the content of eye fixation time. It is important for such analysis to understand the logic of a task as a whole and of its performance during a particular step in task performance.

It should be noted that qualitative analysis is of fundamental importance in psychology in general. Indeed when we talk about individual cognitive processes, it is necessary to bear in mind that distinction can only be made on the basis of qualitative analysis. The information processing underlying human performance includes the integration of various cognitive processes that do not have a clear-cut border between them. For example, sensory processes of signal detection include decision making, thinking is inseparable from memory, and the perception process also includes mechanisms of memory.

Human information processing is a unitary process. Dividing this process into separate stages provides a useful framework for analyzing separate cognitive processes.

In conducting such analysis, it is very useful to have information about the execution time of some standardized perceptual actions. Eye movement research conducted by Yarbus (1965) demonstrates that average fixation lasts between 200 and 500 ms. According to Goldberg and Kotval (1998), the average time for simultaneous perception is 250–300 ms; similar data were obtained by Zinchenko (1981).

An increase in the perceptual complexity of the task leads to an increase in the duration of fixations, reaching 340–380 ms, and in some cases even higher. In MTM-1, there are two eye time elements: eye travel time and eye focus time. They are basic elements for the description of the simplest mental elements in MTM-1. Eye focus element is the time required for recognizing a simple signal or object. Its duration is about 0.3 s. When it comes to the perception of a well-discernible signal or feature of an object, the duration of perception is also about 0.3 s.

Let us assume that the duration of fixation based on the analysis of eye movement registration was 1 s. Therefore, the time for a mental action is

$$T_{ment} = 1 - 0.3\ \mathrm{s} = 0.7\ \mathrm{s}$$

Content of cognitive activity during this period of time can be determined using qualitative analysis of mental actions performed by a subject (Lomov, 1982; Zarakovsky, 2004). An illustration of some information about the duration of cognitive components of activity, in particular the duration of perceptual actions (in seconds), follows:

1. Perceiving (indentifying) a simple symbol:

$$M(T_{per}) = 0.3 \pm 0.1$$

2. Perceiving information from a window display:

$$M(T_{wdisp}) = 0.2 \pm 0.7$$

3. Average time for receiving information from one aviation display:

$$M(T_{avdisp}) = 0.5$$

4. Perceiving information from a pointer display:

$$M(T_{pd}\mathrm{i}_{sp}) = 1.0$$

5. Perceiving a seven-digit number:

$$M(T_{didg}) = 1.2$$

6. Detecting a moving signal in the field of vision (field of vision ≈ 40°):

$$M(T_{field}) = 0.15 - 0.17$$

7. Searching and detecting a target on the screen:

$$M(T_{scr}) = 0.37 \pm 0.15$$

8. Perceiving one out of four operative OUI in a 40° field of vision:

$$M(T_{field}) = 0.6 \pm 0.07$$

9. Actualization of one object (word, signal) from memory:

$$M(T_{act}) = 1.2 \text{ s}$$

10. Checking logical conditions—OR and AND types:

$$M(L_{or}) = 0.3 \text{ s}$$

$$M(L_{and}) = 0.7 \text{ s}$$

There are other temporal characteristics associated with more complex components of activity. A disadvantage of data associated with more complex components of activity may be related to the components not having a clear description. The more complex mental elements of activity are and the more their duration is, the more precise these elements' description should be. Usually, such data should be described in a standardize manner and their start and end points should be presented. Anyone who reads the description of specific mental elements of activity should clearly understand what type of cognitive activity is performed during a particular period of time. Unfortunately, psychology does not provide recommendations for clear description of activity elements with a predetermined time standard. Therefore, presented above types of time standards are of limited applicability. The first ones who realized the importance of standardized descriptions of activity elements with duration were Gilbreth and Gilbreth (1920). These authors proposed some principles of the description of elements of activity. Currently, there are far more accurate methods for the classification of the elements of activity and for determining their duration (see, e.g., MTM-1 system). Such principle of cognitive elements description can be useful when analyzing the cognitive component of work.

For determining the duration of higher-order cognitive actions, it is important to know the duration of perceptual actions. In most cases, receiving of information can be defined as simultaneous perceptual actions. They can be easily understood even when we have only short and not standardized descriptions of such actions. Much more difficult to identify and describe are successive perceptual actions that are involved in the recognition of unfamiliar and complex stimuli. For this purpose, it is necessary to conduct a more detailed analysis of the stage of receiving of visual information. If we know duration of eye fixation and duration of perceptual actions, we can define time performance of higher-order mental actions.

Let us consider an example. If a pilot looks at aircraft instruments and the longest fixation takes 3 s, then the duration of complex mental processes is calculated as

$$T_{ment} = 3 - 0.5 \text{ s} = 2.5 \text{ s}$$

Specifics of cognitive actions and operations involved in reading a display would be determined based on subsequent qualitative analyses. In our example average time for receiving information from one aviation display is equivalent to 0.5 sec.

When a subject perceives a complex unfamiliar object, he/she has to construct its image and interpret meaning of the perceived data. Recognition and construction of a perceptual image of such an object can be considered a system of perceptual actions. The purpose of such actions is detecting the object and creating its perceptual image. The time taken for the construction of the perceptual image is the boundary that separates the complex perceptual process from higher-order cognitive processes. The final stage in the recognition and construction of the image includes categorization and decision making at the sensory-perceptual level (Bedny and Meister, 1997). Therefore, at the eye fixation stage, the perceptual process is completed when the subject performs simple binary yes/no decisions about the perceived object or its features (the perceived object or its features may or may not belong to a specific perceptual category).

For example, an in-flight board operator first receives information from a display and makes a decision about a perceived complex object based on an analysis of its individual features. Only then does he/she make a conclusion about the perceived object. In one experimental study, an operator made a conclusion about the perceptual image of an object utilizing basic features of the object and integrating the obtained data (Zarakovsky, 1976). An operator extracted and analyzed three basic features of an object, where perception of separate features took 0.3 s each, resulting in a total perception time of 0.9 s. In this example, the complex perceptual action consisted of three simple perceptual actions united by a higher-order goal. After the completion of the complex perceptual action, the operator starts to perform higher-order actions including decision making and thinking actions.

This is in agreement with data obtained in cognitive psychology. We can see that the perceptual process includes such stages as extracting and analyzing sensory features, decision making about the final perceptual image, etc. From an activity perspective, sensory features can be considered as perceptual units of activity that are used by a subject when he/she performs perceptual actions. By analogy with motor actions, they serve as a material entity that is utilized during the performance of such actions. Each action is directed to achieve its goal. Sometimes, simple actions can be integrated by their higher-order goals. Simple perceptual actions are components or operations of more complex perceptual actions. The capacity of working memory limits the number of simple perceptual actions that can be integrated into a single, more complex one. Achieving the goal of a perceptual action is the boundary that separates the perceptual stage from higher-order cognitive stages.

In some cases, a researcher is faced with very complex perceptual tasks when their performance time should be determined experimentally. Such perceptual tasks can be transformed into task-problems, which involves thinking actions and the process of categorization at the thinking level. Such tasks can be considered as perceptual only by their final objectives. In fact, such task-problems are more likely to be classified as thinking tasks, since they contain a large proportion of mental actions and decisions at the thinking level.

There are also situations when researchers observe *blind fixations* (fixation unrelated to the task at hand) and it is necessary to identify their causes and cognitive actions and operations associated with them. In cognitive psychology, eye movement is usually attributed to attention processes, or perception (Rayner, 1998; Smith et al., 2000). Blind fixations are considered to be discrepancies between eye movements and attention. However, a blind fixation also can be explained by a subject's involvement in the mental manipulation of information associated directly with the task being performed: Eyes are fixed on a specific object but a subject does not see it. There is a need to conduct more experiments in order to identify the objective reasons for the occurrence of blind fixations, because some blind fixations may be based on individual characteristics of subjects or their current mental state. We will consider eye movement data as sources for determining duration of cognitive components of activity in more detail in Section 9.2. Thus, in AT, more attention is paid to the relationship between eye movements and higher mental functions. The eye movements–hand movements relationship in AT has been studied from the perspective of unity of cognition and behavior.

Hence, AT studies of eye movement and mouse movement are based on different theoretical data. SSAT advocates the use of the principle of unity of cognition and behavior as key in the study of human–computer interaction (HCI) (Bedny et al., 2001). It is recommended that both eye and mouse movements are used in applied research because this helps to interpret eye movement data. It should be mentioned that in cognitive psychology, scientists sometimes also utilize the combination of eye and mouse movements (Hornof, 2001; Smith et al., 2000). However, these studies are based on totally different theoretical data. We will discuss principles of eye movement interpretation when users perform computer-based tasks. These principles can be used for the analysis of any task that utilizes visual information and cognitive components of activity make up the bulk of the task.

6.3 Description of Motor Actions and the Time of Their Performance

Technological revolution has led to increased computerization and automation and to a corresponding trend toward reducing manual components of work, thereby requiring advance conceptual knowledge and cognitive skills. Although automation can make physical aspects of work easier, it has new demands for workers such as increasing processing uncertainty and decreasing predictability of the situation or task at hand. In abnormal situations, workers do not just monitor the state of automatic process but also take part in controlling the system. Such interventions involve development of hypotheses, performance of motor actions, and evaluation of their consequences.

Hence, manual control still has important functions in contemporary, complex sociotechnical systems that are not restricted to the executive aspects of work, but also tightly connected with the cognitive components of performance. Due to the interaction of motor and cognitive actions, a worker has great knowledge about what influence she/he exerts on the system. Manual control is a critical factor in quick recognition and correct interpretation of an unexpected situation. Similarly, methods of modern production processes vary based on how they are performed. Contemporary manufacturing operations and computer-based tasks do not have the same sequence of performance every time they are performed. Cognitive and motor actions have a logical organization and their sequence can be changed according to specific rules. The specifics of contemporary work are not taken into account by MTM-1, which studies manual work. In this regard, SSAT offers a method of combining algorithmic descriptions of human activity with microelement analyses utilized by MTM-1 (Bedny and Karwowski, 2007; Bedny and Meister, 1997). The purpose of algorithmic analysis of activity is to describe the logical organization of human cognitive and behavioral actions. Human algorithms depend on prescribed instructions and on how a subject chooses strategies of task performance. The combination of algorithmic analysis of activity with temporal analysis, including MTM-1 method, gives an opportunity to describe very flexible time structures of human activity.

In SSAT, motor actions are performed by individuals through their skeletal-muscular system and can change the state of material objects in the external world according to the goal of an action. The duration of motor actions has a significant impact on the duration of a task. Therefore, when analyzing motor actions, special attention is paid to their duration. Motor actions can be described as a combination of standardized motions that are integrated by a single action's goal, which allows using MTM-1 for standardized descriptions of motor actions. According to SSAT, MTM-1 utilizes psychological units for the analysis of motions. Thanks to such methods behavioral actions that are described by utilizing technological units of analysis (typical elements of task) can be described by using psychological units of analysis (typical elements of activity). Therefore, at the first stage, we apply a traditional method of motor actions descriptions by using technological units of analysis and then transfer them into psychological units of analysis. Psychological units of analysis describe elements of activity in a standardized manner, which allows for unified and unambiguous interpretation of what a performer does. We want to stress that MTM-1 does not use such terms as motor actions and psychological or technological unit of analysis.

Let us briefly consider the concept of motion in SSAT. Motions also are called motor operations in SSAT. A motor operation is a relatively homogenous act that lacks a conscious goal. Motor actions integrate a set of motor operations around a conscious goal. We define standardized motor actions as a complex of standardized motions (usually no less than two or three motions) performed by a human body, unified by a single goal and a constant set of objects and work tools (Bedny, 1987). Under standardized motion, or motor operation, we understand a single motion of a body, leg, hand, wrist, and fingers that has a definite purpose in a work process and also corresponds to the rules of standardized description. One can clearly describe a motor action only if she/he defines standardized motions imbedded in motor actions. Motor actions that are performed by different parts of a body cannot be integrated into one action. For example, two motions simultaneously performed by both the left and right hand cannot be considered as one motor action. Actions can be combined with supplemental motions. For instance, actions performed by hands can be combined with supplementary motions of other parts of the body. Sometimes motion can be associated with a corresponding conscious goal in a work process. On the one hand, they have only one homogeneous motor act and thus should be considered as a motion. On the other hand, they have a conscious goal and can also be regarded as a simple motor action. For the purpose of standardized task description, we, as an exception, call them conscious motions.

We utilize MTM-1 for describing standardized operations or motions, which are components of motor actions. For example, "move arm and grasp lever" is considered as a standardized motor action that is comprised of two standardized motions, "move arm" and "grasp." MTM-I provides the most useable description of standardized motions but it ignores the concept of action. In MTM-1, the concept of cognitive and motor actions does not exist. A specialist in MTM-1 begins task analysis by dividing activity into discrete motions during task performance. In contrast, in SSAT, motor components of a task are divided into actions and then each action is, in turn, divided into operations (motions). For an algorithmic description of activity we use, in addition, such units of analysis as members of an algorithm. A member of an algorithm can combine several motor or similar type of cognitive actions that are integrated by a higher-order goal. A brief introduction to MTM-1 is given, for example, in Barnes (1980). A detailed description of this system can be found in MTM Association (see UK MTMA, 2000). For the practical use of MTM-1 a specialist must be trained and qualified under the MTM Association. The minimum course duration is 105 classroom hours (see UK MTMA recommendations). MTM-1 is based on ideas of Gilbreth but it is fundamentally different from their Therbligs system. Critical comments on Gilbreths' system cannot be directly applied to MTM-1. Analysis of MTM-1 from an SSAT point of view will be presented further in the chapter.

MTM-1 should be considered first of all as a language for describing activity elements and can also be considered as a tool for determining performance time. It is instrumental for describing motor components of activity.

From an SSAT perspective, it is critically important that each basic motion in MTM-1 be clearly classified and described. The starting point and end point of each basic motion should be identified. There are such basic motions as reach, grasp, move, position, release, etc. Later, a table of body motions was also introduced to this system. The combination of these motions allows to describe any holistic motor activity. Depending on the specifics of implementation, these basic movements are divided into classes. The time of a basic movement depends on the type of movement, its class, and the distance of motion. Description of basic motions and their classes is supplied in drawings, examples and method of extraction and time of motion calculations which enables highly accurate classification of motion. One of the main criteria for classification of motions is their purpose, which is close to the notion of goal in AT. Hence, from an SSAT perspective, basic motions can be considered as motor operations. Motor operations are important types of units of analysis of motor activity (Leont'ev, 1978; Rubinshtein, 1957). Classifications of basic motions (case and description) take into consideration the level of concentration of attention during their performance. This means that MTM-1 with some approximation takes into consideration cognitive mechanisms of motions' regulation. There are also some rules of combining basic motions. The possibility of performing them simultaneously depends on the level of concentration. MTM-1 also has such elements as eye action. There are eye travel time (ET) and eye focus time (EF) that are associated with cognitive elements of work. All this demonstrates that MTM-1, to some degree, takes into consideration cognitive components of work and cognitive aspects of movement regulation. Thus, SSAT proposes to use MTM-1 for describing motor actions. Anyone familiar with MTM-1 can, with some approximation, describe human motions and motor actions during task performance. One should bear in mind that no real motion can exactly match a basic motion in MTM-1 because our activity is very flexible and the same motion when repeated multiple times varies by temporal and spatial characteristics. However, such variations can be ignored if they are within the given range of tolerance. Similarly, in mass production, a worker produces the same part multiple times and each individual part has unique size and shape. However, if the size and shape vary within a given range of tolerance, these parts are considered to be identical. MTM-1 follows the same principle. It ignores the concept of strategy and flexibility of human performance, ignores the concept of action, and does not take into account principles of combining movements with elements of cognitive activity. The rules of combining motions do not cover all practically important situations. There are no principles of classification of activity elements according to their complexity. All these deficiencies are eliminated in SSAT.

Some psychologists superficially criticize time study in general. For example, Schultz and Schultz (1986, p. 410), when criticizing time study, refer to some conflict situations that can arise between workers and trade unions on the one hand and experts in the field of time study on the other hand. Such conflicts can be explained by different interests of these parties, or not sufficient professional skills of time study specialists, etc. However, time study is an essential element of management that facilitates proper organization of the production process, design of efficient work method, planning, correct payments and other important aspects of management. Time study can also be used for the analysis of cognitive work that requires usage of special chronometric methods for mental components of work. Therefore, industrial psychologists should not deny time study but participate in the development of science-based time standards. In summary, time study and MTM-1 can be used to describe tasks and job demands. Time study can range from relatively broad work periods measured in minutes to very detailed elements of tasks measured in fractions of seconds.

Let us consider an example of describing motor actions by using technological and psychological units of analysis with a physical model of production operation "installation of 30 pins into 30 holes of pin panel" especially designed for this purpose. The subject had to move two hands simultaneously to a pin box in front of them; grasp two pins and put them in holes of the pin board also located in front of them but closer. There are two motor actions that are performed simultaneously by two hands: one action is "move left hand to the pin box and grasp the pin" and the same time "move right hand to the pin box and grasp the pin." The description within quotation marks are examples of action descriptions by using technological units of analysis, which is just a description of one element of task. In addition, we can use psychological units of analysis. In our example it will be *R32B+G1C1*. This is an MTM-1 description of a motor action performed by the subject. In addition, we use the concept of motor action that is not used in MTM-1. The goal of the action is to grasp the pin. This goal of action integrates two motions R32B and G1C1. R32B means—reach a single object in a location that may vary slightly from cycle to cycle, distance of movement is 32 sm. Such motion requires an average level of attention concentration. The purpose of the next motion is to grasp an object when there is – interference with Grasp on bottom and one side of nearly cylindrical object which has diameter larger than 12 mm. The MTM-1 manual has a much more detailed description of these motions. Thus the description "move left hand to the pin box and grasp the pin" (technological unit of analysis) in combination with *R32B+G1C1* (psychological unit of analysis) give us a precise description of this motor action. We will use a combination of such units of analysis for describing activity time structure and for quantitative evaluation of task complexity. A list of motor action descriptions using MTM-1 specifications for motions is presented in the following. This list contains very brief descriptions of motions. For detailed descriptions of basic motions, we recommend the use of the MTM-1 manual.

Reach (*R*)—basic motion (operation), the predominant purpose of which is to move a hand or a finger toward a destination.

There are five classes of reach.

Reach A—reach an object in a fixed location or an object in the other hand or on which the other hand rests (low level of attention concentration is required).

Reach B—basic motion to a single object in location that varies slightly from cycle to cycle (average level of concentration of attention is required).

Reach C—reach an object jumbled with others objects in a group so that search and select occurs (high level of concentration of attention is required).

Reach D—reach a very small object or where accurate grasp is required (high level of concentration of attention is required).

Reach E—Reach an indefinite location to get a hand into a position for body balance or the next motion or out of the way.

Symbolic description R30B means "reach an object by method B, distance of motion is 30 sm."

Move (*M*)—basic motion (operation) with the predominant purpose to transport an object to a destination.

There are three classes of move:

Move A—move an object to another hand or against stop (low level of attention concentration is required).

Move B—move an object to an approximate or indefinite location (average level of concentration of attention is required).

Move C—move an object to an exact location (high level of concentration of attention is required).

Symbolic description M30B means "move an object by method B, distance of motion is 30 sm."

Turn (*T*)—basic motion (operation) to turn a hand, either empty or loaded by a movement that rotates the hand, wrist, and forearm about the long axis of the forearm. The time performance of this element depends on two factors: (1) degree of rotation and (2) weight.

Examples:

T45S—turn a small object (up to 2 lb) to 45°

T90M—turn a medium weight object (between 2.1 and 10 lb) to 90°

T180L—turn a heavy object (between 10 and 35 lb) to180°

A turn can be performed independently or can be combined with reach or move.

Grasp (*G*)—basic motion with predominant purpose to secure sufficient control of one or more objects using the fingers.

Grasp is further divided into five types:

G1A—Grasp a separate object of any size that is can be easily grasped.

G1B—Grasp a very small object or one lying close by against a flat surface.

G1C—Grasp a nearly cylindrical object (has three types depending on the diameter of the object) that is placed in isolation, or is in contact with others objects.

G4—Grasp an object jumbled with others objects so that selection or search occurs (has three types depending on the size of the object).

G5—simple contact (there is also regrasp or transfer type).

Release (*RL*)—basic motion has two types:

RL1—performed by opening fingers

RL2—contact and release

Positioning (*P*)—basic motion to align, orient, and engage one object with another object.

This basic motion is classified according to

1. *The type of symmetry*: S—*symmetric (cylinder*); SS—semi-symmetric; NS—nonsymmetric
2. *The class of fit*: P1—*loose*, no pressure required; P2—*lose*, light pressure required; P3—*exact*, heavy pressure required.

In addition, there are *easy to handle* and *difficult to handle* types.

Example:

P1SE—positioning a cylindrical object with loose contact within a cylindric hole, easy to handle.

Disengage (*D*)—motion to break contact between one object and another.

The time of disengagement is affected by the class of fit (tightness of connection) and how easy it is to handle.

Example:

D1E—loose connection, very slight effort, easy to handle

D2D—close connection, normal effort, difficult to handle

Apply pressure (*AP*) has two types: *AP1*—effort 15–22kg; *AP2*—effort up to 15 kg.

This is just a brief description without the performance time of the elements.

Cognitive actions are also associated with visual activity in MTM-1.

Eye focus (EF) depicts the time required to focus eyes on an object and look at it to determine certain readily distinguishable characteristics within an area that can be seen without shifting the eyes. This element includes mental activity for simple *yes-no* or *if-then* decisions.

EF requires 0.27 s (we use 0.30 s).

Eye travel (ET) reflects the time of eye movement from one position to another. It depends on the distance of eye movement and perpendicular distance from the eyes to the line of travel.

The above description gives just a general idea about MTM-1. For practical use of this system, a specialist should read MTM-1 manual.

In SSAT, these motions are components of motor actions. We need to remember that MTM-1 does not utilize the concept of motor action. Thus, in SSAT, MTM-1 is used for description of motor actions.

6.3.1 Other Method of Action Classification

MTM-1 does not considered motions that can be used in verbal activity. Verbal type of actions performs not only communicative but also regulative functions within human activity. Verbal actions are the minimal verbal expression for transmitting meaningful information aligned with the desired goal. If we segment verbal speech too discretely, the extracted segment loses its meaning and we fail to achieve the goal of expression. Segments of speech, used as verbal actions, should correspond to all requirements of actions that include such features as (a) expression we performed intentionally implying that we were motivated to perform it; (b) prior awareness of what we wish to tell others, which implies that the expression is goal directed; and (c) alteration of prior incorrect expressions that enables voluntary regulation of expression. At the same time, one does not realize how she/he produces separate verbal operations.

Speech can be both external and internal. The latter is sometimes manifested by electrophysiological indices of the articulation muscles, even in the absence of audible speech. For example, it was discovered that increasing complexity of a task increases activity of the articulation muscles (Sokolov, 1960). Verbal performance of a task differs from verbal communication or explanation. In verbal performance, speech clearly emerges as a system of verbal actions that correspond to actually performed actions.

Verbal actions can be extracted in communication, reading, and typing. For example, a verbal motor action determines meaningful typing units and can include a single word or several interdependent words that convey one meaning. This is in line with Vygotsky's (1978) idea about meaning to be one of the basic units of activity analysis. Similarly, cognitive psychology shows the possibility of segmentation of verbal activity in verbal protocol analysis (Bainbridge and Sanderson, 1991). According to SSAT (Bedny and Karwowski, 2007), reading and typing normally require the third level of concentration of attention and have a third level complexity of activity. Sometimes, processing the most subjectively significant units of text can be transferred into a higher level of complexity. If the text is relatively homogeneous, there is no need to extract separate verbal actions. However, the time for the processing the text and level of complexity of the text (according the level of concentration of attention) should be determined.

In AT there are additional methods for classifying actions (Bedny et al., 2000; Lomov, 1986; Zarakovsky and Pavlov, 1987). The basic criteria for classification are both the nature of an object of actions and the method of their performance. This classification includes the following actions:

1. An object-practical action that is performed with real objects
2. An object-mental action that is performed mentally with images of objects
3. A sign-practical action performed with real signs, like receiving symbolic information from different devices, as well as its transformation
4. A sign-mental action performed mentally by manipulating symbols

In order to correctly extract the required actions and develop their classification according to these criteria, we need to identify the means of work, the object, tool, and the goal of the actions. The nature of an action is dependent on the interrelation between these components in any specific situation. The first step is to distinguish between the concepts of means of work and tool. The means of work is a broad concept that includes a variety of tools and equipment and cannot be used as a synonym for tools. In SSAT, the concept of tool is, from a psychological point of view, closely associated with the concept of action. Outside a specific task, we cannot determine what a particular tool is for. In this sense, the computer certainly cannot be classified simply as a tool. Rather, the computer should be considered a means of work, which can present or create various tools for a subject that can be used during performance of actions in a computer-based task. Moreover, the computer as a means of work can present not only tools for performance of actions, but it also creates artificial objects that can be transformed according to the goals of actions. The concept of tool in AT also has different meanings. According to Vygotsky (1962), tool can be not only material but also mental. Vygotsky called tools that mediate mental activity *sign*. When individuals performed mental actions they use signs as tools in the same way they use physical tools to perform external material actions. Signs fulfill the role of internal psychological tools. Language is a major system of signs that mediates human mental activity. This theoretical discussion helps us to understand how we can utilize the last method of actions classification. Let us consider the simplest example. A user tries to correct the misspellings in a text. Here, the task is "correct spelling." This is a deterministic task, requiring a well-defined sequence of actions. In order to classify actions according to considered method, we have to identify the object, tool, and goal of each action. The nature of action depends on the interrelation between these components. The required actions of this task are as follows:

1. Reach and grasp the mouse with the right hand.
2. Move the cursor to the initial position preceding the misspelled word and depress the left mouse button with the index finger.

3. With the mouse button depressed, highlight the required word by dragging the cursor to the end of the word; release the mouse button.
4. Move the pointer to the spelling icon on the toolbar, depress the left mouse button with the index finger, then release.
5. Examine the list of options presented by the dialog box.
6. Decide on the most suitable spelling option.

We will not describe the entire task here. Let us classify these actions according to the last method. When the user performs the first action, the mouse is the object engaged by the subject. The conscious goal of this action is the mental understanding of the future result "grasping the mouse." This is an object-practical action. In the second action, the mouse becomes a tool through which the subject implements the movement of an object, the cursor, to the start position. The object for this action is the cursor. The goal of the action is to bring the cursor to the required position. The pointer is a symbol on the screen. However, as the meaning of the sign is not important here, this also regarded as an object-practical action. In a similar way the fourth action should be classified as object-practical action.

In the fifth action, the list of options in the dialog box become the object. In executing this action, the subject does not employ any external tools. This is a sign-practical action because it is performed with real externally presented signs. In the sixth action, a particular item in the list of object is the object. In a similar way we can classify the other actions. This is a sufficiently difficult method of action classification. However, sometimes we need to consider a specific action in a more detail manner, for example, the usability of the mouse such as its *graspability* and *clickability* during the performance of the first action. Similarly, the complexity of the sixth action, (the decision-making action, can be considered. We can simplify the action performed by transferring it from a complex class of actions to a simpler one. For example, by introducing an externally presented symbol, we can transfer a sign-mental action into a sign-practical action.

Usually, specialists have to use the method of standardized description of actions we described earlier in the chapter. The purpose of classifying actions is to present activity as a systemically organized structure. All actions are organized as a system due to the existence of the general goal of activity and mechanisms of self-regulations (Bedny and Karwowski, 2004). One of the most important characteristics of actions is their duration. If we know the time of actions' performance, the duration of cognitive and specifically motor operations, and the rules for combining the elements of activity in time, it becomes possible to construct a temporal structure of activity during task performance. This is an important step in activity design.

6.4 MTM-1 and Strategies of Activity Performance

Motor components of human activity are not eliminated in HCI tasks. A mouse, a keyboard, and a joystick are the objects that dominate the motor components of computer-based tasks. Even the multitouch screen technology that allows manipulating virtual objects on the screen also requires motor activity. Users can push, pull, and grab various virtual objects. Motor actions have to be flexible and precise. Such actions do not require physical efforts. MTM-1 can be useful for standardized descriptions of motor activity during interaction with the computer and other mobile devices.

One of the disadvantages of MTM-1 is the fact that it ignores variability of human activity and the possibility that a user may use various strategies of task performance. Therefore in this chapter we consider the relationship between variability of human performance and the ability to utilize MTM-1, which was originally not adapted to study flexible human activity.

Motor components of activity are important in traditional man–machine systems, where we can distinguish manual, semiautomatic, and automatic controls. Moreover, even in automatic systems, manual control can be important. For example, in case of malfunction, efficient transition from automatic to manual control a is critical factor. In aviation, the principle of joint control was introduced (Beregovoy et al., 1978), which allows effective transition from automatic to manual control of aircraft. Moreover, total elimination of manual control has a negative effect on an operator performance. For example, in automatic flight control, the visual control of pilots during flight was disturbed. Elimination of manual components of work may produce monotony. This is particularly specific for vigilant tasks. We need to pay attention also to the fact that in many production operations manual components are the major part of the work. Therefore, motor components of activity always will be important in human work. The nature of motor actions during task performance has changed significantly. These changes are related to the fact that heavy physical work has significantly reduced. This has led to motor actions, in most cases, not requiring much physical effort. However, the accuracy requirements of motor actions, the ability to coordinate them in time and space, significantly increased. The problem of regulation of motor actions and their coordination with cognitive actions makes adequate pace of performance central in studying motor actions. However, in engineering psychology and ergonomics there is no precise understanding of the concept of motor action. In AT, motor action consists of motor operations that are integrated by the goal of an action. Motor action is considered as a self-regulated element of the motor component of activity. In contrast, in engineering psychology and ergonomics, instead of motor action the term motor response is used very often. Human being is viewed as reactive system. Such methods as the reaction time measurement and Fitts's law are used for the description of motor components of activity. Such methods can be used only in situations when an operator reacts to isolate signals, using discrete actions in highly predictable situations. When an operator performs a sequence of mental and physical actions in response to the appearance of different signals, the speed of performance of these actions is lower than that of isolate reactions. Our study demonstrates that the pace of motor activity in such situations is approximately equivalent to the pace that was described in MTM-1 (Bedny, 1979; Bedny and Karwowski, 2007). Moreover, such methods as the reaction time measurement or Fitts's method consider human activity as a summation of independent responses or reactions. A critical analysis of such method was performed by Salvendy (2004) when he considered time motion study. However, as we demonstrate further in this chapter, this method ignores the mechanisms of activity self-regulation and the strategies of task performance derived from them.

According to SSAT, motor actions should be described in a standardized manner. Because motor actions consist of motions, we can describe them by using MTM-1.

The MTM-1 system is a powerful tool for the analysis of behavioral actions. Time-and-motion economy is not a useful method when we study motor components of activity. This method considers human motor activity as mechanical aggregation of independent motions. However, human activity is a system, which has a logical and hierarchical organization. MTM-1 can be a powerful tool when it is combined with such concepts as motor actions, goal, self-regulation, and structure. MTM-1's rules for the selection of basic elements for tasks description are necessary but insufficient. Only after the analysis of the strategies of task performance and the study of logical and hierarchical organization of motor and cognitive actions can professionals use the rules of the MTM-1 effectively. In this chapter, we will demonstrate that MTM-1 can be used for the analysis of behavioral components of activity only after analyzing real strategies of task performance.

We present the analysis of MTM-1 from an SSAT perspective. The task can be presented as a logically organized system of cognitive and motor actions. Psychological operations are constituent elements of actions. Any motor action includes a number of motor operations. Motions are considered as motor operations. Study of motions and motor actions and their relationship with cognition is an important area of research in SSAT (Bedny et al., 2011). The unity of cognition and behavior is considered a critically important principle of activity study. Motions in MTM-1 are classified according to their purpose. This makes it possible to use motions as motor operations during the analysis of motor activity. However, it is possible only in combination with methods specially developed in SSAT. These methods involve the analysis of strategies of performance based on analyzing mechanisms of activity self-regulation.

MTM-1 allows utilizing a standardized description of movements. They, in turn, must be considered as a component of motor actions. Thus, MTM-1 may be used for the standardized description of motor actions. This is important for analytical methods of task analysis. However, this system ignores such concepts as behavioral actions and activity strategy. In SSAT, the concept of strategy is tightly connected with the concept of self-regulation (Bedny and Meister, 1997). The concept of self-regulation and algorithmic analysis of activity demonstrates that the principle *only one best way of performing a task* is not adequate in contemporary task analysis. Purely manual work is significantly reduced while cognitive components of work are increased. The task becomes variable and is performed in various ways. Such tasks have a logical, hierarchical, and probabilistic organization. Thus the prime objectives of this study were as follows:

1. To gain a deeper understanding of principles of activity regulation and formation of preferable strategies of task performance. According to SSAT, performance time of motions and motor actions depends on the structure of activity and specifics of its regulation by the performer.
2. To evaluate MTM-1 and the efficiency of its application by using traditional methods. It was hypothesized that the existing method of MTM-1 analysis contradicts with the principles of systemic organization of human activity as a goal-directed system. Only after psychological analysis of activity strategy would it be possible to select the basic elements of task performance.
3. To demonstrate that the rules of MTM-1 are necessary for the selection of basic elements or motions for task analysis and description but are not sufficient to determine the performance time of motions and motor actions.
4. To demonstrate that work activity has a hierarchical organization and therefore MTM-1 analysis cannot be reduced to the description of human motions during task performance. SSAT contains the following units of analysis of motor activity: motions, motor actions, and their combinations (members of the algorithm). The latter case is the integration of several motor actions by higher-order goals.

A person actively selects and develops various strategies of task performance based on the mechanisms of self-regulation. Such strategies include conscious and unconscious components. Therefore task analysis and time study cannot be performed without the analysis of the strategies of activity. In the following experiments, particular attention is paid to the study of strategies of activity utilized by subjects in performing isolated and sequentially executed motor actions.

The description of standardized motions (sometimes called *microelements*) is given according to MTM-1. In our experiment, we have chosen the basic motion or element *Reach* (*R*) as an object of study. *Reach* is used when the predominant purpose is to move a hand or a finger to a specified destination

(Barnes, 1980; Karger and Bayha, 1977). We have developed a special device that permitted identifying subjects' preferable strategies of activity and measuring performance time of separate actions including the basic motion or element *reach*. The five versions of the element *reach* (Karger and Bayha, 1977) have been already discussed under Section 6.3.

The main criterion for such classification is the level of concentration during performance of a basic element. Therefore, the basic element *reach* varies in complexity according to the level of concentration of attention. For example, no or minimal visual control is required when a subject performs *Reach A*. From the definition of *Reach B*, it follows that some visual control is required and its level is greater in comparison to *Reach A*.

Reach B requires more time for performance than *Reach A*. The most complex element is *Reach C*. It includes a high level of visual or muscular control and decision making. Therefore, performance of *Reach C* requires even more time than *Reach B*. All other basic elements are classified according to such principles. Hence, MTM-1 deals with manual motor motions and simple mental processes that are involved in their regulation.

There are three types of the element *Reach*:

1. Hand does not move at the beginning or end of *reach*
2. Hand moves at either the beginning or the end of *reach*
3. Hand is in motion at both the beginning and the end of *reach*.

There are some other rules of applying the basic element *reach* in MTM-1. However, for this study this description of *reach* is sufficient. In our analysis, we also used basic elements of MTM-1 such as *release* (*RL*), *grasp* (*G*), and *apply pressure* (*AP*). These basic elements are briefly described in the following (Karger and Bayha, 1977). There are two versions of the basic element *release*: (1) normal release (*RL1*), which requires simple opening of fingers, and (2) contact release (*RL2*) that does not require time for performance because the following element *reach* starts simultaneously. According to the rules of standardized description of behavior, element *RL2* should be utilized in the description of performance in our study. Basic element *grasp* also has different versions or cases. We use only one version of the element *grasp—G5*, which does not require time for performance either. In our study, the subjects' index finger was in contact with the button and, therefore, no time was needed for performance. According to the rules of MTM-1, this element should be used for the description of behavior. In MTM-1, there are three versions of the basic element *AP*. Usually only two of them are utilized: *AP1* and *AP2*. The third one (*AP3*) requires more forces to be applied and, therefore, more time for execution. A detailed description of this system can be found in the publications of the MTM-1 Association.

Our brief analysis of the basic element *reach* demonstrates that there are some limitations in the description of MTM-1, and in the description of the element *reach* in particular. For example, the nature of the relationship between subsequent and previous motions cannot be reduced to the position of a worker's hands before and after the basic element is completed (state of rest or motion), as it is presented in MTM-1. This is a biomechanical analysis, which takes into consideration only the initial acceleration or reduction of inertia of the hand. The psychological regularity of integrated strategies of activity utilized by the worker is virtually ignored. The method and time for task performance depend on the strategies chosen by the worker and therefore depend on the entire structure of activity during task performance. In order to prove that the use of MTM-1 should not begin from the decomposition of activity into basic elements, but rather from a qualitative analysis of the holistic structure of activity and the preferred strategies of task performance, we carried out a specifically designed experiment.

For analyzing the entire structure of activity when the basic element *reach* was used and different strategies of activity were applied by subjects, we designed and developed a special device. There are panels for the subjects and the experimenter on the opposite sides of the device. There are horizontal and vertical panels on the subjects' side. On the vertical panel there are red bulb 1, green bulb 2, white bulb 3, and yellow bulb 4. The red and green bulbs were placed at the upper horizontal positions and the white and yellow bulbs at the lower horizontal positions. In the central position of the vertical panel, digital indicator 5 was placed. All the bulbs and indicators were placed in the optimal field of vision. On the horizontal panel there is a start position 6 for the index finger. Also on this panel was an intermittent button 7 (black color) in the middle position and buttons 8 (green color) and button 9 (red color) were in the right-edge position (see Figure 6.2).

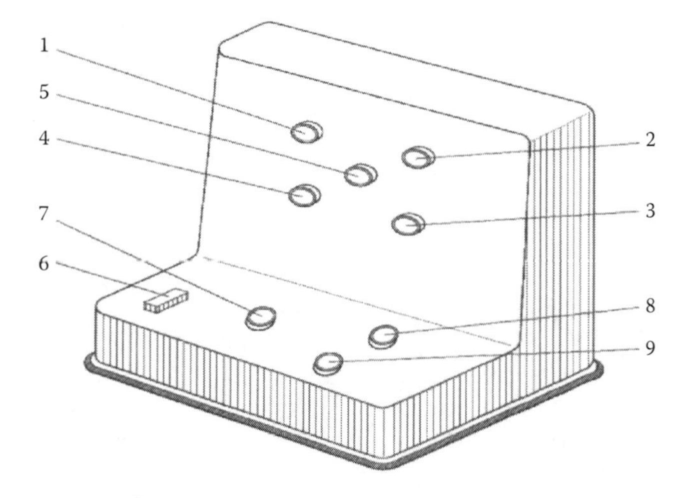

FIGURE 6.2
Experimental device for measuring performance time of two motor actions that are performed sequentially. 1—red bulb, 2—green bulb, 3—white bulb, 4—yellow bulb, 5—digital bulb, 6—start position, 7—intermittent button, 8—green button, 9—red button.

The distance between the start position 1 and the intermittent button 7 and between the intermittent button 7 and two edge buttons 8 and 9 is 12 cm.

Two lines between button 7 and buttons 8 and 9 make a 60° angle. Therefore, after pressing intermittent button 7, the hand moves 30° up to button 8, or 30° down to button 9. The experimenter's panel has switches, buttons, and two timers. The electronic stopwatches measure time in 1/100th of a second. This allowed the experimenter to set up the desired system of signals to the subjects and record the execution time of all tasks, as well as the execution time of individual motor actions.

Ten undergraduate male students of the Ukrainian Industrial and Civil Construction University participated in the experiment. Their mean age was 20.5 years. All of them were right-handed and used their dominant hand in the experiment. All participants were in good physical condition and did not use glasses. Five students participated in the first series and another five in the second series of experiments. Individual reaction times of the participants did not vary significantly. Each group was pretrained to perform the corresponding task the day before the actual experiment. The training was continued until subjects showed a relatively stable time of execution of tasks. Subjects were informed about their performance time of two actions together, but not about the time of each action measured separately.

The first series consisted of two sets (sets 1 and 2) while the second series consisted of three sets (sets 3, 4, and 5). The subjects were informed that they had to perform 30 trials in each set. We chose 30 trials for each task to ensure that the subjects did not change their strategy and that the time of task performance was relatively stable. A brief description of each set is presented in Table 6.1.

The general hypothesis of this study was that performance time of separate motions and motor actions depends on the strategies utilized by the subject during task performance, which is not taken into account in MTM-1. This hypothesis can be proved by our experimental study if a complex basic element requires less time than a simpler basic element in MTM-1. This implies that the rules of MTM-1 should be applied only after discovering preferable strategies of task performancc, which are derived from activity self-regulation mechanisms. We have chosen various types of the basic element *reach* for our study.

In the experiment, the pace of execution of two actions is higher than those provided by MTM-1. Therefore, we have performed a comparative analysis of the performance time for the first and the second action. If one of the actions has more complex basic elements as per MTM-1, its performance time should be greater. Otherwise, we can conclude that the real performance strategies have not been taken into consideration during analysis of tasks with MTM-1.

Let us consider independent and dependent variables in all sets of experiments. In set 1, the condition when the green bulb was turned on/off was an independent variable. Performance times of the first action moving the hand from the start position to the intermittent button and the second action moving the hand to the green button were used as dependent variables.

TABLE 6.1

General Plan of the Experiment

Number of Sets	Description of Sets
	First Series of Experiments
Set 1	If green bulb 2 turns on → press intermittent button 7, then press green button 8
Set 2	If red bulb 1 turns on → press intermittent button 7, then press red button 9 If green bulb 2 turns on → press intermittent button 7, then press green button 8
	Second Series of Experiments
Set 3	If red bulb 1 turns on → press only intermittent button 7
Set 4	The same as in the set 2
Set 5	a. The first situation with *white* bulb 3: if white bulb 3 turns on → press intermittent button 7, then digital bulb 5 turns on with *even* number, then press green button 8; b. The second situation with *white* bulb 3: if white bulb 3 turns on → press intermittent button 7, then digital bulb 5 turns on with *odd* number, then press red button 9; c. The first situation with *yellow* bulb 4: if yellow bulb 4 turns on → press intermittent button 7, then digital bulb 5 turns on with *even* number, then press red button 9; d. The second situation with *yellow* bulb 4: if yellow bulb 4 turns on → press intermittent button 7, then digital bulb 5 turns on with *odd* number, then press green button 8

In set 2, two conditions when the green or red bulb was turned on were used as an independent variables. Performance times of the first action moving the hand to the intermittent button and the second action moving the hand to the green or red buttons were used as dependent variables. In the set 3 the dependent variable was performance time of the first action (see Table 6.1). In set 4 independent and dependent variables are similar to set 2.

Set 5 has more complicated independent variables than the prior sets. The combination of signals from white or yellow bulbs (bulbs 3 and 4) with even or odd numbers presented on digital bulb 5 creates four alternative variations of information. These four alternatives are independent variables. Performance time of the first and the second actions are dependent variables.

In the second experimental study, we used data from three sets of experiments that had three conditions for the performance of the first action. The the first condition was when subjects performed only the first action (set 3). The second condition was when subjects performed two actions in sequence and the second action had only two alternatives (set 4); the third condition was when subjects performed two actions in sequence and the second action had four alternatives (set 5); Performance time of the first action in these conditions was a dependent variable (see Table 6.1). In all experimental sets, the subjects were instructed to act quickly and precisely.

In this study, we utilized a mixed design experiment comprising of within-participants and between-participants comparisons. In all cases, we utilized a combination of signals presented to participants on the panel as the independent variable. Table 6.1 depicts all sets of experiments. Tables 6.2 and 6.3 show performance times of motor actions by all participants. The average response time over 30 trials is listed for each participant. The average time across participants is the mean of the within-participants average.

A statistical analysis of the difference between performance times of different actions was performed using within- and between-subjects *t*-tests. For the multi-mean comparison, we used a one-way repeated measures analysis of variance (ANOVA). Let us consider the obtained data.

In set 1 experiments, subjects were informed that only green bulb 2 would be used. The subjects placed their index finger on the start position 6 (Figure 6.2). As green bulb 2 turned on, the subjects had to move their index finger and press the intermittent button 7, and then move the finger to the green edge button 8 and press it (task 1). When the subjects press the intermittent button 7, the first electronic stopwatch turns off, while the second electronic stopwatch turns on automatically. Then, as the subjects press the green button 8, the second electronic stopwatch turns off.

The device measures performance time of the first and second actions separately in combination with their corresponding cognitive components. Summation of the performance times of these actions provides information about the performance time of the whole task.

In Table 6.2, we present the mean (average) time for each participant over 30 trials. The means of these means are presented at the bottom Table 6.2. Standard deviation across participants is presented in parentheses.

In set 1 (task 1), we were interested in comparing the execution time of the first and second actions, where the main component that influenced performance time was the motion to move the index finger from the start position to the intermittent button, and then to the green button (see Table 6.1). This movement can be identified as the basic element *reach*. Therefore, differences in performance time of the first and second motor actions depend on the version of the basic element *reach* in these actions. Hence, we should also find

TABLE 6.2

Performance Time of Motor Actions in the First Series of Experiment (in Seconds)

	Set 1		Set 2	
Subjects	**First Action**	**Second Action**	**First Action**	**Second Action**
1	0.33	0.25	0.6	0.29
2	0.42	0.23	0.49	0.30
3	0.39	0.28	0.48	0.39
4	0.37	0.24	0.47	0.37
5	0.45	0.32	0.66	0.55
Average time	0.39	0.26	0.54	0.38

out what versions of this basic element are considered as components of two motor actions. In this regard, it is necessary to conduct an analysis of these motor actions from the MTM-1 perspective.

According to MTM-1, the first motor action (movement from start position 1 to the intermittent button 7) consists of the following basic elements or motions: *Contact Release* (*RL2*), *reach case A* (*R12A*), and *grasp* (version *G5*). For basic elements *RL2* and *G5* no performance time is needed in MTM-1. *RL2* is simply an interruption of a contact. The motion *grasp* (version *G5*) is simply a *contact* with the button. Such basic motions or elements are used for precise description of a worker's activity. To reach an object in a fixed location requires basic element *RA* because the intermittent button has the same position during the entire experiment. This is the simplest *reach* movement in MTM-1. After contacting the intermittent button, subjects had to press it. This basic motion is called *apply pressure* (*AP2*) or, sometimes it is designated as APB. It takes 16.2 TMU (a special Time Measurement Unit that is equivalent to 0.036 s).

After performance of *AP2*, the first motor action is completed. Let now us consider the second motor action. *RL2* and *G5* are also basic elements of this action. That is, in the comparison of the performance time of the first and second motor actions, we can ignore basic elements *RL2* and *G5* because they do not require time for performance.

In the set 1 experiment (task 1), after green bulb 2 is turned on, subjects had to move their index finger from the start position to the intermittent button 7 and press it. After that, the subjects had to move their finger only to green button 8 and press it. When subjects pressed the intermittent button the first stopwatch turned off and at the same time the second stopwatch turned on. In this set of experiment, subjects had to reach the intermittent and the red button both of which were in the same fixed locations. Hence, not only in the first motor action, but also in the second motor action the basic element *reach* (*R12A*) should be used. Upon contacting green button 8, subjects had to press it. This is the basic element *AP2*. Basic motions *AP2* occur in both motor actions. The method for their execution (apply pressure—*AP2*) remained the same throughout the experiment for both actions. The execution time of *AP2* can be regarded as the time constant.

From this it follows that *AP2* has no effect on performance time comparison of two actions. We did not use the basic element EF because perception of the signal from the bulb is a highly automated mental operation overlapped by movement.

According to the rules of MTM-1, the basic element EF was not being used in such situations. In general, this basic element is rarely used in MTM-1. It means that the difference in execution time of two motor actions depends only on the *version* or *case* of the basic element *reach* utilized in these actions.

The main hypothesis of this experiment was as follows: the first motor action should take more time than the second motor action because it involves program formation process for both actions. According to the rules of MTM-1, the performance time of these two actions should be similar. So, MTM-1 cannot adequately describe the strategy of a subject's task performance.

From this it follows that two motor actions (from the start position to the intermittent button and from the intermittent button to the green button) include the same basic elements *R12A* + *AP2*, where the distance is 12 cm. Only these two basic elements influence on the performance time of the considered motor actions. Because the performance times of the first and second motor actions are determined by the same basic elements *R12A* + *AP2*, their execution time should be the same. However, the results of the experiment suggested otherwise (see Table 6.2, performance time of the first and the second actions in the first set of experiments).

There is another rule in MTM-1 that effects performance time and therefore it should be considered. The same basic element *reach* may have different execution times only if the hand is in motion before and after completing the first movement.

Another factor that should be considered is the *type* of *reach*.

According to MTM-1, in all of our experimental sets we had the same type of the motion *reach*. The reason is that this basic element was always performed from the rest position and always ended with the element apply pressure in both actions (first action—move hand from start position to the intermittent button and press it; second action—move hand from intermittent button to the red or green button and press it). According to the rules of MTM-1, the hand which a subject uses to perform the *AP* is considered as being at rest and the two motor actions should be considered as independent. So the basic element or movement *reach* in the first and second motor actions is independent and its performance time should be the same in both actions, according to MTM-1 rules. Indeed, in our experiment after performance of the motion *reach* in the first action, the subject stops the hand for a moment and then begins to move it again. According to MTM-1 rules, using the hand movement acceleration from the first motor action to perform *reach* motion in the second action is impossible. Thus, the performance time of the first and second motor actions is determined by two basic elements *R12A* + *AP2*. Moreover, these motor actions are considered as independent (they do not influence each other). It means that according to MTM-1 their performance times should be the same.

However, as it is shown in Table 6.2 (see set 1, task 1), their times are different. According to MTM-1, subjects perform two exactly similar motor actions. However, the first action required 0.39 s and the second action required 0.26 s. The performance time for the first action is 33% more than that of the second action. This difference was statistically significant (within subjects, *t*-test $t(4) = 7.11$, $p < 0.01$). The result of this experiment is presented in Table 6.2 (set 1).

Van Santen and Philips (1970) described a system for the time study of mental components of work. However, the normative data of this system is not fully published, which makes it very difficult to evaluate the system. On another hand, an analysis of this publication shows that the system has too many detailed. For example, it uses nerve impulses for the calculation of components of work. According to this system, the time of the mental component of the reaction on one signal in case of uncertainty of such signal appearance is 0.1 s.

From this it follows that the time for motor component of the first action in our experiment is 0.39 – 0.1 = 0.29 s. The second action according to MTM-1 is independent from the first one. From this it also follows that the motor component of the second action is 0.26 – 0.1 = 0.16 s. This calculation demonstrates that the time for the first and second motor actions is different and therefore two motor actions could not be considered as independent. Two motor actions are integrated into a system when the first and second actions influence each other. Data analysis allows us to conclude that the time for performing the first action is longer than the time for completing the second one due to the time difference in receiving and interpreting information and formation of unitary programs for the execution of the actions. Therefore, the first and second actions influence each other, which, in this particular case, contradicts MTM-1 rules. In the absence of deviation from the developed program of performance, current information about the interim and final results of the actions does not have a significant impact on the execution time. Let us consider the experiments in set 2 next.

The same subjects took part in set 2 experiment. Task 2 was performed in this set of experiments. Subjects were informed that either red bulb 1 or green bulb 2 would be used. Subjects placed their index fingers on the start position 6. As red bulb 1 turns on, the subject had to move their index finger and press the intermittent button 7, and then move the finger to the red edge buttons 9 and press it. As green bulb 2 turns on, the subjects had to perform motor actions in the same way they did in the first set of experiments. In set 1 experiments, the subjects pressed intermittent button 7 and then green button 8. In set 2, the subjects can press green button 8 or red button 9 (see description of procedures in Table 6.1). In this set of experiments, each subject performed 30 trials as in set 1.

Moving hand from the intermittent button to the red or green button requires an alternative decision to move the hand to one of these places. In other words, the second motor action is a movement to location, which varies from cycle to cycle. According to MTM-1 rules, this is the element *reach R12B* (case *B*). In the previous experiment, when the subject had to move the index finger only to the green button, that basic motion was *R12A*. According to MTM-1, to perform the movement *R12B* requires more time. At the same time, the first movement was the same as in the first experiment, and is considered as *R12A*. Therefore, the first motor action includes movements *R12A* + *APA*, and the second motor action includes movements *R12B* + *AP2*. We recall that basic elements *RL2* and *G5* were not considered in our experiment because they did not require time for performance. The result of this experiment is presented in Table 6.2 (set 2).

The main hypothesis of this experiment was that the first motor action requires more time because it includes a program formation process for both actions and decision making for the selection of the second action from two alternatives. According to MTM-1, the time for the second motor action should be greater than that for the first one because the first motor action includes basic element *R12A* and the second one includes basic element *R12B*. Therefore, MTM-1 ignores real strategies of performance of these two actions.

The first action, which in average includes more simple basic motions for all subjects, required 0.54 s. The second action, which includes a more complex basic motion, was performed in 0.38 s. This contradicts MTM-1 rules. The difference was statistically significant (within subjects *t*-test $t(4) = 3.85$; $p < 0.05$).

The data obtained can be explained in the following way. The subject forms the program and makes their decision of how to perform the second action not after pressing the intermittent button 7, but during the execution of the first action. It was observed that if the performance time of the first action was decreased, then the performance time of the second action increased. However, the performance time of the first action was always more than that of the second action. Thus, the second action, which included the more complex basic element *R12B*, took less time. This contradicts the rules of MTM-1. The second action can takes less time only in a situation when the two actions are interdependent and the first action has an effect on the second one. In such situations, the subjects decided to press the green or red button during performance of the first action. This explains the differences in the performance times of the considered actions. Under the rules of MTM-1, two actions are separated by the basic element *AP2* and they should be independent. Therefore, the decision to move the finger to the green or red button should be performed during the second motor action. Only in such situations would the second action take more time than the first one.

Our experiment demonstrates the opposite result, and in contradiction to rules of MTM-1, the first action took more time than the second one.

It is interesting to compare the time performance of the first action in the set 1 experiment with the time performance of the first action in the set 2 experiment. In both cases, subjects had to move their finger from the starting position to the intermittent button and press it. According to MTM-1, these identical actions require the same execution time. However, for the first action in set 1, the performance time was equal to 0.39 s, while the same action in set 2 required 0.54 s (Table 6.2). The difference was statistically significant (within subjects *t*-test $t(4) = 3.85$; $p < 0.05$). It can be explain by the fact that in set 1, the subjects developed a program to perform two identical actions, whereas in set 2 they had to develop a different program for two actions, where the second action was different and more complex than the first one.

The analysis of data in the second series of the experiments is described next.

The second group of five subjects participated in the second series of the experiment (set 3, task 2; set 4, task 3; and set 5, task 4). The subjects received some preliminary training before the actual experiment. The training was conducted according to the same principles as described earlier. Each set

TABLE 6.3

Performance Time of Motor Actions in the Second Series of Experiment (in Seconds)

	Set 3	Set 4		Set 5	
Subjects	**First Action**	**First Action**	**Second Action**	**First Action**	**Second Action**
1	0.31	0.44	0.30	0.42	0.51
2	0.27	0.42	0.27	0.72	0.80
3	0.29	0.51	0.29	0.54	0.48
4	0.34	0.47	0.36	0.64	0.75
5	0.26	0.35	0.26	0.42	0.53
Average time	0.29	0.44	0.30	0.55	0.61

included 30 trials. In this series too the basic elements *contact release* (*RL2*) and *grasp* (*G5*), which did not require time for their performance, were not used. The first action ended with *AP2*, and therefore the first and the second motor actions during their sequential performance should be considered as independent according to MTM-1 rules. Basic motions *AP2* occur in both motor actions and therefore the time for their performance was considered as the time constant.

The results in Table 6.3 demonstrate mean (average) time for each subject over 30 trials and average time across all subjects is presented at the button of Table 6.3 (means of the means across five subjects).

Let us consider the set 3 of experiment when subjects performed only the first action. Time performance of action in this set is presented in Table 6.3. This group of subjects performed only the first action in response to the red bulb being turned on. They performed it 30 times as all other subjects did in the other sets. The performance time of the first action was 0.29 s. The set 4 experiment was the same as set 2. The main hypothesis of this experiment was similar to that in set 2. In the set 5 subjects have four alternatives for selecting the second action. They also have to move a hand from the start position to the intermittent button 7, and then to the green button 8 or the red button 9. However, the information presented to the subjects was more complex (see Table 6.1). Hence, all sets (sets 3, 4, 5) include the same first action. The next step was to compare the subjects' execution time for the first motor action in sets 3, 4, and 5 (see Table 6.3).

The main hypothesis of this comparative study was that the performance times of three externally similar motor actions were different because they required different performance strategies. According to MTM-1, the performance times for these actions should be similar because they involve the same basic element *R12A*.

Upon analyzing the results of the experiment, we could see that even though all three actions were carried out in response to similar signals with the same motor response, the execution time for the action "move hand to the intermittent button" was different (Table 6.3). In set 3, time was 0.29 s, in set 4 it was 0.44 s, and in set 5 it was 0.55 s.

The difference was statistically significant. To assess the statistical significance for within-group comparisons, we used one-way within-subjects ANOVA ($F(4,16) = 16.35$; $p < 0.01$). Post-hoc *t*-test (set 5, movement 1 versus set 3, movement 1, $t(4) = 6.44$; $p < 0.005$); (set 4, movement 1 versus set 3, movement 1, $t(4) = 1.9$; $p > 0.05$); (set 5, movement 1 versus set 4, movement 1, $t(4) = 4.33$; $p < 0.05$).

Set 4 was exactly the same as set 2 of the first series. In both sets, subjects performed the same task (task 2). Hence, the result obtained in this set was approximately the same as in the second one. The time for the first action for this group was 0.44 and for the second action it was 0.30 s (Table 6.3). The difference was statistically significant across subjects (within subjects *t*-test $t(4) = 6.11$; $p < 0.01$). From this it follows that the second group of subjects used the same strategy of task performance as the first group did in set 2. Let us consider the set 5 experiments.

The procedure for experiments in set 5 is described in Table 6.1 (task 4). Hand movements in this set were similar to the movements in the preceding set.

According to the rules of MTM-1, using an additional digital bulb does not require introduction of the element *EF*, since the perception of one-digit numbers is a highly automated perceptual action, and is therefore overlapped by the time of motor action.

Moreover, in this set, white bulb 3 or yellow bulb 4 turns on before the first motor action and digital bulb 5 turns on before the second motor action. Hence, this factor has the same effect on the performance time of the first and second actions. In this task, subjects had four alternatives for performance of two motor actions (see Table 6.1, set 5).

The main hypothesis of this experiment was the following. In this task, subjects could not know in advance to which extreme button (red button 8 or green button 9) they had to move their hand to. They could know this only after pressing the intermittent button 7. Therefore the first action would be simple (includes *R12A*) and the second action would require more time than the first one (includes *R12C*). According to MTM-1, the second action should include *R12C* because it requires a high level of visual and muscular control or mental decision in order to select one part jumbled with others. In our example the considered action requires mental decision.

As will be seen later, a more complex strategy was used by the subjects. Let us consider this strategy. For this purpose, we have to compare the performance times of the first and second actions in the fifth set (task 4). The second action included the basic element *reach* accompanied by a mental action for the selection of the required button. Decision making was performed primarily on the basis of the information stored in memory.

Hence, according to MTM-1, the basic element *reach* in the second action should be considered as *R12C*. The basic element *RC* is much more complicated than *RA* and even *RB*. Hence, according to MTM-1, in this situation the difference in the performance time of the first motor action (0.55 s) and second motor action (0.61 s) should be statistically significant. Indeed, the basic element *RC* for the second action requires much more time than the basic element *RA*, which is needed for the first action according to data from MTM-1. However, our experimental data demonstrate that the difference according to criterion *t* was not statistically significant (within subjects *t* test $t(4) = 2.06$; $p > 0.05$). Professionals would not be able explain this result by using MTM-1. The first action, which can be considered as *R 12A* according to MTM-1, should take less time than the second one, which is considered as *R12C*.

Let us compare the performance time for the first action in sets 1 and 4. When we compared the time of the first action in set 1 (Table 6.2) with the time of the first action in set 4 (Table 6.3), we could see a difference, even though according to rules of MTM-1 the actions are almost identical. In set 1 the average time for the group was 0.39 s and in set 4 it was 0.55 s (between subjects *t*-test, assuming unequal variances $t(5) = 2.52$; $p = 0.05$).

At the same time, as we compared the execution time of the first action in set 2 (Table 6.2; 0.54 s) with the execution time of the first action in set 5 (Table 6.3; 0.55 s), we could see that the time difference was not statistically significant (between subject *t*-test $t = (7) = 0.14$; $p > 0.05$).

According to MTM-1, the second action in set 4 and the second action in set 5 involve similar versions of basic element *reach*. However, the second action in set 5 requires more time (within subjects *t*-test $t\ (4) = 5.02$; $p < 0.01$). Similarly, according to MTM-1, the second action in the set 2 (see Table 6.2) and the second action in set 5 (see Table 6.3) involve the same version of *reach* and should therefore take approximately the same performance time. However, the second action in set 5 required more time than the second action in set 2 (between subjects *t*-test, assuming unequal variances is $t(7) = 2.87$; $p < 0.05$). Hence, the fact that the differences between performance time of the same type of actions in one situation is statistically significant and is not statistically significant in a different situation, cannot be explained using the MTM-1system rules.

Let us consider data obtained from the activity self-regulation theory perspective.

For the explanation of the statistically insignificant differences in the performance times of the first and second actions in set 5 (task 4) of experiment we had to consider the strategies utilized by subjects during their performance of these actions in a more detailed manner. In set 5, subjects had not only two but four alternatives. They had to make a choice based not only on presented information (color of bulbs and digital numbers), but also on the memorized rules. That is, they had to make their decisions not only on the externally presented information, but also on the information extracted from their working memory. An analysis of the performance time of all actions, as well as observation of the subjects' behavior and discussions with them, clearly demonstrated that subjects could not conclude what the second action would be during performance of the first action. They were able to do this only after the completion of the first action. The subjects' strategies, in most cases, were as follows: when, for example, the yellow bulb 4 was turned on, the information stored in working memory of what should be done when the white bulb 3 turns on was eliminated. It is important to note that such types of decisions (elimination of redundant information from working memory) were made during the performance of the same external motor action "move hand from start position 6 to intermittent button 7." Upon completion of the first action, the subjects held in their working memory only the information related to the implementation of the second action when yellow bulb 4 was activated. Hence, the first decision was made during the execution of the first action and the second decision was made during the execution of the second action. Indeed, there were four alternatives. Prior to any bulb being turned on, the subjects were able to store four alternatives in their working memory. As soon as one bulb lights up, two nonrelated alternatives were immediately eliminated from working memory as unnecessary information. The subjects kept information in working memory about two remaining alternatives only, which were used for the performance of the required version of the second action. Therefore, despite the fact that the subjects moved the hand to the same intermittent button, they still performed their first motor action simultaneously with the preliminary mental act of choice, which is a prerequisite for the next decision for the selection of the second appropriate version of motor action. It enabled subjects, first, to optimize their strategies of task performance, second, to reduce the quantity of information held in working memory, and, third, to increase their processing speed. Such strategies were formed gradually on the basis of subjective evaluation of the motor actions performed. This assessment and the development of adequate strategies sometimes were not clearly realized by the subjects.

One subject explained his strategy in the following way: "I basically memorized the instructions about what to do when the white bulb lit; and when yellow one lit, I acted in the opposite way." The subject's statement clearly demonstrate that during the performance of the first action—to move his hand to the same intermittent button—he still performed the mental decision in order to eliminate unnecessary information from his working memory and store the required information in memory for the performance of the next motor action. According to MTM-1 rules, when the movement of hand involves decision (or choice), it becomes the basic movement *RC*.

An analysis of the strategies used during the execution of the first and second actions shows that they were based on the performing of basic element *reach*, which included mental decision making. If mental decision was needed in order to make the choice, it was a case of *RC* in MTM-1. It means that not only the second, but also the first action included the basic element *RC*. It explains why the difference in the performance of the first and second actions in set 5 is not statistically significant.

However, if we do not consider real strategies of task performance, then the first action would not include the basic element *reach*, which requires mental decision and should be considered as *R12A*. This is explained by the fact that MTM-1 ignores the possibility of the subject eliminating unnecessary information from working memory for the following second and independent version of the action during performance of the first action. At the same time, the second action includes the basic element *reach,* which needs to be accompanied by a mental decision and therefore should be considered as *R12C*. This version of *reach* is much more complicated than *R12A*. Therefore, the differences in the performance time of two motor actions should be statistically significant according to the rules of MTM-1, when real strategies of task performance are ignored. An analysis of the real strategies of the sequentially performed actions clearly demonstrates that both motor actions included the basic element *reach,* which needs to be accompanied by a mental decision. These actions according to MTM-1 could not be viewed as independent because the first one is ended by basic element *AP*.

An analysis of the real strategies of performance of these two actions demonstrates that they are interdependent and influence each other. The first motor action involves a mental decision that requires elimination of unnecessary information from working memory during the performance of the second action. The second action requires a mental component in making a decision for the correct selection of the required version of the second action. According to MTM-1 rules, motor motions that require a mental component in making a decision should be related to *RC*. Thus, after the analysis of the strategies of actions performed in sequence, it becomes clear the basic elements *reach* in both actions should be related to the category *RC* and therefore similar time is needed for performance of the actions. This was proved by the described experimental data. This also explain the fact that the performance times of two motor actions in set 5 requires more time than two actions in set 4.

Let us consider a possible strategy when two sequentially performed actions can be considered as independent (set 5, task 4). Unlike the data obtained in sets 1, 2, and 4, the subjects in set 5 could not make the final decision about what the second action was during the performance of the first one. We can therefore assume that processing of information and formation of programs for performing the of two motor actions must be exercised consequentially and that these actions are considered to be independent of each other. In such situations the strategy of performance should be the following: Subjects perform the first action (moving hand to the intermittent button) without analyzing the information that can be used for performing the second action. After completion of the first action, subjects make a decision to select one out of four existing alternatives in working memory.

This strategy could be used when two actions are *independent*.

However, as was demonstrated by experimental data, subjects utilized another strategy.

They eliminated unnecessary information about two unrelated alternatives from working memory while performing the first action. When subjects started to perform the second action, they kept only needed information about two alternatives in working memory. This strategy demonstrates interdependence of two actions as they are integrated into a structurally organized system. Such a strategy allows a subject to reduce difficulty of task performance.

It should be understood that we have described strategies used by subjects who were suited to the conditions presented in the experiment. Under the conditions of our experiment, the subjects were pretrained for each task before they performed the 30 trials in the actual experiment. In a production environment, for example, in the conditions of an assembly line, workers perform the same task multiple times. In contrast, an operator, who controls and monitors complex equipment, performs many different tasks.

Each task has its own probability of occurrence during the working shift. This means that a subject's readiness for perceiving and processing information and for implementing strategies of performance of a particular task among many other possible tasks is different compared to performing the same task multiple times.

Let us take, for example, task 4 in set 5. Set 5 contains four decision-making actions.

Under the conditions of our experiment, subjects may hold in advance in their working memory all necessary procedures and rules of tasks performance. They do have a high level of readiness to perceive and process information and to implement the required actions. Under such conditions, mental actions are performed in combination with motor actions. In accordance with MTM-1 rules, the time to perform cognitive and motor actions separately is not required in the considered situation. Such situations may only lead to increased time for performing the motor actions that are combined with cognitive components of activity. However, when the same task appears among many others, the performing strategy changes. The reason is that in such situations the needed information is being extracted from the long-term memory. The subjects' readiness to perform adequate actions is lower and so is his/her level of skill to perform a particular task. These factors allow us to predict that motor actions will not be executed simultaneously with the cognitive components of activity. First, executing motor actions requires retrieving information from the long-term memory. Then, a decision-making action will take place, and only after that the first motor action will be performed. The second part of the task will be performed next. This means that without a preliminary analysis of the possible strategies of performance, it is impossible to correctly utilize MTM-1 or any other similar system. In some cases, the strategy of performance can be very complicated and experimental methods may be required to analyze it.

The experiment was designed to analyze application of MTM-1 from the self-regulation point of view. To simplify the description of our experiment, we discussed only the basic strategies without referring to the functional blocks of activity self-regulation. This experiment was constructed on the basis of the theoretical principles that derive from the analysis of the models of self-regulation (see Section 3). In this regard, as an example, let us consider how some strategies should be explained on the basis of analysis of the most relevant functional blocks of activity self-regulation. Here we consider strategies for sets 1, 2 and 4. With some modification, this discussion can be relevant for set 3 as well.

Based on the given instructions and visual perception of conditions for task performance, the subjects developed a conceptual model of task (function block 13 *conceptual model*). On the basis of these factors, the subjects perceived the task as consisting of two sequentially performed motor actions. However, in the process of task performance, the conceptual model of task (block 13) was transformed into a more adequate dynamic model (block 9) when two actions were no longer perceived as independent. Such a model of task (mental representation of task by subjects) contradicted the objectively given instructions. This modification happened primarily under the influence of block 8 (assessment of task difficulty) during task performance. Subjects perceived execution of two independent and sequentially performed motor actions as being a more difficult task, which resulted in a subjectively negative evaluation of the performance result according to the speed criterion (block 17 *negative evaluation of result*). The reason is that the increase in the difficulty of task performance reduces the speed of actions. New mental representations of a task reduces difficulty of the task and this factor, through block 18 (positive evaluation of result), influences block 11 (making a decision about correction), which is involved in the modification of the subjectively relevant task conditions or dynamic model of task (block 9). Thanks to the interaction between block 9 (dynamic model or subjectively relevant task conditions), block 8 (assessment of task difficulty), block 10 (formation of a program of task performance), and block 11 (making a decision about correction) the dynamic model of task and the real program of task performance are also changed. Two independent actions are subjectively integrated into a coherent structure, which leads to the reduction of the difficulty of task performance (block 8) and a positive evaluation of result (block 18) because actions are performed faster. The program of activity performance (block 14) is reconstructed and simplified. The subjects do not form two programs for two separate motor actions. They form a unified program for two sequential actions. This allows subjects to refocus attention on the elements of the task and the program's execution elements. The program of two actions is formed under conscious control. The first simpler action is mainly regulated by a subprogram at the unconscious level. This allows subjects to make a decision about the second action, when they perform the first one. Thus, subjects optimize activity performance mainly based on such subjective factors as accuracy and difficulty.

The material presented in this chapter demonstrates that in some studies, we can restrict ourselves to the description of only the real strategies of task performance without specific consideration of the functional blocks of activity self-regulation. However, we actually consider their function in our study. For example, in the description of the subjects' strategies, we analyze the goal of the task, consider how the subjects developed program of activity performance, how they perform decision, and how they optimize activity in terms of task difficulty. In fact, all of the described strategies can be correlated with the corresponding functional blocks of activity self-regulation described in Sections 3.2 and 3.3. The method of describing the process of self-regulation in the analysis of the considered task depends on the specifics of the task and research purposes.

One important theoretical conclusion of this study is that human activity is a complex self-regulative system. The structure of such a system depends on utilized strategies of performance. New methods are required to complement traditional task analysis and time study. In modern forms of labor, with increasing role of mental processes, interaction of motor components with mental components of activity becomes much more complex. This is a major reason work activity unfolds over time as a complex structurally organized system. In the analysis of such systems, it is important to know the performance time of certain cognitive and motor components of activity and determine their logical and hierarchical organization and probability of their occurrence. One powerful method of time study of the behavioral (motor) components of activity is MTM-1. This system divides manual tasks into basic motions (basic elements) and assigns a predetermined time standard to each motion, which depends on the nature of motions and conditions under which they are made. However, the nature of motions and their organization also depend on preferable strategies of activity during task performance and the hierarchical and logical organization of activity. This is particularly relevant for tasks that consist of a complex combination of cognitive and motor components of activity. In such situations, MTM-1 rules for the selection of basic elements for tasks description are necessary but insufficient. Hence, it would be incorrect to begin analysis of manual components of work by dividing motor components of activity into motions. Activity consists not only motions but also other hierarchically organized units. The experimental material presented demonstrates that task analysis should not begin with a decomposition of tasks into motions, but with the psychological analysis of the whole structure of activity and analysis of preferable strategies of performance. Only after analysis of strategies of performance, including description of logical organization of motor and cognitive actions and members of algorithm, can professionals use MTM-1 rules effectively. This is explained by the fact that the choice of adequate basic elements for the description of a particular task depends not only on the existing rules in MTM-1, but also on the preferred strategies of activity that subjects use. This factor should be taken into consideration when professionals try to use MTM-1. The system is particularly important for the design of time structure for holistic activity and for determining the time performance of motor components of activity. Activity has a hierarchical organization. Motions are components of motor actions and the last components are the subsystems of activity, which are called members of algorithm. MTM-1 can be used for describing motor components of activity (motor operations) at the analytical design stage followed by experimental verification and correction of the data obtained. This principle is also used in engineering designs. MTM-1 presents a standardized language for the description of motor components of activity and, therefore, can be used for the creation of models of work activity which are necessary for ergonomic designs.

7

Morphological Analysis of Work Activity during Performance of Human–Computer Interaction Tasks

7.1 Introduction to Morphological Analysis of Activity

Simon (1999) wrote that the evaluation of any complex system is connected with the analysis of its structure. Activity is a complex, multidimensional system requiring the use of systemic principles of its analysis and therefore the concept of structure is critically important in the study of human work activity. The system may have different organizations of its elements. There are three types of organizations: linear, logical, and hierarchical. These can be combined in different ways in the system. The organization of the elements of activity determines the structure of activity during task performance. In the study of activity as a system, particular attention should be directed to the units of analysis, to the relationship between the elements of the system, and to the stages and levels of analysis. Any specialist can represent the same activity in terms of different models describing the structure of activity from various perspectives. Consequently, we may have different representations of the same activity. There are various types of the systems. Activity is the structural system that consists of different elements that are interrelated. The element of the system cannot be understood if it is considered in isolation from the whole and, therefore, the system is more than the sum of its elements. Activity is a situational system because it is constructed and adapted to a situation according to mechanisms of self-regulation. It includes flexible reconstructive strategies (situated components) and preprogrammed or preplanned components. Finally, activity should be viewed as a functional system that mobilizes, forms, and disappears upon the achievement of the desired result. The activity system has a loop-structure organization having continual feedback about the progress of performance. The structure of activity unfolds in time as a process. Cognitive psychology ignores the concept of activity structure and considers cognition only as a process. It should also be noted that in systemic-structural activity theory (SSAT), the focus is not only on the cognitive (informational) aspect of activity study but also on the energetic or emotional-motivational aspects of activity analysis. As will be shown further in the chapter, the consideration of activity as a structure that includes informational and energetic components is a basis of the task complexity evaluation. Some general aspects of the relationship between informational and energetic components of mental activity was also considered in Chapter 2 of this book. Morphological analysis is an important stage in the study of any complex systems. Morphological analysis has been used in many fields of science for the discovery and description of structural interrelations between elements of complex systems. Such analysis precedes quantitative evaluation of task complexity (Zwicky, 1969). The term *morphological* comes from the Greek *morph,* which means shape or form. Thus, the morphological method of study describes arrangement of different elements of a holistic object under investigation. This method describes the object being studied as a structurally organized system. The object can be physical such as anatomy of an organism, or mental such as a concept or an idea. Morphological analysis is a general method for description and analysis of diverse complex objects. This method is not quantitative but very useful for the description of an object's structure. Analysis and synthesis are important principles of morphological analysis. Complex phenomena or social problems can be analyzed into any number of nonquantified elements. Similarly, sets of nonquantified elements can be synthesized into well-defined relationships or structures. This is a formalized method that gives possibility for the various solutions, including quantitative analysis (Ritchey, 1991). Alternating between analysis and synthesis is the fundamental scientific method that is being used in various fields of science such as mathematics, economics, psychology, etc. Analysis is defined as the procedure by which a complex whole is broken down to elements or components. Synthesis is quite the opposite procedure; it combines separate elements or components into a structurally organized system. Analysis and synthesis as scientific methods always complement one another. Analysis and synthesis are important concepts in activity theory (AT). For example, Rubinshtein (1959) suggested that the thinking process is first of all analysis and synthesis. When a subject is trying to understand a problem situation, he/she decomposes the whole situation into elements. However, it is worth noting that the features or elements do not exist in isolation. Therefore the subject tries to discover relationships between the elements—this is synthesis. Analysis and synthesis always exist in integration. This example demonstrates that analysis and synthesis methods can be applied not only for the study of material and structurally organized systems but also for the study of the systems' functions. In the analysis of functions, a specialist breaks down the system to the identified functional processes or activities, which the system carries out in order to perform for what it was created for (Ritchey, 1991). Poorly defined parameters of the problems being studied become evident immediately when they are described as a structurally organized system. The major purpose of morphological analysis is to transfer problems that are not clearly defined or described into clearly defined and structured ones. Morphological analysis is a general method for nonquantitative modeling of an object. In a complexity field, this method precedes quantitative evaluation procedures. This is an important stage in task analysis. This stage of task analysis is also required for the following quantitative evaluation of activity structure including the evaluation of task complexity. The original method of morphological analysis of activity was developed in SSAT. The basis of this method is an algorithmic description of an activity and the development of its time structure. This is the psychological aspect of morphological analysis.

According to Simon (1999), any complex system is made up of a large number of parts that have many interactions. Hence, such system cannot be reduced to the sum of its parts. Some subsystems of activity are hierarchically organized. Such subsystems are restricted by the span of control and by subordination of a higher-order goal of this subsystem. Simon introduced the term *span* of hierarchy understanding it as the number of subordinates beneath each element. Hence, a system can be divided into subsystems or modules and not all structured systems are hierarchical. The concept of hierarchy allows scientists to choose the level of analysis of functioning complex systems. In nonhierarchical systems, modules should be organized logically. Activity also is a complex structural system and therefore the concept of *hierarchical levels* can be applied to studying activity as a system. Activity can be considered as logically organized units, where each unit is a subsystem that can, in turn, be considered to be hierarchically organized. In algorithmic analysis, activity during task performance is divided into algorithmic elements (members of algorithms) such as operators and logical conditions. These elements have a logical organization. They are involved in informational (mental) and material (physical) transformation of elements of situation for achieving the goal of an activity. Members of algorithms include actions and operations and have a hierarchical organization. The hierarchical organization of activity can be presented as follows: member of algorithm → action → operation. Each member of a human algorithm is a hierarchical subsystem that includes one or more actions integrated by a higher-order goal in comparison to the goal of an individual action. The simplest member of an algorithm can include a single action. One important characteristic of members of an algorithm is having logical but not hierarchical relationship between each other. The concept of span of hierarchy can be applied only to separate members of an algorithm. The span of hierarchy of a member of an algorithm is restricted by the specificity of the logical organization of activity and by the capacity of working memory. For example, if a member of an algorithm includes only one cognitive or behavior action, this action can be divided into hierarchical units such as operations and function microblocks. If a member of an algorithm includes several actions, then it should be divided into hierarchical units such as actions, operations, and functional microblocks. The level of decomposition depends on the specificity of the study. Therefore, the span of hierarchy for each member of an algorithm is determined by the number of levels of decomposition. The whole activity during task performance is integrated by one task goal. Bertalanffy (1962) introduced the notion of *functional equivalence*, which means that different systems might lead to the same result. If a strategy of activity, which one can consider as a system, produces some deviations inside this strategy, it can be considered as *functional equivalence inside the strategy*. However, if a subject applied a different strategy that produced a similar result, it would be considered as *functional equivalence between different strategies*.

The logical and hierarchical organization of activity elements determines the structure of activity during task performance. This structure depends on strategies of activity. If the strategies of activity change due to self-regulation, then the structure of activity changes as well. The structure of activity unfolds in time as a process. Such understanding of activity makes it much easier to apply the concept of complexity to job evaluation. Cognitive psychology ignores the concept of activity structure while considering cognition only as a process. Such approach makes complexity evaluation impossible. In a real work situation, a subject can switch his/her attention from one task to another.

Such understanding of activity allows to apply quantitative measurement procedures during task analysis much easier. For example, only after morphological analysis is completed does it become possible to quantitatively evaluate the activity's complexity during task performance. Cognitive analysis, which is utilized in SSAT as a parametric method of study, describes activity as a process. For complexity evaluation of a job, the job should be divided into tasks. For example, in manufacturing, the work process should be divided into different tasks or production operations and then the activity complexity during the performance of each task should be evaluated.

Models of activity that are obtained during morphological analysis can be developed by utilizing analytical methods or simulation methods or a combination of both. Sometimes cognitive aspects of activity can be modeled only by utilizing a simplified experiment in laboratory conditions at the preliminary stage of analysis. The purpose of such experiments is not to compare data gathered during the experiment, but rather to obtain some empirical data about the strategies of human performance, time performance of some elements of activity, etc., for further development of activity models. A model that is developed based on the empirical data obtained from this type of an experiment is a simulation model. Models of activity that are created by utilizing analytical procedures are analytical models. In SSAT, morphological analysis includes an algorithmic description of the activity and the development of the activity's time structure. These two stages present the activity as a structurally organized system that can be evaluated quantitatively by using objective measurement procedures. According to SSAT, activity is a multidimensional structurally organized system and therefore multiple methods of activity analysis should be utilized. SSAT includes various methods for the study of human activity as a system, which we will look at next.

The *parametric method* allows to concentrate on the study of different parameters of activity that are treated as relatively independent. The cognitive approach is an example of the parametric method. The systemic approach includes also morphological and functional analysis of activity.

Morphological analysis is involved in the description of constructive features of activity. This method uses such major units of analysis as cognitive and behavioral actions and operations (cognitive acts and motions). Based on morphological analysis, one can describe activity structure in terms of logical and temporal-spatial organization of actions. Morphological criteria entail representing activity as activity–action–operation. In SSAT, morphological analysis includes algorithmic analysis of an activity and the analysis of its time-structure.

Functional analysis includes the description of activity as a self-regulative system. This method allows to utilize such unit of analysis as the function block *or* mechanism of self-regulation. Describing the specificity of the functioning of each block and its interaction with other function blocks helps to understand possible strategies of activity. Motive–goal conditions are a critical factor that determines strategies of activity performance. Basic models of activity self-regulation were described in Chapter 3. The chapters of this book contain examples of the application of functional analysis, when activity is described as self-regulative system. This allows us adequately to describe an activity's strategies in performing specific tasks. Only then it is possible to adequately describe the structure of activity during task performance.

Qualitative methods deal with verbal descriptions of activity. This method can be used for traditional objectively logical analysis, which is the verbal description of work, for analysis of work space organization, for description of work conditions, etc. A second qualitative method is the sociocultural aspects of the study of activity (Vygotsky, 1978). Culture is regarded as a mediator between the user and the technology. It includes beliefs, attitudes, values, social norms, and standards. A third qualitative method for studying human activity; it has to do with the individual style of performance. This method considers individual characteristics of personality in relation to the objective requirements for job performance (Bedny and Seglin, 1999; Klimov, 1969).

Quantitative methods are methods where mathematical procedures are used for activity description. In SSAT, basic quantitative methods include evaluation of task complexity and reliability assessment.

Genetic analysis is used to describe the major strategies of activity at different stages of activity acquisition. We need to describe the activity structure while subject are acquiring the method of task performance. If we know how the structure of activity changes and what the final structure of activity is during task acquisition, then we can evaluate the usability of equipment more efficiently. This method becomes useful because it is difficult to discover the differences in design solutions when task performance occurs on the automatic level of performance. In cognitive psychology this phenomenon is known as automaticity. Schneider and Shiffrin (1977) have distinguished between controlled processing and automatic processing. Controlled processing requires conscious thoughts and attention, working memory and, mental efforts. Automatic processing occurs with little or no conscious attention or effort. The more intermittent strategies are utilized during activity acquisition, the more complex the activity becomes.

All methods of study of work activity are organized into four stages of analysis: (1) qualitative analysis, which can be parametric or functional; (2) algorithmic analysis, which describes activity as a logical system of human actions; (3) time structure analysis, which describes the duration of different elements of activity and their unfolding in time; (4) quantitative method of activity evaluation, which measures the complexity or reliability of. The quantitative method of evaluation of task complexity is a systemic method of analysis because it evaluates activity structure (system of activity) quantitatively. The stages of analysis can be broken down to various levels of analysis. All stages and levels of analysis have a loop-structure organization, implying that the result of analysis of one level and stage may require reconsideration of preliminary levels or stages of the analysis. The logical and hierarchical organization of the activity elements determines the structure of activity during task performance. This structure depends on strategies of activity. If, due to self-regulation the strategies are changed, the structure of activity also changes. The structure of activity unfolds in time as a process. Such understanding of activity makes it much easier to apply the concept of morphological analysis to the study of human work activity.

The material presented demonstrates that the basic concepts of SSAT are system and structure. However, these concepts can be applied not only to the study of man–machine systems but also to the study of work activity. Man–machine systems or human–computer interaction systems can be optimized based on the analysis activity structure. This optimization can be performed based on qualitative, formalized, and quantitative analysis of activity structure.

7.2 Algorithmic Task Analysis versus Constraint-Based Approach

In cognitive psychology, there are no methods for describing flexible activity. From this follows the basic assumption that there are unpredictable external disturbances acting on the system and therefore there is no one right way of getting the task done. Human behavior is dynamic, requiring workers to adapt to moment-by-moment changes in the context. The basic conclusion from this discussion was that the existing principle of discovering *one best way of task performance* is incorrect. Hence, there is the necessity to use constraint-based approach to task analysis. The basis of this principle is an assertion that performers can do the task utilizing any chosen method within the specified constraints. Vicente (1999, p. 72) wrote that workers can independently decide how exactly the task should be performed. The constraint-based approach contradicts with the instruction-based approach is due to the fact that in cognitive psychology there is no method for describing flexible human activity. In fact, the author refrains from solving various issues in such important areas of study as task analysis and design. Any design solution has to take into consideration constraint-based principles. For this purpose, it is necessary to identify the most effective strategies to accomplish a particular task in specified constraints conditions (Bedny and Karwowski, 2007). Flexible human activity can be described by using algorithmic description of task performance.

Until the mid-1950s, the notion of an algorithm was considered purely mathematical. Later, with the advance of computers, the notion of algorithm began to be used in computer science. The notion of human algorithm was introduced in AT. Human algorithm can be viewed as a system of logically organized mental and motor actions that can be used for solving specific classes of problems or perform similar type of tasks. Computer algorithm, similar to human algorithm, guides the process of solving problems. However, computer algorithm does not necessarily model problems the way in which humans solve them. Elementary computer operations are static, while mental operations and actions utilized by humans are dynamic. The same algorithm can be used by different computers, but this is not true for humans. The building blocks of human algorithm are cognitive and behavioral actions.

Algorithmic analysis of activity is a particularly powerful morphological approach. It consists of subdividing activity into qualitatively distinct psychological units and determining the logic of their sequential organization. Each member of the human activity algorithm consists of tightly interdependent homogeneous actions (only motor, only perceptual, or only decision-making actions, etc.), which are integrated by a higher-order goal into a holistic system. Subjectively, a member of such algorithm is perceived by a subject as a component of his/her activity (mode), which has a logical completeness. Usually, the amount of actions for one member of an algorithm is restricted by the capacity of short-term memory. While motor actions can be performed simultaneously, mental or cognitive actions are usually performed sequentially. Cognitive actions can be combined with motor actions according to the rules described in SSAT (Bedny and Meister, 1997; Bedny and Karwowski, 2007). The members of an algorithm are called *operators* and the units of activity analysis as *logical conditions*. Operators represent actions that transform objects, energy, and information. For example, we can describe operators that are implicated in receiving information, analysis of a situation and its comprehension, shifting of gears, levers, etc. Logical conditions include the decision-making process and determine the logic of selecting the next operator. Each member of the algorithm is designated with a special symbol. For example, operators can be designated by the symbol O and logical conditions by the symbol l. All operators that are involved in the reception of information are categorized as afferent operators, and are designated with the superscripts α, as in O^{α}. If an operator is involved in extracting information from the long-term memory, the symbol μ is used, as in O^{μ}. The symbol $O^{\mu w}$ is associated with keeping information in working memory, and the symbol O^{ε} is associated with the executive components of activity, such as the movement of a gear. Operators with the symbol O^{ε} are depicting efferent operators. From this description, one can see that, for example, O^{ε} cannot include any cognitive actions. Similarly O^{α} can include only perceptual actions. If an operator is involved in extracting information from the long-term memory (only mnemonic actions), the symbol μ is used, as in O^{μ}. After receiving the information (performance of O^{α}) it is impossible to use this information immediately. Worker keeps this information in memory and therefore the symbol $O^{\mu w}$ is used. This symbol describes element of activity that involves keeping information in working memory.

In the algorithmic description of an activity, there are situations when a combination of mental actions and operations should be described. For example, when a person makes a decision, sometimes it is necessary to keep that information in memory until a decision is made. In other situations, decision making requires extraction of information from memory for performing decision-making actions. Combination of decisions with memory function complicates decision-making actions and this should be taking into consideration during algorithmic description of tasks. In such situations, we use symbolic description of logical conditions as l^{μ}, where μ designates the memory function that complicates decision making. Let us consider another example with the following decisions: "if a red bulb is turned on a worker has to press a red button, and when a green bulb is turned on he has to press a green button." For the description of such decisions the symbol l is used. However, for the condition "if an even number is displayed a worker has to press a red button, and when an odd number is displayed he has to press a green button" a worker uses information extracted from memory to make a decision and we use the symbol l^{μ}. Simple logical conditions include *and*, *or*, and *if-then* rules. They have two outputs *yes* and *no* with values 0 and 1. More complicated logical conditions are usually designated by L and have logical conditions with more than two output with each output having various probabilities where the sum of probabilities of all events is 1.

The symbols "l" for a logical condition has to include an associated arrow with a number on top that corresponds to the number of a logical condition associated with it. For example, logical condition l_1 is associated with the number on top of the arrow, $\uparrow^1$. A downward arrow with the same number, $\downarrow^1$, has to be presented in front of the corresponding member of the algorithm to which the arrow refers. Thus the syntax of the system is based on a semantic denotation of a system of arrows and superscripted numbers. An upward pointing arrow of the logical state of the simple logical condition "l", when "l" = 1, requires skipping all following members of the algorithm until the next appearance of the superscripted number with a downward arrow (e.g., $\downarrow^1$). So the operation with the downward arrow with the same superscripted number is the next to be executed. If "l" = 0, then the following member of the algorithm should be executed.

Complex logical conditions have multiple outputs. For example, $L_1\uparrow^{1(1-6)}$ indicates that this is the first complicated logical condition that has six possible outputs: $\uparrow^{1(1)}, \uparrow^{1(2)}, \uparrow^{1(3)}, \ldots, \uparrow^{1(6)}$. Arrows after logical conditions, $\uparrow^{(1)}$, demonstrate transition from one member of an algorithm to another ($\uparrow^{1}\downarrow^{1}$). This means that the logical condition according to the output addressed from the upward to the downward arrow is associated with the particular member of the algorithm. Therefore human algorithm can be deterministic as well as probabilistic (Bedny and Meister, 1997). A deterministic algorithm has logical conditions with only two outputs with values 0 and 1. A probabilistic algorithm has logical conditions which have more than two outputs with various probabilities or logical conditions which have two outputs that can have any value from 0 to 1.

In some cases, logical conditions can be a combination of simple ones. These simple logical conditions are connected through rules such as *and*, *or*, and *if-then*. Logical connections between simple ones are designated with the standard symbols such as, "&," "∧," "→," etc. For example, complicated logical conditions, comprised from simple ones, may be designated as L_1 (l_1^1 & l_1^2 & l_1^3). This symbol means that it is the first complex logical condition. The symbol with a lower case "*l*" designates that it is a simple logical condition that belongs to L_1. Numbers 1–3 used as superscripts designate the number of logical conditions. Complicated logical conditions are particularly important in diagnostic tasks. In the example, the complicated logical condition L_1 is comprised of three simple ones that are combined via the logical conjunction *and*. This complicated logical condition can be used for example to determine whether a particular phenomenon belongs to a certain category, particularly when the phenomenon attributes are connected via conjunctions. Three simple logical conditions can answer the following questions: Is feature 1 present? Is feature 2 present? Is feature 3 present? Only when all three questions receive the response "yes" can one conclude that a phenomenon belongs to a particular category. In contrast, if in our example simple logical conditions are combined via the logical disjunction *or*, it will be sufficient when any one simple logical condition has the required attribute. Sometimes, a complicated logical condition includes different logical connections.

Thinking actions are often based on externally provided information (e.g., mental manipulation of externally presented data), or made with reliance on the information held by or retrieved from memory (manipulation of data in memory), or thinking action require keeping intermittent data in memory. In this case, we describe thinking operators as $O^{\alpha th}$ or $O^{\mu th}$ (α denotes the thinking process of an operator based on external, e.g., visual information, and μ means that the operator requires complicate manipulation in memory). Such symbolic descriptions are used when in describing thinking it is important to distinguish if thinking is based on externally presented information or information extracted from memory.

Sometimes, similar members of an algorithm follow each other. For example, there may be several afferent members, O_1^{α}, O_2^{α}, and O_3^{α}, or efferent members, O_1^{ε}, O_2^{ε}, and O_3^{ε} of an algorithm. In this case, experts can use some approximate rules to extract different members of the algorithms.

If the sequence of the performed actions can be kept in working memory, then the number of actions in one member of an algorithm should be no more than three to four.

If actions are simple and performed sequentially, and their order does not need to be kept in working memory, then their integration into separate members of an algorithm is determined by logical completeness of parts of the activity. Such actions are integrated by higher-order goals and have a limited number of interdependent work tools and objects. The limited capacity of short-term memory can also influence strategies of the grouping of these actions.

Let us consider a simple hypothetical example of an algorithmic description of a task performed by a human. The task is "A driver bypasses a car in front of his/her car." This is a real scenario that every driver is familiar with (see Table 7.1).

The algorithm should be read from the top to the bottom. A symbolic description of a member of the algorithm in a standardized form in the column on the left is an example of psychological units of analysis because they have clearly defined psychological characteristics. A verbal description of a member of the algorithm in the right column is an example of technological units of analysis. These units of analysis describe elements of work and they do not possess a clearly defined psychological description. Driving a car is a familiar task for most people. Therefore, verbal description of the members of the algorithm in the right column of the table does not require much explanation. However, very often it is quite difficult to describe human

TABLE 7.1

Algorithmic Description of Task "Bypassing the Car in Front"

Member of Algorithm	Description of Algorithm Member
$\overset{3}{\downarrow}\overset{2}{\downarrow}\overset{1}{\downarrow}O_1^{\varepsilon}$	Continue driving
$O_2^{\alpha th}$	Mental order or command "to bypass the vehicle ahead of my car"
$O_3^{\alpha th}$	Look at the speedometer and evaluate speed
$l_1 \underset{1}{\uparrow}$	If speed makes bypass possible ("Yes") go to O_4^{α}. If "No," go to O_1^{ε}
O_4^{α}	Look in front
O_5^{th}	Do the positions of the cars in front allow bypassing?
$l_2 \overset{2}{\uparrow}$	If "No," continue driving (go to O_1^{ε}); If "Yes," go to $O_6^{\mu\alpha}$
$O_6^{\mu\alpha}$	Keep information in memory about position of car in front and look back
$O_7^{\mu th}$	Do the positions of the cars behind also allow bypassing?
$l_3^{\mu} \overset{3}{\uparrow}$	If "No," continue driving (go to O_1^{ε}); If "Yes," go to O_8^{α}
O_8^{ε}	Bypass (turn the wheel of your car left and then right and go ahead)

work activity clearly with everyday language. Such a description of the task is a major challenge to the expert who tries to evaluate the activity of a person during task performance not through observation, but by reading the documentation. Algorithmic analysis is an important method of formalized description of a task. It represents not only the logic of the transition from one element to other elements of activity, but also with the aid of symbolic notation gives an idea of the basic psychological characteristics of each member of the algorithm. Algorithmic analysis is essential for the description of variable human activity. Probabilistic algorithms have a special role in such situations.

The presented material allows to make some basic conclusions. One, the critical stage in the description of flexible human activity is the algorithmic description of task performance. Human algorithm should be distinguished from a mathematical or a computer algorithm. The main units of analysis in human algorithm are cognitive and behavioral actions. In algorithmic analysis of task performance, human activity is divided not only into actions, but also into algorithmic elements and members such as operators, and logical conditions. Members of an algorithm can be considered as a subsystem of activity, which has a higher-order goal that can integrate the same type of actions into such subsystems. A logical condition is a decision-making process that determines which member of the algorithm and, therefore, which actions are selected. An algorithm can be probabilistic as well as deterministic. Deterministic algorithms have logical conditions with only two outputs, 0 or 1. If an operator's activity is multivariate and it can be described by a probabilistic algorithm. In such cases, logical conditions can have more than two outputs and vary from 0 to 1. Hence, logical conditions can transfer activity flow from one member of the algorithm to another with various probabilities. Each member of an algorithm is designated a special symbol, which demonstrates the psychological meaning of each member. Algorithmic analysis of human activity eliminates contradiction between the so-called one best way of performance and the constraint-based approach. We have described the principles of algorithmic analysis of human work activity in a brief manner in this chapter. We will discuss this method during the analysis of computer-based tasks in the third section of the book.

Section III

Quantitative Assessment of Computer-Based Task

8

Quantitative Assessment of Task Complexity Computer-Based Tasks

8.1 Analysis of Existing Method of Complexity Evaluation of Computer-Based Tasks

The study of complexity stimulated the development of an interdisciplinary field, *science of complexity*, due to various aspects of complexity being studied in different fields of science.

The main purpose of the study in this field is the development of formalized and quantitative methods of evaluation of a complex systems structure. In different fields, scientists emphasize different aspects of complexity. In mathematics, there is the computational complexity theory. The study of complexity attracts the attention of economists, biologists, and other scientists. For such areas of science as psychology, ergonomics, economics, and human–computer interaction (HCI), psychological complexity is especially important (Bedny and Karwowski, 2007; Thomas and Richards, 2012). It is important to understand that what is objectively complex for some may not be equally complex for other subjects.

Under complex system, we understand any system that consists of a large number of parts that have many interactions, and the whole is more than the sum of its parts (Simon, 1999). Task complexity evaluation is an important approach to enhancing work efficiency, improving design solutions, and optimizing work in general. Moreover, quantitative method task complexity assessment in the field of HCI is not developed. Therefore, in this chapter, we extend the task complexity evaluation approach to the field of HCI. Optimization of work performance according to complexity criteria enhances the efficiency of HCI. In the systemic-structural activity theory (SSAT), task complexity is considered from two perspectives: functional analysis perspectives when activity is represented as a goal-directed self-regulative system and morphological analysis perspectives where activity is seen as a complex multidimensional structure. Functional analysis considers how a subject evaluates the difficulty of a task and how this evaluation influences strategies of task performance. In functional analysis complexity is an objective characteristic of a task and difficulty is its subjective characteristic. Relationship between objective complexity of a task and its subjective difficulty is critically important in selecting strategies of task performance. Therefore, the concept of self-regulation plays a critical role in task complexity/difficulty relationship. A subject can select various strategies of task performance depending on its perceived difficulty. A task can even be rejected by a subject if perceived as an extremely difficult one. We considered this aspect of complexity/difficulty relationship in Chapter 3. Attention should be drawn to the fact that from an ergonomic point of view, a technological system can be excessively complex but its interaction with a human can be simple, and vice versa. We will analyze only the psychological aspects of complexity, that is, those aspects of designing systems that are associated with human performance in the HCI system. Various aspects of this issue have been studied by different authors (Jacko, 1997; Jacko and Ward, 1996; Jacko et al., 1971, etc.).

In this chapter, we consider task complexity from morphological perspectives as a multidimensional, integral characteristic of a task. The more complex a task is, the higher the cognitive demands for its performance. Components of task complexity impose demands on the mental efforts of users. A user cannot experience complexity by itself but rather perceives it as a subjective difficulty. When the complexity of a task increases, the probability that performance will require more cognitive effort and motivational mobilization increases as well.

Complexity as a multidimensional phenomenon cannot be evaluated by one measure. Multiple measures for task complexity should be selected for task complexity evaluation. Based on complexity measures, we can optimize human performance or design solution.

So, for quantitative analysis of task performance, we have to utilize not one but multiple measures of complexity. It should be noted that most scientists agree that multiple measures are required instead of an integral one (see, e.g., Kieras, 1993). Quantitative measures of complexity suggest the procedures for their calculation. The term *measures* is always associated with principles of measurements or calculations.

Analytical quantitative methods of task analysis are not sufficiently developed in traditional engineering psychology. Most of such methods have only some theoretical meaning, but not sufficiently developed for practical applications. Quantitative methods become especially important when it comes to analyzing computer-based tasks. In this chapter, we investigate the complexity of the activity during computer task performance because one of the main purposes of ergonomic design is to reduce cognitive demands of a given task. Task complexity is an important characteristic of not only cognitive but also motor components of activity because motor actions include cognitive components, which are responsible for the regulation of motor actions and motions (Gordeeva and Zinchenko, 1982; Sanders, 1980; Sternberg et al., 1980). For example, motor programming is a cognitive component of motor activity. The duration of this stage depends in large degree on complexity of movement. The evaluation of complexity of computer-based tasks is based on the same basic principles described earlier in Bedny and Meister, 1997; Bedny and Karwowski, 2007. However, computer-based tasks have their own specifics. In this regard, we have to consider the specifics of these issues, which relate to the assessment of the complexity of computer-based tasks. It also requires consideration of the publications that have been in this field.

First of all, we have to analyze the GOMS (goals, operators, methods, and selection rules) method, which was developed for the evaluation of complexity of computer-based tasks. The GOMS method was developed by Card et al. (1983). After that, various GOMS methods have emerged in literature in the last two decades (Kieras 1993, 2004; Kieras and Polson, 1985). There are some superficial similarities between GOMS methods and the approach of complexity evaluation of computer-based tasks in SSAT. These two approaches differ significantly. Kieras (2004) states that the GOMS method can be used only after a basic task analysis has been carried out. This is due to the fact that this method does not have its own psychologically sound theoretical basis for the analysis of human activity or behavior. For example, goal plays an important role in the GOMS method. However, goal concept is not clearly defined in GOMS. In activity theory (AT and SSAT), the term *goal* has a totally different meaning. Goal in SSAT is a desired future result of activity. It includes imaginative and verbally logical components. It always includes conscious components and is always associated with motives creating vector *motives* → *goal*. Goal is a cognitive component and motives are energetic components of activity. We have to distinguish goal of task from goal of subtasks and goal of actions. Logical rules in SSAT are used as external or internal tool for transformation of internal or mental and external or material object according to the goal of activity (Bedny and Meister, 1997). In GOMS, goal concept includes both motivational and cognitive components. Because our behavior is polymotivated, it is difficult to understand what the real goal of the task is (Bedny and Chebykin, 2013). In GOMS, the goal is something that the user tries to accomplish. It is a ready-made end state of the system to which behavior is directed. In SSAT, there are external requirements of the task that should be transferred into a subjectively accepted goal. A goal should be perceived, interpreted, accepted, and accomplished through behavior. An objectively given goal and a subjectively accepted goal are not the same. This idea is particularly important for computer-based tasks, where users often have to independently formulate the goal of given tasks. This factor is also important when users have to correctly interpret an objectively given goal (Bedny and Bedny, 2011). According to Diaper and Stanton (2004), the goal in cognitive psychology is not a theoretically grounded concept and it is very difficult to use in practice. As a result, these authors suggest to abandon the concept of goal in task analysis. On the contrary, this concept is of fundamental importance in AT and SSAT. The goal is understood in SSAT in a different manner. Without goals there is no task.

In SSAT and GOMS, the concepts of actions and operators have totally different meanings. In SSAT, actions can be cognitive and behavioral. Actions are goal-directed, self-regulated subsystems of activity. From AT perspectives, behavioral actions include cognitive regulative mechanisms. Behavioral and cognitive actions include psychological operations. Actions in SSAT can be classified according to particular principles and extracted from activity flow based on specific procedures. In GOMS, a concept of action does not have any theoretical justification and is used as an empirical term. GOMS does not define actions and methods of their extraction from activity flow and uses it just as an intuitive terminology. SSAT uses the concept of a human algorithm, where major units of analysis are cognitive and behavioral actions. GOMS describes human behavior in terms of computational concepts such as logic algorithms of performance, as GOMS resembles a computer algorithm. Let us consider Kieras' example (1993, pp. 138–139):

Accomplish the goal of < goal description >
Report < goal accomplished >
Decide: If < operator… > Then < operator >
Else < operator >
Forget that < WM-object-description >

This is an artificial computer-like algorithm, which completely opposed the concept of a human algorithm in SSAT. For example, the last operator is not a voluntary human action. Forgetting is an involuntary process and cannot be considered as a voluntary, goal-directed action. In AT, "Report < goal accomplished >" can be introduced only in a situation when a subject should provide information about goal accomplishment. The "Else < operator >" also cannot be considered as an element of activity and cannot be used as a unit of analysis for algorithmic description of human performance. SSAT offers scientifically proven principles of designing human algorithms. GOMS measures complexity by analyzing the number of rules (productions) of a computer-like algorithm. However, the level of behavior decomposition, which determines the quantity of productions, is not precisely defined.

Each production in GOMS is formed based on conditions and actions. Each condition has only two outputs (0 and 1). Human actions rarely are so simple and clear-cut. Real-life conditions can have more outputs and probabilities. In SSAT, there are deterministic and probabilistic algorithms. Probabilistic algorithms cannot be described by *if-then* rules containing 0 or 1 value.

We demonstrated that an algorithmic description of human activity is not sufficient for quantitative assessment of task complexity (Bedny and Karwowski, 2007; Bedny and Meister, 1997). Activity is a process, and for its complexity evaluation during task performance, a time structure of activity should be developed and measures of complexity should be created. It is also critically important to determine which types of activity elements can be performed simultaneously and which can be performed sequentially only.

These aspects of complexity assessment are not discussed in the GOMS analysis method. GOMS describes behavior as only a sequence of behavioral elements and cognition as a simple sequence of production units. However, some elements of activity can be performed simultaneously, which affects task complexity. The possibility to perform elements of activity simultaneously or sequentially depends on the level of complexity of these elements. GOMS suggests the following formula to calculate task performance time (Card et al., 1983; McLeod and Sherwood-Jones, 1992):

$$T = T_p + T_c + T_m,$$

where T_p, T_c, and T_m are performance times for perceptual, cognitive, and motor components of activity. This formula would only work when all elements of activity are performed in sequence and have a probability of 1. More often than not, elements of activity are combined in time and thus have a probability less than 1. Hence, analytical methods of task execution time calculation cannot be reduced to a simple summation of execution time of separate elements of a task.

The GOMS methodology (Kieras, 1993) reduces measures of complexity to determining task execution time, time needed to learn the system, and counting quantity of productions. However, these are examples of indirect data of complexity analysis that is unrelated to the actual quantitative (mathematical) estimation of complexity. It is known that more complex tasks are in some cases executed in a shorter period of time because a user combines some elements of activity in time. Kieras and Polson (1985) suggest such possible measures of complexity as the number of productions in task representation, the maximum number of goals in working memory during task performance, and the number of conditions and actions in production, but they do not specify how such units of measures should be selected and how these measures can be calculated. Later, Kieras (1993) suggests using a number of production rules or units of roughly equal *size* to measure complexity. However, units of roughly equal *size* are not commensurable units of measure and attempts to use them contradict all principles of measurement in physics and mathematics.

The GOMS method has been under serious criticism even by its followers. Karat (1993) wrote that the GOMS method misses many key components of behavior that should be considered in interface design. According to the GOMS method, cognitive operations are of equal difficulty and problem-solving aspects of behavior are practically eliminated from analysis. This makes it impossible to use units of roughly equal *size* or production rules in complexity assessment (Bedny et al., 2012). In contrast, in SSAT, cognitive and behavioral units can be evaluated according to a five-point scale of complexity. GOMS describes behavior as only a sequence of elements when in fact some elements of activity are performed simultaneously.

This factor influences task complexity. The possibility of performing elements of activity simultaneously or sequentially depends on the level of complexity of these elements. Diaper (2004, p. 23) wrote that GOMS' models undoubtedly misrepresent human psychology. According to him, this method also incorrectly predicts task performance time by simply summing the time associated with each primitive operator.

For complexity evaluation of computer-based task, we use a five-point order scale for evaluation of activity elements complexity and determining possibility to perform elements of activity simultaneously or sequentially (Bedny and Karwowski, 2007).

It should be noted that the existing methods of complexity assessment that utilize measures of complexity based on evaluation of human behavior do not use the order scale for the gradation of activity elements standing on any objective ground. GOMS does not distinguish levels of complexity for various elements of behavior and offers no procedures to do it. They simply ignore differences in levels of complexity because there were no developed procedures for this purpose. For example, in GOMS, the assumption is that all cognitive operations are of equal difficulty, or production rules should be of roughly equal *size*. Human activity has a complicate structure and includes cognitive, behavioral, and emotionally motivated components. Some components are not obvious and are hidden from the expert analysis. Hence, it is not possible to evaluate activity structure based on purely subjective judgment.

We will not examine in detail the principle of the development of measures of complexity in this chapter. These questions will be discussed in subsequent chapters. In this chapter, we will consider a specific example of computer-based task and measures of complexity, which can be used for such purpose.

8.2 Theoretical Principles for Evaluating the Complexity of the Computer-Based Task

Complexity is a multidimensional concept that requires multiple measures for its evaluation. For each specific task, some measures might be more important than others. Some of these measures can be of zero value if they are not important for a particular task. Having multiple measures allows enhancing and redesigning tasks based on each measure and their comparison. This is done by getting a clear understanding of the task's design shortcomings and various aspects of difficulty in task performance. We will start our analysis with a consideration of the relationship between mental efforts and level of concentration of attention required for performance of a particular element of activity.

Such concepts as level of attention concentration, level of activation of neural centers, and level of wakefulness are important as a theoretical basis for evaluation of mental efforts. Kahneman (1973) considered attention as a mechanism responsible for coordination and regulation of mental efforts during performance. The same idea is presented in the self-regulative model of attention suggested by Bedny and Karwowski (2011). The more complex the task is, the more mental efforts are required and the higher is the level of attention concentration. Usually, time to perform actions increases if a higher level of attention is required for their performance. However, a more complex task can be performed in less time because the subject can mobilize his/her efforts and combine some of the task elements. In the works of Bloch (1966) and Lazareva et al. (1979), it is shown that the complexity of performance is associated with the level of brain activation. There are specific and nonspecific levels of activation. Nonspecific or global level of activation is tightly connected with the functioning of the reticular activating system. Works of the aforementioned authors point out that nonspecific forms of activation are connected with the difficulty of performing particular tasks. Therefore, during evaluation of complexity as an objective characteristic of difficulty, the level of specific activation can be neglected. Only a particular range of wakefulness associated with nonspecific activation of neural system should be considered during the evaluation of task complexity. Nonspecific level of neural centers activation is a continuous process. The higher the level of neural centers activation is, the higher is the attention concentration. Complexity of activity can also be viewed as a continuum that allows to present complexity as an ordered scale. Based on analysis of the literature and our own study, we have developed an order scale for assessing the complexity of separate elements of activity according to attention concentration criterion (Bedny, 1987; Bedny and Meister, 1997; Bedny and Karwowski, 2007).

Motor components of activity can be evaluated according to three categories of complexity. The motions that requires a low level of attention concentration belong to the first category of complexity. For example, MTM-1 motion *Reach* (*RA*) (reach object in fixed location) requires a minimum level of control and attention. Element *Reach* (*RB*) is more complex. It involves reaching an object in a location that can vary and requires an average level of attention. This element is related to the second category of complexity. Element *Reach* (*RC*) is the most complicated. Its purpose might be to reach an object mixed with other objects. This last element requires a high level of control or concentration of attention and is equivalent to the third category of complexity. The simplest cognitive action requires the third category of complexity. An example of this category of complexity is a simple *yes–no* or *if...then* decision. Such decision-making actions are performed with a high level of automaticity and performance time for such elements approximately 0.30 s. Hence, even the simplest cognitive operations and actions should be related to the third category of complexity. All cognitive actions and operations performed with a high level of automaticity are related to the third category of complexity. This is the simplest category of cognitive actions.

There are more complicated cognitive components of cognitive activity. Decision-making when the required response is not known in advance, or decision-making performed in an ambiguous situation, is more complicated than decision-making where the required response is already known. Therefore, this type of decision-making, the same as the whole group of more complex cognitive actions, should be related to the fourth category of complexity in comparison to the action considered previously. This group of cognitive actions includes mental actions that are associated with overloaded attention, recognition of unclear signals, decision-making and performance of actions in contradicting situations, etc.

Some motor actions should also be in this category. For example, the operator performs motor actions such as when a signal on the screen moves forward, the operator is required to move a control backwards in exact position. This kind of scenario requires remembering instructions, a greater level of concentration of attention, etc. Then, the motor action, which was of the third category of complexity, becomes more complex (fourth category of complexity) (Bedny, 1987; Bedny and Karwowski, 2007). If an operator performs a task under stress, the level of complexity increases. Therefore, a five-point order scale for motor and cognitive activity complexity was developed. This ordered scale distinguishes three categories of complexity for cognitive elements of activity. Usually, the simplest is the third category and the more complicated is the fifth category. Motor elements can be related to the high-order category (higher than the third category of complexity) in some specific conditions described earlier. The presented five-point scale of complexity can be applied with sufficient precision to the complexity evaluation of various elements of activity. Thus, motor and cognitive scales partly overlap each other. The qualitative content of cognitive and motor elements of activity, probability of their occurrence, and possibility of their performance, not just sequentially but also simultaneously, should be considered. Multiple measures of complexity should be used instead of just a single one.

Another important issue of task complexity analysis is the development of adequate units of measure for its evaluation. As shown, there were multiple attempts to develop a quantitative method of task complexity evaluation. However, adequate units of measurement for this purpose were not developed. Suggested measures such as task solving time, number of transitions, and total number of system's states are inadequate from a mathematical point of view because they are noncommensurable units. It is like mixing apples and oranges. Suggested measures do not always correlate with complexity. For instance, a complicated task can be performed at the same time as a simpler one. The subject can spend more mental effort performing task in a short time or with fewer transitions during performance. Manipulation of one control can be more complex than manipulation of several controls. Similarly, one cannot just calculate the amount of actions performed by an operator during task performance for task complexity evaluation because one motor action or decision-making can be more complicated than several simple ones. The examples listed earlier demonstrate an attempt to utilize noncommensurable units of measure. As was similarly suggested by GOMS, *units of roughly equal size* cannot be used as units of measures. Existing methods of complexity ignore energetic and emotionally motivational aspects of activity. Increasing activity complexity leads to an increase in concentration of attention and to a higher level of emotional-motivational components of activity. Considering human beings as working computers ignores these factors. All the earlier examples demonstrate utilization of noncommensurable units of measure (Bedny et al., 2012).

Activity is a process, and therefore, its complexity evaluation during the performance of a computer-based task can be performed only after a time structure of activity is developed. In order to describe the time structure of the activity, it is necessary to distinguish its constituent elements. In cognitive psychology, there are no clearly defined principles of extraction of such elements for task analysis. For example, such concepts as cognitive and motor actions are not clearly defined. According to Vicente (1999, p. 4), action is a goal-directed behavior of an actor and an actor is a worker or an automation. Therefore, human behavior is not distinguished from the machine one. Moreover, in AT, cognitive and behavioral actions are elements of human activity. Actions can be cognitive and motor. They are directed to achieve a conscious goal of actions. Actions consist of operations and so on. Activity during task performance can be presented as a logically organized system of cognitive and behavioral actions that are directed to achieve a goal of the task. In SSAT, cognitive and motor actions are described in a standardized manner, which is of fundamental importance when analyzing a structure of human activity and solving design issues in ergonomics.

The concept of *goal* is used in various fields of science and practice. In his definition of goal-directed behavior, Vicente utilizes a concept of goal that is used in psychology, cybernetics, engineering, management, philosophy, and so on. The concept of goal is interpreted differently in various fields of psychology and directions of science. Human goal cannot be considered in a similar way as a goal of automotive system. In SSAT, a goal is a psychological reflection of a desired future result of our own activity. It represents the most important form of anticipation and includes verbally logical and imaginative components. Elimination of the task's goal and goals of separate actions during task performance reduces human work activity to a chain of reactions or responses. Generally, such behavior entirely depends on external, environmental stimulation. We also need to distinguish between the *overall or terminal goal of task* and *partial or intermittent goals of actions and subgoals of a task*. The goal cannot be presented to the subject in a ready form but rather as an objective requirement of the task. However, these requirements should be conscious and interpreted by the subject. At the next stage, these requirements should be compared with past experience and the motivational state, which leads to the goal acceptance process. Subjectively accepted goal does not always match the objectively presented goal (requirements). Moreover, very often subjects can formulate the goal independently. During an ongoing activity, the goal can become more specific and corrected if needed. So, we can conclude that the goal does not exist in a ready form for the subject and cannot be considered simply as an end state to which the human behavior is directed. Thus, during time structure development, we always take into consideration the concept of goal. In some fields of psychology, goal and motives are considered as unitary mechanisms (Lee et al., 1989). In SSAT, goal is a cognitive component and motives are energetic components of activity. Before we develop a time structure activity, we need to extract cognitive and behavioral actions that are elements of activity. The determining factor in the selection of actions in the structure of activity is the goals of actions. In general, we can see that such concepts as goal, action, and activity have a completely different meaning in SSAT compared to how it is understood in cognitive psychology.

In the process of describing activity time structure, it is necessary to determine which elements of the activity can be performed simultaneously and which can be performed only sequentially.

The possibility to perform elements of activity simultaneously or sequentially depends on the level of complexity of those elements (concentration of attention during performance of these elements). In SSAT, there are rules that determine the possibility of performance of cognitive and behavioral elements in sequence or simultaneously (Bedny, 1987; Bedny and Karwowski, 2007). Time structure of activity can be presented in table or graphical form. If time structure of activity is complex, then it is recommended to utilize not only a table but also a graphical form of time structure description.

During motor activity, the major user tools are the mouse and keyboard. There is a problem with segmentation of motor activity while the user works with a keyboard. The main criteria for segmentation of this kind of motor activity are the existence of the goal of motor action and the principle of rhythmic organization of repetitive motor motions. However, in the analysis of this type of motor activity, there is usually no need to extract separate motor actions. For example, the same level of concentration of attention should be assigned to typing homogeneous text. Usually, this period of time should be related to the third category of complexity. Therefore, in order to assess the complexity of this component of work, it is sufficient to know the duration of typing a particular text. If such period of time is not homogeneous and some text is considered by a user as more significant, this period of time should be related to the fourth category of complexity. The complexity of reading a text can be evaluated similarly. We present the fragment of algorithmic and time structure description of computer-based drawing task in later text (Sengupta and Jeng, 2003). This work was performed under our supervision.

The study procedures included qualitative cognitive analysis, eye movement and mouse movement registration, video registration, an algorithmic description of task, and time structure. All methods can be presented as three stages of analysis (from existing four stages of analysis in SSAT). In all cases, when an algorithmic description is accompanied by performance time of individual members of an algorithm and of its elements, this is a representation of a time structure of activity in a tabular form. Table 8.1 depicts a fragment of an algorithmic description of activity and its time structure for a computer-based drawing task in a tabular form.

Some members of the algorithm are prone to abandonment. This is explained by the fact that during performance of HCI tasks, a user often does not know in advance the sequence of actions he has to perform. In such situation, the user explores the possibility of different actions. This exploration can be performed in internal and external plane. Internal explorative activity manifests itself in the delay of external actions. It is important to conduct

TABLE 8.1

Algorithmic Description of Activity and Its Time Structure in Tabular Form (Fragment)

Symbol	Member of Algorithm	Classification of Actions	Time (in ms)
O_1^{α}	Visual perception of the given figure.	Simultaneous perceptual action	300
$l_1^{\mu}\overset{1}{\uparrow}$	Mental selection of required figure.	Decision-making at the perceptual level (including the following operations):	250
		1. Actualization of information from memory	370
		2. Decision-making	
$\overset{1}{O}{}_2^{\mu}$	Sustain actualized information about selected circle in working memory (abandoned option).	See description below	—
$\overset{1}{\downarrow}\overset{2}{O}{}_2^{\mu}$	Sustain actualized information about selected square in working memory during performance of O_3^{ε} (correct action).	Mnemonic action of maintaining information in working memory while performing O_3^{ε} (direct connection action)	300
O_3^{ε}	Move pointer from the starting position to the drawing toolbox.	Motor positioning action under visual control (required coordination of eye and arm movement)	300
O_4^{α}	Examine area with related tools.	Perceptual actions	200
$l_2\overset{2}{\uparrow}$	Selection of required tool.	Decision-making at the perceptual level	200
$\overset{1}{O}{}_5^{\varepsilon}$	Move to circle drawing icon and click (abandoned option).	See below	—

chronometrical studies and determine the time associated with such a delay. Contents of this internal activity can be determined with some approximation based on its qualitative analysis utilizing observation, interview, and expert's analysis. It is important to determine the mathematical mean of the delay time of the task and the probability of its occurrence. Explorative activity can be performed in external form. A user can observe the result of externalized explorative motor activity. In such situation, time for performance of explorative activity (including cognitive and behavioral components) is measured together. In the production environment, some explorative components of activity can lead to corruption of database. Explorative actions with significant time of performance and those that can lead to corruption of database should be eliminated during the design of HCI tasks. We will discuss this type of activity in more detail in Chapter 12. In Table 8.1, there are members of algorithms that are underlined by two lines. This means that such members of algorithms are performed simultaneously with other members of algorithms. Time performance of such member of the algorithm is not considered when we determine the execution time of the whole task.

In Table 8.1, the left column presents psychological units of analysis in symbolic standardize form. For example, symbol O_1^α clearly demonstrates that this member of algorithm relates to perceptual activity. The second column of Table 8.1 presents technological units of analysis in common language terms. These units of analysis may also be called typical elements of the task. The third column presents psychological units of analysis. These units of analysis present description of cognitive and behavioral action in a standardized manner. In the last column, time performances of algorithm members were presented in milliseconds. The algorithm and time structure data were derived from qualitative analysis that includes retrospective protocol analysis, which consists of observation and expert analysis and chronometrical method of study. It is also derived from instrumental analysis as eye movement registration and analysis of video. We present the same fragment of activity time structure in graphical form in the following (see Figure 8.1).

In this model of activity performance, individual elements of activity are presented by horizontal lines. The length of each horizontal line segment depends on the duration of its performance. In some cases, elements of activity can be performed simultaneously depicted by horizontal line segments located one under the other. Activity is described as a process that has a complicated structure. When developing the temporal structure of activity, it is necessary to use not only technological but also psychological units of analysis because time structure of activity demonstrates how elements of activity unfold over time. Strategies of task performance can be described with great precision by activity time structure utilizing psychological units of analysis. Comparison of time structure of activity with equipment configuration or interface demonstrates efficiency of design solutions. If, for example, time structure of activity is very complex, then redesign of interface is required. Psychological units of analysis are also instrumental in the assessment of task complexity.

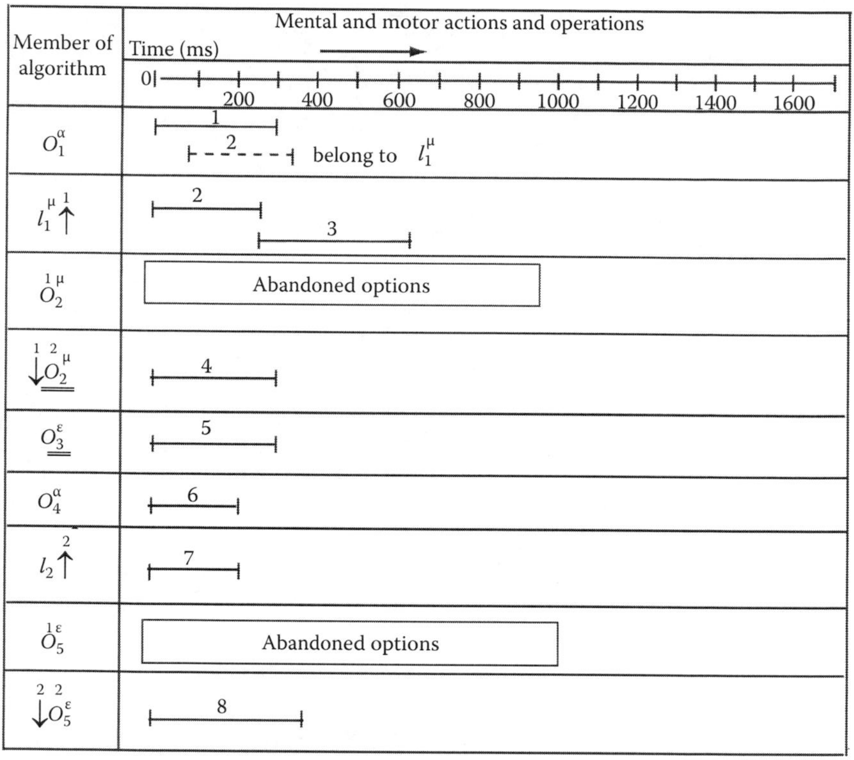

Member of algorithm	Mental and motor actions and operations Time (ms) → 0, 200, 400, 600, 800, 1000, 1200, 1400, 1600
O_1^{α}	1 2 belong to l_1^{μ}
$l_1^{\mu}\uparrow^{1}$	2 3
$\overset{1}{O}{}_2^{\mu}$	Abandoned options
$\downarrow^{1\,2} O_2^{\mu}$	4
O_3^{ε}	5
O_4^{α}	6
$l_2\uparrow^{2}$	7
$\overset{1}{O}{}_5^{\varepsilon}$	Abandoned options
$\downarrow^{2\,2} O_5^{\varepsilon}$	8

FIGURE 8.1
Time structure of activity during performance of HCI task in graphical form (fragment).

These topics will be discussed in Section 9.5. In the example considered earlier, we can outline three stages of task analysis and description: qualitative stage, algorithmic analysis, and time structure analysis. All these stages precede quantitative evaluation of task complexity. Time structure of activity should be distinguished from a time line chart that describes distribution of elements of work in technological terms.

Let us consider another fragment of time structure of activity. This helps us to understand the basic principles of task complexity evaluation. A model of a manual production operation was studied under laboratory conditions. Each element of activity is relatively simple. However, the logical structure of activity is sufficiently complex. The task requires installing 30 pins into the holes according to existing rules. Some pins had a flute. When subjects grasp pins with the flute from the box, they had to use the following rules:

1. If a fluted pin is picked up by a subject's left hand, it should be placed in the hole so that the flute should be inside the hole.
2. If a fluted pin is picked up by a subject's right hand, it should be placed in the hole so that the flute should be above the hole.

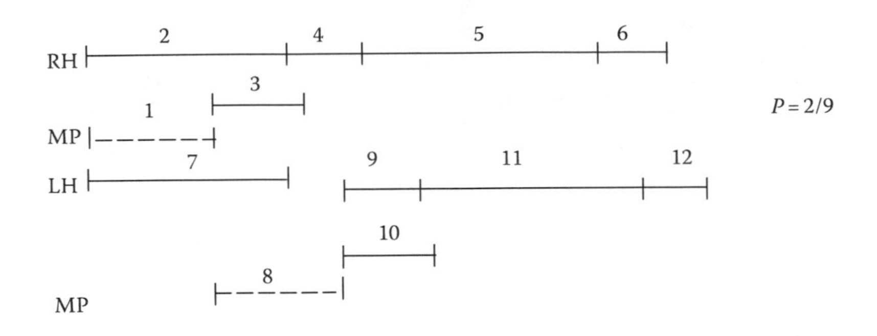

FIGURE 8.2
Graphical time structure of activity during installation of two fluted pins (one simultaneous installation by two hands).

Graphically, time structure of activity during performance of this fragment of task when subject grasps two fluted pins is presented in Figure 8.2.

In this figure, RH means right hand, LH means left hand, and MP means mental processes (dashed line). P means the probability of appearance of these elements of activity during task performance. As can be seen, some elements of activity are performed simultaneously and some in sequence. The following is a verbal description of these elements:

1—Receive information about the pin's flute shape in the right hand and decide to turn it into the required position

2 and 7—Simultaneously while receiving information and making decision, move the pins with left and right hands into the approximately correct position

3—Simultaneously while moving the pins, turn one in the right hand 90° so that the flute should be above the hole after installation

4—Without interruption, move the pin with the right hand in the exact position (above the hole)

5—Install the pin with the right hand to the hole

6—Release the pin in the right hand

8—After the first decision is made, perceive information about the flute in the left hand and make the decision to turn the pin 90° so that the flute should be inside the hole after installation (according to SSAT rules, cognitive actions cannot be performed simultaneously, and therefore during this decision-making, the left hand movement is interrupted because decisions cannot be performed simultaneously, and elements 1 and 8 are performed in sequence)

9—Move the pin into the exact position

10—Simultaneously turn the pin into the required position by the left hand
11—Install the pin by the left hand into the hole
12—Release the pin in the left hand

The presented example illustrates that elements of activity can be performed simultaneously and cognitive elements can be performed in parallel only with motor activity. This means that mouse movement can be combined with simple decision-making. In addition, some components of activity may appear in the structure of the task with a different probability. For example, considering elements of activity appears in the structure of the task with the probability $P = 2/9$.

The material presented here clearly shows that the activity is a process that has a complex structure and evolves in time. Its time structure has various cognitive and motor components. If we know the duration of individual elements, the probability of their occurrence, and their nature of being combined in time, it is possible to calculate the complexity of the analyzed activity structure. The units of complexity measures that are classified according to standardized principles are intervals of time when various elements of the activity are performed.

Let us briefly consider a few examples of task complexity evaluation for some elements of activity during task performance. These examples demonstrate the importance of time structure analysis. Suppose two elements of activity are performed simultaneously (Figure 8.3 depicts two

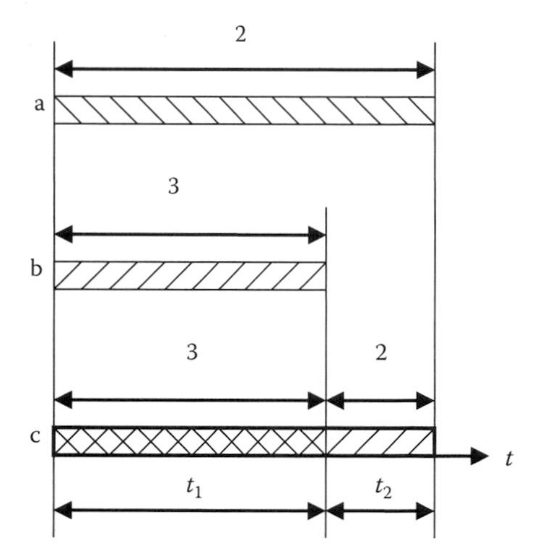

FIGURE 8.3
Example of graphical interpretation of complexity motor components of activity, which are performed simultaneously (element *a* has the third and element *b* has the second level of complexity according to attention concentration). c illustrates complexity of an interval of time t_1 when two activity elements are performed simultaneously and t_2 when element "a" is performed independently at the final time period.

elements of activity *a* and *b*). These segments' length depicts their performance time. Element *a* requires an average level of attention concentration and therefore, according to SSAT rules (Bedny and Karwowski, 2007), is related to the second level of complexity. Element *b* requires the third level of attention concentration and therefore is evaluated as a motor element of the third category of complexity. Therefore, according to SSAT rules, an interval of time t_1 belongs to the third category of complexity. The element on the right with performance time t_2 is related to the second level of complexity.

The time interval t_1 when two elements are performed simultaneously has the third category of complexity and the remaining time interval t_2 is of the second category of complexity.

Later, we present the description of the utilized rule: a time period, when two elements of activity with different categories of complexity are performed simultaneously, should be evaluated by complexity of a more difficult element. Let us consider another situation. If two motor components of activity are of the third category of complexity, they can be performed simultaneously only within the optimal visual field. Then, a period of time when these elements are performed simultaneously should be related to the fourth category of complexity. All formalized rules are described in detail in Bedny and Karwowski (2007).

Once we have performed an algorithmic description of task performance and developed a time structure of an activity (morphological analysis of activity), we can move to the stage of quantitative assessment of task complexity. For this purpose, we can define the general performance time for each step of algorithm of task performance. In an algorithmic description of activity during task performance, such steps are called member of algorithm. They include one or several actions integrated by a high-order goal. After that, an algorithm (task) execution time can be calculated according to the following formula:

$$T = \sum_{j=1}^{n} P_i t_i$$

where

P_i is the probability of occurrence of the *i*th member of the algorithm (step of algorithm performance)

t_i is the duration of the *i*th member of the algorithm

At the next step, we calculate what fraction of time is spent on receiving information, making decisions during task performance, and utilizing a five-point scale in the evaluation of the complexity of decision-making

(Bedny, 1987; Bedny and Karwowski, 2007). Let us consider, as an example of calculation, a measure as a fraction of time for logical components of work (fraction of time for decision-making process during task performance).

Time performance of all logical conditions (steps of decision-making process during task performance that determines the logic of the transition from one member of an algorithm to another) can be evaluated according to the following formula:

$$L_g = \sum_{i=1}^{k} P_i^l t_i^l$$

where

P_i^l is the probability of occurrence of the *i*th logical condition (*i*th decision-making stage)

t_i^l is the duration of the *i*th logical condition

The next step would be to determine the relationship between the time spent on logical conditions (all steps of the decision-making process) and the time spent on execution of the whole task (the fraction of time for the logical components of work or decision-making process):

$$N_l = \frac{L_g}{T}$$

Here

L_g is the time for performance of logical conditions

T is the time for the entire task performance

This measure characterizes the complexity of the decision-making process during the task performance. There are several other measures that characterize various aspects of the decision-making process. It is important to understand that activity elements do not follow a strict sequence. These elements have a certain logical organization and different probability of appearance in the activity structure. They can be performed in sequence or simultaneously. This was taken into consideration when we calculated measures of task complexity. In Table 8.2, we present all measures of complexity and their psychological interpretation, which can be utilized in the study of computer-based tasks. Practitioners can select, in any particular situation, the more informative measures of task complexity.

TABLE 8.2

General List of Complexity Measures for Computer-Based Tasks and Their Psychological Meaning

Name of Measure	Formula for Calculation	Variables	Psychological Meaning
1	2	3	4
Time for algorithm execution (total time of task performance)	$T = \Sigma P_i t_i$	P_i is the occurrence probability. t_i is the occurrence time of i-th member of algorithm.	Duration of activity during task performance
Time for performance of logical conditions (decision-making)	$L_g = \Sigma P_i t_i$	P_i is the occurrence probability. t_i is the occurrence time of i-th logical conditions.	Duration of decision-making component of activity
Time for performance of afferent operators (sensory–perceptual actions)	$T_\alpha = \Sigma P^\alpha t^\alpha$	P^α is the occurrence probability. t^α is the occurrence time of r-th afferent operators.	Duration of perceptual component of activity
Time for performance of efferent operators (motor activity)	$T_{ex} = \Sigma P_j t_j$	P_j is the occurrence probability. t_j is the occurrence time of j-th efferent operators.	Duration of executive components of activity
Time for discrimination and recognition of distinctive features of task approaching threshold characteristics of sense receptors	$'T_\alpha = \Sigma P_{r'} t_{r'}$	$P_{r'}$ is the occurrence probability. $t_{r'}$ is the occurrence time of r'-th afferent operators, characteristics of which approach threshold value.	Duration of sensory–perceptual components of activity connected with processing of threshold data
Total time for performance in goal area (cognitive component)	$T_{gol} = \Sigma P_{gol} t_{gol}$	P_{gol} is the occurrence probability. t_{gol} is the occurrence time in goal area.	Duration of cognitive components of activity in goal area
Total time for performance in object area (cognitive component)	$T_{obj} = \Sigma P_{objl} t_{obj}$	P_{obj} is the occurrence probability. t_{obj} is the occurrence time in object area.	Duration of cognitive components of activity in object area
Total time for performance in tool area (cognitive component)	$T_{tool} = \Sigma P_{tool} t_{tool}$	P_{obj} is the occurrence probability. t_{tool} is the occurrence time in tool area.	Duration of cognitive components of activity in tool area

(Continued)

TABLE 8.2 (*Continued*)

General List of Complexity Measures for Computer-Based Tasks and Their Psychological Meaning

Name of Measure	Formula for Calculation	Variables	Psychological Meaning
1	2	3	4
Proportion of time for cognitive activity in goal area to total time of task performance	$N_{gol} = T_{gol} / T$	T_{gol} is the total time for performance in goal area. T is the total time of task performance.	Relationship between cognitive activity in goal area and total time of task performance (complexity of goal interpretation or goal formation stage)
Proportion of time for cognitive activity in object area to total time of task performance	$N_{obj} = T_{obj} / T$	T_{obj} is the total time for performance in goal area. T is the total time of task performance.	Relationship between cognitive activity in object area and total time of task performance (complexity of comprehension initial stage of situation)
Proportion of time for cognitive activity in goal and object areas to total time of task performance	$N_{golobj} = T_{gol} + T_{obj} / T$	T_{gol} and T_{obj} are the cognitive activity in goal and object areas, respectively.	Complexity of creation of mental model of situation
Proportion of time for cognitive activity in tool area to total time of task performance	$N_{tool} = T_{tool} / T$	T_{tool} is the total time for performance in goal area. T is the total time of task performance.	Relationship between cognitive activity in tool area and total time of task performance (complexity of executive stage)
Proportion of time for logical conditions to total time for task performance	$N_l = L_g / T$	L_g is the time for performance of logical conditions. T is the total time for task performance.	Relationship between decision-making process and total time for task performance (complexity of decision-making stage)
Time for performance of operators associated with thinking process	$T^{th} = \Sigma P^{th} t^{\alpha th}$	P^{th} is the occurrence probability. t^{th} is the occurrence time for thinking components of activity.	Duration of thinking components of activity largely associated with manipulation of information presented through interface elements

(*Continued*)

TABLE 8.2 (*Continued*)

General List of Complexity Measures for Computer-Based Tasks and Their Psychological Meaning

Name of Measure	Formula for Calculation	Variables	Psychological Meaning
1	2	3	4
Time for performance of operators associated with thinking process based on external features presented through interface elements	$T^{\alpha th} = \Sigma P^{\alpha th} t^{\alpha th}$	$P^{\alpha th}$ is the occurrence probability. $t^{\alpha th}$ is the occurrence time for thinking components of activity whose operational nature is predominantly governed by information presented externally.	Duration of thinking components of activity largely associated with manipulation of information presented through interface elements
Time for performance of operators associated with thinking process based on data extracted from memory	$T^{\mu th} = \Sigma P^{\mu th} t^{\mu th}$	$P^{\mu th}$ is the occurrence probability. $t^{\mu th}$ is the occurrence time for thinking components of activity whose operational nature is predominantly governed by information extracted from memory.	Duration of thinking components of activity largely associated with manipulation of information in memory
Proportion of time for performance of operators associated with thinking process based on external features presented through interface elements	$N^{\alpha th} = T^{\alpha th}/T$	$T^{\alpha th}$ is the time for performance of operators associated with thinking process based on external features presented through interface elements. T is the total time of task performance.	Relationship between thinking process depending largely on external features presented through interface elements and total time for task performance
Proportion of time for performance of operators associated with thinking process based on data extracted from memory	$N^{\mu th} = T^{\mu th}/T$	$T^{\mu th}$ is the time for performance of operators associated with thinking process based on data extracted from memory. T is the total time of task performance.	Relationship between thinking process depending largely on data extracted from memory and total time for task performance
Proportion of time for thinking components of activity to total time of task performance	$\Delta T\text{th} = T^{th}/T$	T^{th} is the time for performance of operators associated with thinking process. T is the total time of task performance.	Relationship between thinking components of activity and total time for task performance

(*Continued*)

TABLE 8.2 (*Continued*)

General List of Complexity Measures for Computer-Based Tasks and Their Psychological Meaning

Name of Measure	Formula for Calculation	Variables	Psychological Meaning
1	2	3	4
Proportion of time for logical components of work activity depending largely on information selected from long-term memory rather than external features presented through interface elements	$L_{ltm}=l_{ltm}/L_g$	l_{ltm} is the time for logical components of activity whose operational nature is predominantly governed by information retrieved from the long-term memory.	Level of memory workload and complexity of decision-making process
Proportion of time for retaining current information in working memory	$N_{wm} = t_{wm}/T$	t_{wm} is the time for activity related to storage in working memory of current information concerning task performance.	Level of workload of working memory
Proportion of time for discrimination and recognition of distinct features of task approaching threshold characteristics of sense receptors	$Q = T_{\alpha}/T$	T_{α} is the time for discrimination and recognition of different features of task approaching threshold characteristics of sense receptors.	Characteristics of complexity, sensory, and perceptual components of activity
Proportion of time for efferent operators (motor activity)	$N_{mot} = T_{ex}/T$	T_{ex} is the time required for efferent operators (motor activity).	Relationship between motor components of activity and total time for task performance
Proportion of time for afferent operators (sensory–perceptual activity)	$N_{\alpha} = T_{\alpha}/T$	T_{α} is the time required for afferent operators.	Relationship between sensory–perceptual components of activity and total time for task performance
Scale of complexity a. Algorithm b. Member of algorithm	Xr -level of complexity (1,2,..5)	Level of concentration of attention during task performance (1, minimum concentration; 5, maximum).	Level of mental effort during task performance and performance of different elements; unevenness of mental effort and critical points of task performance

Note: Proportion of time refers to the ratio of the time of the element to the total time required.

9

Complexity Evaluation: Practical Example

9.1 Basic Principles of Morphological Analysis of Computer-Based Tasks

As was demonstrated in Chapter 7, morphological analysis is the basis for further application of quantitative design methods. Activity is a process and the main purpose of morphological analysis is to describe the structure of activity as it unfolds in time. Changes in human–computer interface characteristics influence the methods of task performance in a probabilistic manner. Each method of performance is associated with a specific structure of activity. Therefore, through analysis of the relationship between structure of activity during task performance and interface characteristics, it is possible to evaluate complexity of computer-based tasks and usability and reliability of human–computer interfaces. Precisely developed units of analysis and formalized methods of activity description facilitate the creation of formalized models of activity during task performance. Morphological analysis that includes algorithmic and time structure description can also be used as an independent from quantitative stage of analysis method of study.

The first step of morphological analysis of activity is extraction of cognitive and behavioral actions that are involved in task performance. Visual information is the main source of information in human–computer interaction (HCI) system. Most computer-based tasks can be compared with solving chess problems on a computer screen with a chess board. Analysis of chess players' thinking in the applied activity theory is carried out applying Yarbus' method (Pushkin, and Nersesyn 1972; Telegina, 1975, Zinchenko and Vergiles, 1969). The development of the study of eye movement followed mainly by way of eye movement registration. Methods of interpretation of eye movements remained almost unchangeable from that time on. A new method of eye movement analysis has been developed in the systemic-structural activity theory (SSAT) framework. We suggest utilizing eye movement data for the analysis of computer-based tasks. This is a new method that offers basic principles of extracting cognitive and behavior actions during performance of computer-based tasks.

Some general principles of extraction of cognitive actions based on eye movement analysis are presented in Section 6.2. Eye and mouse movement data are very useful in the analysis of any task that utilizes visual information. By combining obtained data with observation of subjects' behavior and verbal protocol analysis, analyzing preceding and subsequent actions of subjects, a specialist can extract various cognitive actions including thinking actions. Such combination of the earlier discussed methods is a reliable source of information about not only motor but also cognitive activity. The obtained data are critically important for farther algorithmic description of task performance where cognitive components of activity play an important role in completing a task at hand. The obtained data are also very important for complexity evaluation of task performance. Increasing the number of fixations and their duration correlates with increasing task complexity. Eye and mouse movement data are very useful for the description of preferable strategies of task performance. HCI activity is very variable. In the case of manufacturing operations, a standard operating procedure is often fixed. In computer-based tasks, strategies of task performance are much more flexible. Therefore, eye and mouse movement analysis can identify multiple versions of algorithmic description of task based on actual user performance. However, only the most representative strategies of task performance should be described algorithmically. Moreover, the same probabilistic algorithm has multiple versions of its realization. Thus, we can select the most optimal strategies of task performance and describe them algorithmically. Selection of the most representative strategies of task performance can be accomplished by analyzing mechanisms of activity self-regulation.

The algorithmic description of a task should be developed keeping in view all constraints of operating with the interface and the requirement of the task. We recommend distinguishing between real, ideal, and optimum algorithms of task performance. Real algorithms of task performance describe data obtained based on experimental analysis of eye and mouse movement. Ideal algorithm describes the best strategy of performance. Usually, real strategies can only approach the best one. In the design process, we focus on the optimal versions of the algorithm that approach the ideal one. An optimal algorithm can be more precisely defined based on its complexity evaluation. However, sometimes we can utilize less precise criteria. An algorithm with the least number of clicks or the least number of mental actions can be used as such criteria. It is expected that using less clicks (motor actions) or less mental actions will result in a lower number of members of an algorithm. For our study, we developed a model of the hypothetical computer-based task (Bedny et al., 2008). The position of the interface elements and their symbolic coding system are depicted in Figure 9.1.

Table 9.1 presents symbols description of different task elements that are used in Figure 9.1.

The triadic schema of activity includes the following elements of activity: Subject → Tool → Object → Goal → Result (Bedny and Harris, 2005). This schema utilizes the Vygotsky (1978) concept of a psychological tool. Direct interaction of the participant with the object does not imply an absence of a

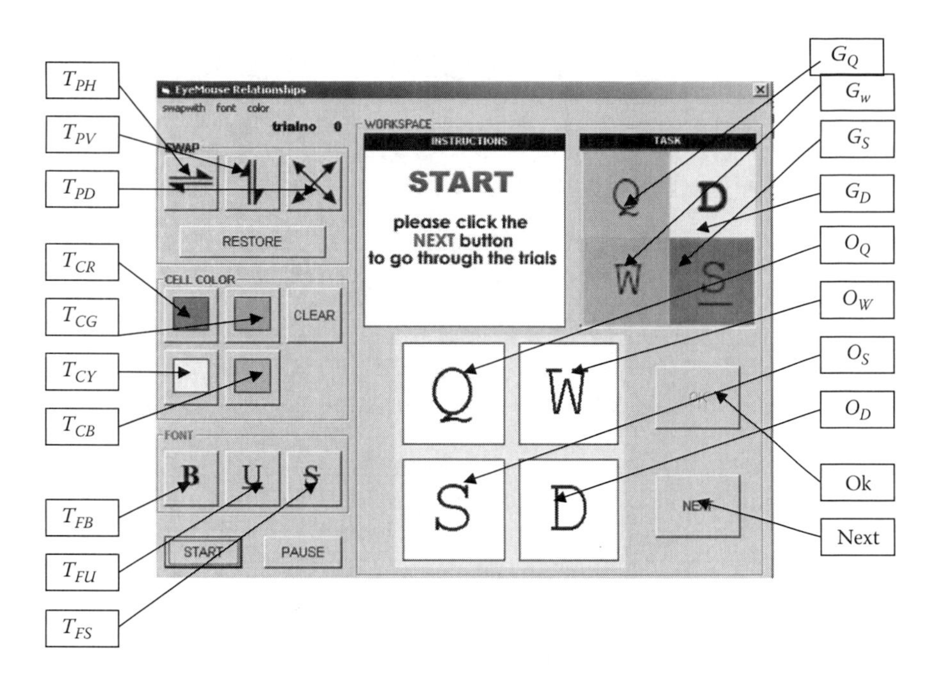

FIGURE 9.1
Position of the interface elements and their symbolic coding system.

TABLE 9.1
Symbols Description of Different Task Elements That Are Used in Figure 9.1

Task	Tools									
Types	Position			Color				Format		
Elements	Horizontal	Vertical	Diagonal	Red	Green	Yellow	Blue	Bold	Underline	Strike
Symbols	T_{PH}	T_{PV}	T_{PD}	T_{CR}	T_{CG}	T_{CY}	T_{CB}	T_{FB}	T_{FU}	T_{FS}
Task	Objects				Goal					
Elements	Q	W	S	D	Q	W	S	D		
Symbols	O_Q	O_W	O_S	O_D	G_Q	G_W	G_S	G_D		

mediating tool. Rather, in such cases, the subject employs *internal* (mental) tools. Based on this schema, the following areas of the screen were identified:

1. *The goal area.* This is the pattern of the desired arrangement the user must impart to the objects in order to successfully accomplish the task. This is the desired result of activity during task performance that in this study is presented in an externalized form. The goal of the task should be distinguished from goals of actions or goals of subtasks. In our example, goal of task is presented externally throughout the trial.

2. *The object area.* This area consists of the elements whose state should be manipulated to achieve the final arrangement given in the goal area.
3. *The tool area.* This area of the interface display consists only of elements or icons, which could be clicked on to impart the desired feature and required arrangement.

Tools presented on the screen are designed as icons that are used to perform specific functions. These externalized tools are presented in material form. They should be distinguished from internal, mental tools.

The features of elements of the situation that were manipulated in the experiment were (a) position, the location of the letters with respect to each other; (b) color, the color of the cell containing the letters; and (c) the format of the letters. Element D in object area has yellow background, Q has blue background, W has green background, and S has red background. In the tool area, the top left cell has a red color and the right cell has a green color. The bottom left cell has a yellow color and the right cell has a blue color. The task was to impart features to the letters so that a given arrangement of elements in the object area according to the goal can be reached. Therefore, alphanumeric characters of the object area that were altered in position, color, and shape according to the goal of the activity are the objects in this task. Therefore, in this task, the goal was presented externally in the goal area and modified for each new task. Any number of actions could be performed by the subject. They were only given an instruction to reach the final arrangement. According to activity terminology, the main focus of the task was to alter the features of the objects using available tools to reach the presented goal. In their initial state, the objects had no specific color or format. The participants had to follow a certain sequence to impart the features of the objects. There is an optimal sequence of steps to perform this task. Any deviation from this sequence would result in an excess number of motor actions (clicks) and hence would reduce the efficiency of the performance. The participants knew initially that there is a preferable sequence of actions but were not familiar with this sequence. They were instructed to figure out the most efficient method of performance while performing the task. Moreover, some sequences of actions can be incorrect. Therefore, participants have some constrains in the possible strategies of task performance. These constrains are not visible to the user just by looking at the interface. The user finds out the constraints only by interacting with the interface. The participants have one final goal presented on the screen that they have to achieve. However, the participants cannot achieve this goal at once. They need to break this task into subtasks. Each subtask has its own subgoal. Participants formulate subgoals independently. These subgoals can change based on the feedback the participants receive while going through the task. Each action provides participants with feedback, through which the participants gradually start to understand how the experimental software works. Participants can change their subgoals when they encounter an undesired output during task performance. Therefore, there is a possibility to explore different strategies of task performance. In the case of computers and computer-based tasks, an initial phase of exploration is often evident (Carroll, 1987). Carroll's research demonstrated that users are motivated to explore the interface when they are confronted with certain task difficulties. From this description, one can see that it is a problem-solving task. Hence, task performance involves not only memorization of instructions but also thinking and decision-making. This makes the task a rather complicated one.

All participants were recruited from the New Jersey Institute of Technology student population. Most of them were graduate students. All of them had computer experience. Participants were given financial incentive to complete the experiment. Eight participants completed 128 tasks (16 tasks each) on this interface. We had eight participants to ensure that we did not miss any major aspects of this task performance and possible strategies used by the participants. In the real production settings, one to three representative users would be sufficient for such analysis. The main purpose of this analysis is to gather the raw data for further development of the analytical models of the operator's activity during task performance. There are no experimental and control groups for further comparison of their results because the performed analysis is based on theoretical models of activity. Analysis and comparison of the data were performed by utilizing theoretical models of activity during task performance. These analytical procedures enable a practitioner to abandon or discard incorrect solutions and develop a more appropriate one based on the comparison of different models of activity performance. In the design process, experimental data, just as any other data, are only a supplemental material for the creation of design models. The analysis and comparison of different models of activity eliminate the need to use comparative analysis between experimental and control groups with the testing of statistical hypotheses.

Analysis of the task's video and the observation showed that the participants use different strategies to obtain the required goal. It was discovered that 55% of the participants completed the task in the following order: changed the position of the letters first, then introduced color, and changed the format as the third feature. There are two other basic strategies. Efficiency of all three strategies is approximately the same. In this chapter, we analyze the most preferred strategy only. The main purpose of this study is to demonstrate the possibility to develop models of human activity during HCI task performance. All strategies are described in the same manner. It is not possible to describe models for all three strategies because of length limitations of the work. The presented method allows us to combine algorithmic description of different strategies into one general algorithm to evaluate in general the task. It also allows us to compare various strategies if it is necessary. For this purpose, we utilized not only deterministic but also probabilistic algorithms.

The goal of the task was to impart the features to the different elements of objects so that the given arrangement is finally reached. Each participant had to select an appropriate sequence of actions to impart the features to the objects. The following methods of data collection were employed in the experiment at the qualitative stage.

1. Eye movement analysis by using eye tracking system and video recording during task performance.
2. Mouse event logging in terms of both the mouse movements and the actions carried out during the task.
3. Debriefing and discussion with the subject and an expert analysis of eye movement scan path.

The combination of these methods and comparison of obtained data enhances the process of the activity description and allows us to develop models of activity during HCI task performance. The reason for the study of eye movement in combination with motor activity derives from the principle of *unity of cognition and behavior* in activity theory. The utilization of this principle helps one interpret the cognitive actions by using the observation of the eye movements—and following that, motor movement—and by evaluating the duration of the eye fixation data. Mouse movements help us to understand the reason behind eye movement during task performance. The purpose of this study was not to develop new methods of eye and hand movement registration but to find a new method of their interpretation. Therefore, traditional methods of eye movement and hand movement registration were used in this study.

The software needed to capture mouse events and coordinate data for the interface has been designed. The collected data were stored in a separate log file. Time of performance in terms of the sampling rate of the coordinates (which was kept at 10 Hz) was based on experimental data of the fastest movement possible with the mouse. The sampling rate was kept at 100 ms to catch every movement made by the user.

The eye movements were analyzed using the dispersion threshold (Salvucci and Goldberg, 2000). The dispersion threshold is measured either in angles or visual arc or in pixels traversed by the point-of-regard data. This is the amount of variation in the coordinates of the point of regard, which will associate a given range of points to a saccade or fixation. The total number of fixations along with the fixation duration is taken as the gaze at the particular area of interest. Based on the different areas of interest and the visual angle of these areas, the average dispersion threshold was calculated based on 200 saccades obtained from video observation and analysis. This value was then used for the algorithm as the dispersion threshold based on transformation, duly applied for normalization of the coordinates for all participants. The duration threshold defines the minimum length of the duration of fixation for which it can be qualified as fixation. As per Yarbus (1969) and Salvucci and Goldberg (2000), the minimum length of the duration of fixation is given a standard of 100–200 ms. Considering the nature of the task, the duration threshold was defined as 100 ms and a 16.67 ms interval was chosen as an appropriate one to understand user's movements and gazes in different areas of the screen.

The equipment that has been used for this study was obtained from the Visual Research Laboratory of the Biomedical Engineering department. A RK 426PCI corneal reflection Eye Tracking system (ISCAN Inc., Burlington, MA) was used in the study. The software used to control the pupil/corneal reflection tracking system, also from ISCAN, is known as the Eye Movement Data Acquisition software. The point-of-regard coordinates of eye movements were recorded. There were inherent difficulties with the analysis of the point-of-regard coordinates because of the equipment restrictions. As a result, eye movement data were also obtained via the analysis of the point-of-regard video. An industrial-grade VCR was used to recode the whole session from the Scene camera of the eye movement registration equipment. Two PCs loaded with Microsoft Windows 98 operating system were utilized, and both of them had the Microsoft Visual Basic development system. The first computer had the necessary software for running the eye registration equipment (experimental workstation), and the other had the software designed for the experiment in Visual Basic (participants' workstation). As can be seen in our study, we utilize the notions of tool, goal, and object areas of the screen. These terms have been associated with corresponding activity theory terminology. Studying eye movement in these areas might provide us the understanding of the performance strategies of the users and hence make an assessment of the usability of the interface in the context of task performance based on activity theory data. Thus, areas of interest on the interface associated with activity theory notions of task performance. Sometimes intermittent goal formation stages of activity associated with specific areas of the interface can be extracted for analysis as goal area on the screen.

9.2 Extraction of Cognitive and Behavioral Actions from Eye and Mouse Movement Data

In this chapter, we describe general principles of cognitive and behavioral action extraction from flow of activity by utilizing eye movement data and developed in the SSAT principle of action extraction. In the next chapter, we applied this approach for extraction of actions in computer-based task described before (see Section 9.1). Action emerges as the primary unit for the morphological analysis of activity (Bedny and Karwowski, 2003). Hence, the continual flow of cognitive and behavior activity should be divided into individual units and presented as a structure. Because users rely on visually perceived interface elements on the screen, eye movements should reflect the participant's activity strategies during task performance. Data obtained by monitoring the eye movement can significantly enrich the analysis of the user's strategies and enhance the extraction of cognitive and motor actions from the continual activity flow. This section provides an SSAT-based foundation for the extraction of cognitive and behavioral actions by utilizing eye and mouse movement data.

During the saccade, an object can only be detected, not recognized (Just and Carpenter, 1976; Yarbus, 1965). Every saccade is contingent on the preceding cognitive process, which is assumed to be a portion of the preceding gaze. This gaze includes among other processes the program of performance before the saccade (Viviani, 1990; Yarbus, 1965, 1969). Not only perceptual but also other mental or cognitive processes are performed during a gaze or series of fixations (Pushkin, 1978; Tikhomirov, 1984; Yarbus, 1969; Zinchenko and Vergiles, 1969). Hence, considering eye movement and gaze time in respective areas gives us the opportunity to study cognitive action durations for the particular activity. According to the activity theory, sensory-perceptual process also includes a decision-making stage at the sensory-perceptual level (Bedny and Meister, 1997). Viviani (1990) suggested that there are three processes that take place during an eye fixation (250–300 ms) before a saccade. These three processes include the analysis of the visual stimulus in the fovea field, the sampling of the peripheral field, and planning of the next saccade.

The following rules are developed based on this study. These rules allow us to separate the eye movement data into movement and gaze pairs:

1. Because saccades are very quick, it is not possible to execute complex mental operations during such short durations.
2. A mental operation performed during a gaze consists of different operations associated with receiving information, interpretation, decision-making, and so on.
3. The final stage of such a gaze also includes setting a performance program for the next saccades. This is the point of separation of two corresponding actions. As a result, one complete eye movement and one complete gaze duration that follows this eye movement is roughly estimated as one complete action. However, the type of action is distinguished on the basis of descriptive analysis and the duration of the action as well as its relevancy to the task at that point in time.
4. In cases where the gaze durations are longer and include multiple fixations, considerations have to be given to the following three aspects: the type of eye–cursor movement at that point, the actions preceding the gaze, and the action following the gaze.

Using these three aspects, the type of action can be estimated fairly accurately.

Therefore, the eye movement registration should be combined with the eye cursor movement registration. The summation of the eye movement and the associated gaze time provides the total approximate time of the cognitive action. However, if the participant performs successive perceptual actions that are involved in extracting information from unfamiliar stimuli that requires the creation of perceptual image, a series of eye movements and gaze pairs can be integrated into one complete perceptual action. In such cases, their components are then considered as operations for this complicated action. In this study, the complex image features and image formations are not encountered, and as a result, an eye movement–gaze pair is used as one complete action. If, for example, the eye movement time is 100 ms and the gaze time is 250 ms, then the total time of action is given by the movement time plus the gaze time, which in this case is 350 (100 + 250) ms. During this time, the user has to locate the tool, then select it mentally and execute the action while gazing at the particular area.

Thus, our method of eye movement interpretation has an important difference from traditional method. We do not use cumulated scan path for eye movement interpretation.

We divide the cumulated scan path of the eye movement into segments that correspond to the individual cognitive actions. One of the most important criteria for division cumulative scan path into segments is the goal of the individual actions.

The symbolic system, which is presented in Figure 9.1 and Table 9.1, is used further for the development of an action classification table.

The action classification table (refer to Table 9.2, fragment that was completed after performance of the first click) is based on the division cumulative scan path into segments and the qualitative analysis of eye movement in the corresponding period of time.

The table shows only one image and an adequate for this image trajectory of eye movements. The full table contains 12 images with their corresponding eye movements. Each image demonstrates eye movements between two clicks. The table displays the most representative strategies of task performance.

The analysis of dwell times that are associated with a particular area on the screen and its corresponding clicks gives us the opportunity to relate these dwell times with the duration and content of the mental actions. A qualitative analysis of each cognitive action involves the following steps: The task is divided into meaningful, logically completed segments of activity; a selected segment of activity is divided into small elements, such as cognitive and motor actions; the goal of each action is determined; the tools that are utilized during this action performance are defined; the transformation of the object of activity during this action performance and how this transformation (result) corresponds to the goal of the action is considered; the purpose of actions that are performed before and after

TABLE 9.2

Action Classification Table (Fragment Before the First Click)

1		2	3	4		5	6	7
Eye Move and Final Position		Activity between Successive Mouse Events (clicks) Mental/Motor Actions Involved	Mouse events	Time (ms)		Total Action Time ($a + b$)	Classification of Actions	Scan Path Generated/Duration
From	To			*a*. Approx. Eye Movement Time to Reqd. Position	*b*. Approx. Dwell Time at Position			
Start	G_Q	Goal acceptance and formation and creation of subjective model of situation; selection of object (OD) for subsequent subtask execution (includes simultaneous perceptual actions, with explorative thinking; comparison of object and goal in relation to the program of performance)		150	180	330	Simultaneous perceptual actions	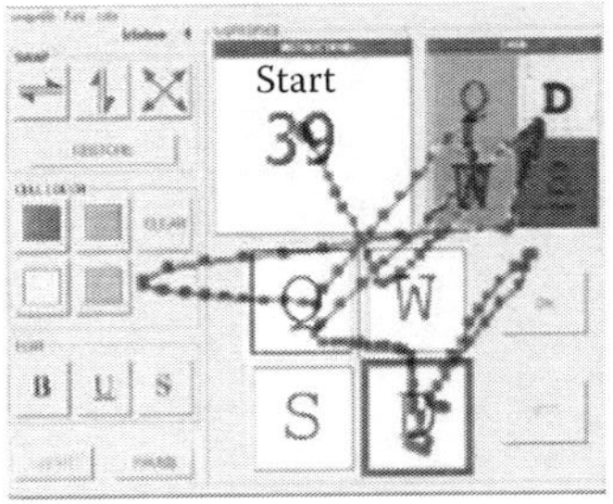
G_Q	O_Q			150	220	370	Simultaneous perceptual actions	
O_Q	T_{CB}			180	150	330	Simultaneous perceptual actions	
T_{CB}	G_S			180	220	400	Thinking action based on visual information	
G_S	G_D			150	190	340	Thinking action based on visual information	

G_D	O_Q			210	220	430	Thinking action based on visual information	
O_Q	O_W			150	330	480	Thinking action based on visual information	
O_W	O_D			150	190	340	Thinking action based on visual information	
O_D	G_D	Decision on program of performance and motor action based on decision		210	190	400	Thinking action based on visual information	
G_D	O_D	Perceptual action with motor action and thinking based on program of performance	Click object element *D* with mouse	210	630	840	Decision-making action at sensory perceptual level, with simultaneous motor action	
O_D	T_{PV}	Motor action of eye and mouse along with selection frcm choice of tools		210	220	430	Simultaneous perceptual action with motor action	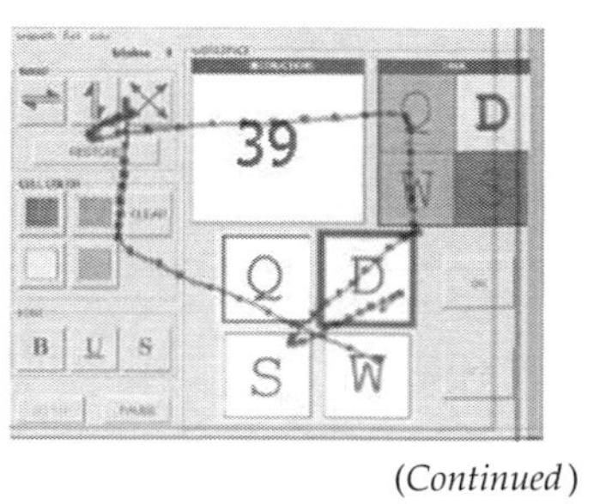

(*Continued*)

TABLE 9.2 (***Continued***)

Action Classification Table (Fragment Before the First Click)

1		2	3	4		5	6	7
				Time (ms)				
Eye Move and Final Position		Activity between Successive Mouse Events (clicks) Mental/Motor Actions Involved	Mouse events	*a*. Approx. Eye Movement Time to Reqd. Position	*b*. Approx. Dwell Time at Position	Total Action Time $(a + b)$	Classification of Actions	Scan Path Generated/Duration
From	To							
T_{PV}	G_D	Eye move to goal area with mouse stationary at tool. Use of peripheral vision for mouse control while focus on the goal area	Click vertical positioning tool	150	420	570	Simultaneous perceptual action; decision-making action during visual assessment; Motor action;	
O_DOK	OK finish	Final comparison of goal and object area and acceptance of completion of task	Finish	120	390	510	Simultaneous perceptual action with motor action.	

the considered action is analyzed; and the duration of the action is determined. As a final step in this analysis, the goal of the motor action that follows the considered sequence of cognitive actions is defined. All of these steps of analysis help us infer what kind of cognitive actions are performed by the participant. Therefore, the basic principles to *penetrate the user's mind during task performance* and uncover mental components of activity are as follows:

1. Break down complex unobservable cognitive processes into more elementary mental actions or operations.
2. Compare the data obtained for eye movement and motor movement registration (based on the principle of unity of cognition and behavior).
3. Use qualitative methods of study to analyze eye and motor movement registration.
   a. Perform concurrent or retrospective verbal protocol analysis during or after task performance.
   b. Cross-examine an expert as to how he or she typically performs task or solves problems; cross-examine a novice about his or her task performance.
   c. Compare novice–expert differences in task performance (differences in strategies, difficulties, typical errors, etc.). Change the conditions of the task performance and measure task performance in new conditions, and question the users about performance in different conditions.
   d. Introduce new elements into task performance or eliminate some of them, increase or decrease the speed of performance, change sequence of task elements, and so on.

Combination of scan path images and dwell times were used and associated with hand movement data. The usage of the eye movement and hand movement data is based on the unity of cognition and behavior principle in the activity theory (Bedny et al., 2001).

The rationale for using mouse events was the fact that every motor action in a computer task is based on the preceding mental processing. To simplify this procedure, the eye movement images were associated with the division of the task into segments that included a logically completed set of actions. Other issues influencing this subdivision were easiness of the eye movement interpretation and the notion of a completeness of the logically related subtask elements. Hence, the method requires the association of eye and mouse movements with the corresponding elements of activity. The duration of dwell time is also an important source of information for classification of mental components of activity. A brief summary of the comparative analysis of eye movement data in cognitive psychology and activity theory is presented next.

The traditional method of eye movement registration extracts the following basic information from the scan path: the frequency of fixations as a measure of the importance of a display, fixation duration as an attribute of the difficulty of information extraction, and the pattern of transitions between display elements as a measure of display efficiency.

In SSAT, the task is divided into meaningful, logically completed segments of activity during which the subject achieves a particular subgoal of the task. Mouse clicks can be used for such divisions. The eye movement scan path related to a segment of activity should be selected. This makes such segment of scan path more understandable during eye movement analysis. Extraction of actions is performed based on the analysis of the considered fragment of eye movement scan path. The following data allow the determination of the cognitive actions performed by the user.

1. Extraction of actions
   a. Each saccade and gaze is considered as cognitive action.
   b. Actions are classified based on dominance in a particular moment cognitive process.
   c. Not only correct, but also incorrect actions should be extracted during this analysis.
2. Classification of actions by
   a. Analysis of logical organization of actions
   b. Analysis of the action purpose
   c. Relation of gaze to visible elements on the screen
   d. Purpose of the following action, and particularly the motor clicks
   e. Duration of the gazes and their qualitative analysis
   f. Analysis of debriefing the subjects and comparison of their reports

The eye movement registration and frame-by-frame analysis of video are related to the detailed, microstructural-level analysis of activity and help the researcher uncover how the user comprehends and interprets the meaning of the task elements and the situation (Bedny and Karwowski, 2004c).

In our eye movement analysis, we divide cognitive activity into small elements, continually relate these elements to the goal, tool, and object areas, and strive to find a relationship between external and internal components of activity. Hence, our classification of cognitive actions is performed with a high level of precision. All actions in Table 9.2 are described according to the SSAT standardized action classification system (see Chapter 6), which is utilized to develop the tabular presentation of activity elements shown in the table. In Section 9.3, we present principles of cognitive and motor actions' analysis and their classification that derives from the data presented in Table 9.2.

9.3 Action Classification Table Analysis

The action classification table is the first formalized method of activity description. In this study, the development of Table 9.2 is the first stage of morphological analysis of activity during the performance of the considered task. This table represents the model of activity during the actual task performance by the most representative user. In some cases, a comparative analysis of strategies utilized by several users is required. Unwanted experimental action classification tables should be removed from analysis. Based on the analysis of the most representative strategies, a general algorithm for the task performance should be developed. Such algorithm would describe the optimal strategy of task performance. Ineffective actions should be eliminated in the optimal algorithm. In the real strategies of task performance, a user can utilize incorrect cognitive and motor actions. In such cases, after their discovery, a performer uses corrective action. In the table where these actions are listed, a specialist would mark them in parentheses as "wrong or abandoned action" or "corrective actions" or designate them by bold lines. Later incorrect actions are analyzed in terms of their causes and effectiveness of their corrections.

Column 1 in Table 9.2 represents the start and the end position of the eye during one complete movement and dwell, which changes the focus of the eye. Similarly, it presents the motor components of activity. Column 2 describes activity elements between clicks. These elements of the task have specific subgoals. For example, the subgoal of the first fragment of the task is "Selects element D in the object area and clicks on it to switch the positions of objects D and W." This goal is not given to the participant in advance. It is formulated by the participant while he or she is exploring the situation. Column 3 represents mouse actions in the same time line. Column 4a presents eye movement time from the start to the final position according to the start and end position in column 1. Column 4b demonstrates dwell time in the required position. Column 5 reflects total action time for cognitive and behavior actions. Column 6 demonstrates standardized description of actions according to the existing SSAT language of description. Column 7 has images of scan path generated by eye movements during the performance of a particular fragment of the task. The comparison of columns 6 and 7 helps us to understand how cognitive elements of activity are associated with the eye movement scan path.

The association of eye movements with the position of the interface elements is based on approximate position of the eye to the nearest element on the screen. The previously explained symbols that represent the interface elements for the designation of the start and the end position (see Figure 9.1 and Table 9.1) have been used. For example, in column 1, the first transition represents the movement of the eye from the *start* position to the element G_Q (goal area–final state of Q). So the total time for the movement of the eye from the start position to the position of G_Q is 150 ms (given in column 4a). The dwell time at the end position (i.e., at G_Q) is given in column 4b and is equal to 180 ms. The summation of the time of these elements, represented in column 4a and 4b, is given in column 5. It reflects the total action time, because the total action time is given by the sum of the gaze time at the particular area or the element and the movement time to the particular element. The performance time of the first action (from the start position to the G_Q) is 150 + 180 = 330 ms. According to column 6, during this period of time, the participant performs simultaneous perceptual actions. In our analysis of task performance, the eye movement and the motor movement registration are the most important methods of study. The sequence of gazes and movements uncover the logical organization of mental actions. The simultaneously performed motor actions are designated by the mouse event data and are used to classify these actions.

Let us consider the first fragment of the task that is associated with Image 1 in column 7. This image demonstrates that after performing a number of cognitive actions, the participant selects element *D* in the object area and clicks on it. As a result, this element is activated and its frame is highlighted. This means that the participant wants to shift this element into another position. In this case, the participant spends most of his or her time on O_D, O_W, O_Q, and G_D and finally selects O_D. The sequence of an eye movement along with time of dwell suggests that the participant is more inclined to act on object *D* and might consider switching the positions of objects O_D and O_W.

The following is the sequence of actions: The eye moves from the start position to the position of G_Q (goal area element *Q*). Dwell time of this movement is 180 ms, and the total time for the action is 330 ms. At this stage, the participant just wants to identify the goal of the task. Therefore, this is a perceptual action. According to the existing SSAT action classification system, it is a simultaneous perceptual action. In visual field $\alpha \approx 10°$, the participant can simultaneously perceive four to six elements. Hence, the participant can perceive not only one letter but all four.

The next eye movement begins from element G_Q (goal area of letter *Q*) to O_Q (see column 1 and column 7, where O_Q means object area of letter *Q*). This is the first shift of eyes into the object area. The duration of the eye movement is 150 ms and the duration of the dwell time is 220 ms. Therefore, the total time of this action is 370 ms. The participant attempts to receive some general information. He moves his or her eyes to the tool area (see column 1 and Image 1), eyes move from element O_Q to element T_{CB} (tool element, color blue, see Figure 9.1). This is also a perceptual action.

At the next stage, eyes shift from the tool area to the goal area again (from T_{CB} to *Gs* where the last symbol means goal area of letter *S*). The purpose of this movement is not only for receiving information (perception). A participant starts to pay attention not to the perceptual features of the situation but to the relationship between the elements of the situation. The relationship between elements of the situation and the task goal is not a perceptual feature of the task but rather is the feature of the task that requires involvement of the thinking process that is performed based on visual information. The duration of the eye movement increases as a result (time of performance of this action is equal to 400 ms). According to the existing classification system of actions, it is a thinking action that is performed based on visual information (Bedny et al., 2000). This is an example of the simplest thinking action. Eyes move from element Gs to element G_D. Similarly, the purpose of this eye movement is to find out the relationship between elements of the situation. Hence, it is also a thinking action performed based on visual information (duration of this action is 340 ms). The next eye movement is involved in the analysis of the relationship between elements G_D and O_Q. This is a thinking action based on visual information. The duration of this action is 430 ms. Similarly, comparisons of positions of elements O_Q and O_W (time performance 480 ms) and then positions of O_W and O_D (time performance 340 ms) are examples of the thinking actions. All these actions are involved not simply in perception of the information but also in analysis of relationships between elements. The relationship between elements is not a perceptual feature of the situation. In our task, these actions are examples of the simple thinking actions that are performed based on visual information. It is interesting to pay attention to the fact that in spite of the short distance of eye movements, the average duration of these actions is more than the duration of perceptual actions. Such actions are more complex than perceptual.

The next thinking action is involved in the formation of program of performance. The eyes move up closely to the goal area. The subject does not need to clearly consider the goal area with element O_D. He remembers the position of the element. In this case, peripheral vision is sufficient. Visual information is just used for the confirmation of correctness of the formulated program of action. The duration of this thinking action is 400 ms. During this time, the subject performed a mental action associated with the production of the program of action performance. This is a thinking action that is performed mainly in the mental plane. This is why the subject simply moves the eye up closely to the goal area.

Let us analyze the last two actions that are associated with the analysis of the first image.

One action is related to the decision to implement the program, and the second motor action is associated with the activation of element O_D (the subject moves the pointer to element O_D and clicks on it). When the subject presses element O_D, it is highlighted by a bold line (see Table 9.2, Image 1, element D). Both actions partly overlap in time. Before activating element *D* in the object area, the participant should perform a decision-making action due to the choice of position in the tool group.

Hence, mental action before the clicking of a vertical positioning tool can be classified as a decision-making action that has been based on visual information. According to the existing system of action classification, this is a simple decision-making action at a verbally thinking level. During this decision-making, the participant starts to move the mouse to the tool area. The scan path does not give the total picture about possible cognitive actions. The dwell time defines how much importance or attention each place requires. The longer the dwell time is, the higher the probability that the thinking process is involved in the task performance at that period of time. The information about the duration of eye movement, dwell time, and cognitive action in general can be obtained from columns 4a, 4b, and 5 that are adjacent to Image 1. In general, it is evident that the subject selects O_D and intends to switch the positions of O_D and O_W.

We cannot consider all the details of eye movements associated with the second image. Moreover, our table reflects only 3 images from 12 (we present only fragment of Table 9.2). Therefore, we briefly consider eye movements associated with this last image. The last image (Image 12) demonstrates that the subject performs final comparison of the goal and object areas and performed motor action that requires pressing the OK button.

The eye movement data in column 7 can be used for the functional analysis of activity during task performance. The basis of functional analysis is the study of mechanisms of activity self-regulation. This stage of analysis is dedicated not so much to separate actions but rather to generalize strategies of performance and their relation to such functional mechanisms as a goal, subjectively relevant task conditions (dynamic mental model), and formation of a program of task performance (Bedny and Meister, 1997). For example, analysis of Image 1 demonstrates that the participant first attempts to receive information about the goal and then about the object area. The eye scan path suggests that the comparisons between the final required state (goal) and the initially given state in the object area is taking place. Thinking actions are required to evaluate the goal in a more specific manner through comparison of subjectively accepted task requirements (goal) with an initial state of the situation. Participants attempt to develop a mental picture or model of initial situation of the task based on comparison of the goal and object areas. Eye movement registration demonstrates how a mental model of a situation is developed. There is a considerable dwell time on the object and the goal areas at the first stage of the task performance (Image 1). In the next image (Image 2), the focus shifts to the tool area to develop a plan of execution and to choose the tools that fit the corresponding actions. Therefore, the general strategy of task performance includes interpretation and acceptance of the goal, development of a mental model of the situation based on comparison of the goal and object areas, evaluation of the tool area, and development of the plan of actions accordingly. By shifting eyes into the goal area, a participant can simultaneously receive information about all four elements located in the goal area (perceptual action). However, while switching to the thinking actions, the participant's attention is concentrated on the individual elements of the goal area and their functional relationship to the elements in the object area. Hence, Image 1 in Table 9.2 demonstrates that a participant at this stage does not simply receive information about a goal and an object area but rather attempts to find out the functional relationship between different elements and comprehend the final goal of performance and the initial state of the object at this stage. Then, a participant formulates a subgoal of the task. So the goal cannot be considered simply as the end state to which behavior is directed—an objectively given goal is subjectively interpreted and accepted. During the functional comparison of the final goal elements with the object area elements, a participant formulates a more specific goal, evaluates initial state, and develops a mental model of the situation. Based on these mental actions and operations, the decision is made about what the intermittent goal should be. Such intermittent goal description and classification requires morphological analysis.

An analysis of the aforementioned eye movements in the action classification table (Table 9.2) demonstrates that during development and interpretation of this data, some combination of functional and morphological analysis elements is used. A functional analysis involves paying attention to eye movements in different areas of interest and strategies of task performance. When a specialist pays attention to action classification, it is a morphological analysis. These methods are often difficult to separate from each other. A functional analysis that is derived from analysis of activity self-regulation helps to discover preferable strategies of task performance.

The mental models demonstrate that the participant disengages himself or herself from such features of the object as color and symbol formation and decides to transform a situation according to the space criterion. This stage in task performance is shown in Image 2 where the scan path reveals more transitions of the eyes to the tool area G_D–O_D–T_{PV}. Here, the participant is more concerned with the completion of the task and decides on a course of actions based on the importance of the assessed tools.

Eye movement registration helps us to discover wrong actions and understand their causes. For example, during analysis of image 10, it was discovered that the subject perform strikethrough action and this action was not correct. Instead of using the underline tool, the subject erroneously used the strikethrough tool. Thus, an error is reported in terms of using the underline tool where the participant mistakenly uses the strikethrough tool.

When the participant issues the commit command that is the OK button (GS-Feedback), the error is subsequently detected and the participant's focus immediately shifts to the object area. Here, most of the dwell time is spent once again in the object and goal areas to detect the difference in the letters' positions. Finally, once the difference is detected, the participant easily furnishes the task using the complete feature. However, before going on to the next trial, the user checks the arrangement in comparison to the given goal. It is an evaluative stage of the subtask performance. It can be observed that eye movements basically follow the natural pace of the task performance.

Thus, the suggested method of eye movement interpretation is totally different in comparison to the traditional method. Cumulated scan path that is presently used in cognitive psychology is not sufficiently informative. Cumulative scan path should be broken into segments that are associated with actions performed by subjects. SSAT suggests a standardized method of action classification. Eye movement method is also important for functional analysis when activity is considered as a self-regulative system. Analysis of the strategies of task performance and classification of actions allows to develop algorithmic description of task and its time structure.

9.4 Algorithmic Description of Task and Its Time Structure Analysis

Algorithmic description of the task and its time structure analysis are the basis of morphological analysis of activity. In SSAT, we use the term human algorithm, which describes logical sequences of human cognitive and behavioral actions. User's activity in HCI tasks is very flexible. Therefore, subjects' algorithm should be developed based on analysis of the most representative strategies of task performance. Sometimes it requires analysis of algorithms of representative subjects. However, the human algorithm gives us a fair idea of user performance of a particular task. In cases where existing designs need to be evaluated for changes, this represents the ideal solution for comprehensive analysis of the existing design of HCI. When we describe the duration of each member of the algorithm, this is a combination of algorithmic description with time structure analysis. In this task analysis, time structure is presented not in graphical but in tabular form. This is explained by the fact that graphical description of time structure of activity is used when some elements of activity are performed simultaneously. In such situations without graphical description of time structure of activity, it is difficult to conduct task complexity evaluation. In this study, the factor associated with complexity evaluation of simultaneously performed actions is ignored because the main elements of activity are cognitive and therefore should be performed sequentially. It should be noted that algorithmic description of task performance consists of subdivision of an activity into qualitatively distinct psychological units with the determination of their logical organization. Such elements are called members of the algorithm. They consist of one or several cognitive or motor actions that are integrated by a higher-order goal or the goal of this particular member of the algorithm. Because of the limit of the working memory capacity, members of the algorithm are usually comprised of one to four integrated actions. A member of the algorithm can be classified according to its qualitative characteristics. Afferent operators are associated with receiving information and are designated by the symbol O^{α}. Operators associated with extraction of information from long-term memory or keeping information in working memory are designated by O^{μ}. Efferent operators are involved in executive components of activity and are designated by O^{ε}.

If a member of an algorithm includes a thinking activity such as a comparison between elements of the situation, discovery of the functional purpose of the symbols on the screen, their relationship, performance of logical actions, and so on, then it is designated by O^{th}. This means that this member of the algorithm describes the thinking activity. Very often, thinking actions that are performed during HCI are based on visual information. Hence, a member of an algorithm can include thinking actions that are performed based on visually presented data. In such cases, the symbol $O^{\alpha th}$ is used.

Logical conditions determine the logic of the selection and realization of different members of the algorithm and include a decision-making process. They can be designated by *l* or *L* (based on the combination of several logical conditions). There are some other symbols that are used in algorithmic description of activity, which were described in previous sections.

An algorithmic description is the model of activity during task performance. Such algorithms closely reflect the real user's performance strategy. At the next step, an expert performs psychological analysis of an algorithm. Each member of the algorithm can be evaluated as a subsystem of activity. At this stage of analysis, preliminary qualitative analysis can be reconsidered based on the new data. The algorithm of the task performance and time structure of activity is presented in Table 9.3. Performance time of some members of the algorithm is not shown because they are performed simultaneously with some others members of the algorithm.

Such concepts as functional equivalence between different strategies, interchangeability, and a range of tolerance demonstrate a possibility of utilizing the term representative strategies of task performance. Most variations in task performance can be reduced to the most representative strategies of task performance.

During experimental study, three representative strategies were discovered. In this study, three representative strategies were discovered. Only one of the most representative strategies of user performance is described. Therefore, in algorithmic description of the task, only one output from logical conditions is utilized. The probability of this strategy is 0.4. The other two strategies can be described similarly. The complexity of each strategy can be calculated. In order to determine the general task complexity, it is necessary to take into account the probability of occurrence of each strategy. In the further discussion, we assess the complexity of one strategy only. Variation in task performance that deviates from the described strategies can be neglected. As per the action classification table (Table 9.2), sets of actions can be attributed to individual algorithms of performance. This is one of the ways to describe the structure of holistic activity in contrast to analysis of separate aspects of activity during task performance. This is an example of the systemic description of activity during task performance.

The algorithmic description gives a fair idea about human performance in a particular situation. The algorithm presented in Table 9.3 can be considered in terms of potential improvement of the task sequence and hence the strategy

TABLE 9.3

Fragment of Algorithmic and Time Structure Description of Task Performance

Algorithm	Description	Actions Obtained from Action Classification Table	Time (ms)
O_1^{α}	Look at the goal area and the initial state of the object area.	Simultaneous perceptual actions (three actions)	1030
$O_2^{\alpha th}$	Find out differences between the goal area and the object area.	Thinking actions based on visual information (four actions)	1650
$O_3^{\alpha th}$	Find out differences between the goal area and the object area and simultaneously perform O_4^{ε}.	Thinking actions based on visual information (two actions)	740
O_4^{ε}	Move cursor closely to the object area.	Simple motor action	
l_1	Decide to click object (element O_D) and simultaneously perform O_5^{ε}.	Decision-making action based on information from memory	840
O_5^{ε}	Simultaneously with l_1, click object element O_D.	Simple motor action	
O_6^{α}	Look at the tool area and simultaneously perform O_7^{ε}.	Simultaneous perceptual action	430
O_7^{ε}	Simultaneously with O_6^{α}, move cursor closer to the tool area.	Simple motor action	
l_2	Decide to click tool (element T_{PV}) and simultaneously perform O_8^{ε}.	Simultaneous perceptual action Decision-making action during visual assessment	570
O_8^{ε}	Simultaneously with l_2, move cursor close to a specific icon and click icon.	Average precision motor action	
$O_9^{\alpha th}$	Evaluate how the object area matched to the goal area.	Thinking action based on visual information	400
$O_{10}^{\alpha th}$	Evaluate intermittent state of the object area.	Thinking actions based on visual information (four actions)	1740
O_{11}^{α}	Look at the goal area and then look at the tool area.	Simultaneous perceptual actions (two actions)	
l_3	Decide to click object (element O_s) and simultaneously perform O_{12}^{ε}.	Decision-making action at sensory-perceptual level	280
O_{12}^{ε}	Simultaneously with l_3, click object element O_s, by using the mouse.	Simple motor action	
O_{13}^{α}	Look at the tool area and simultaneously perform O_{14}^{ε}.	Simultaneous perceptual action with partly overlapping motor action (see below)	1200
O_{14}^{ε}	Simultaneously with O_{13}^{α}, move cursor close to a specific icon (horizontal position tool).	Average precision motor action	
O_{15}^{α}	Look at the object area to evaluate change of position of objects (O_S and O_W—horizontal shift) and perform O_{16}^{ε}.	Simultaneous perceptual action with motor action (see below)	400

(Continued)

TABLE 9.3 (*Continued*)

Fragment of Algorithmic and Time Structure Description of Task Performance

Algorithm	Description	Actions Obtained from Action Classification Table	Time (ms)
O_{16}^{ε}	Click tool to activate action simultaneously performed with O_{15}^{α}	Simple motor action	
O_{17}^{α}	Continue looking at the object area.	Simultaneous perceptual action	
l_4	Decide to click object (element O_Q) and simultaneously perform O_{18}^{ε}.	Decision-making action at sensory–perceptual level	330
O_{18}^{ε}	Click object element O_Q.	Simple motor action	
O_{19}^{α}	Look at goal area to evaluate color of elements.	Simultaneous perceptual	370
l_5	Decide to click blue icon tool.	Decision-making action at sensory–perceptual level	420
O_{20}^{α}	Look at tool area and simultaneously perform O_{21}^{ε}.	Simultaneous perceptual action with motor action	400
O_{21}^{ε}	Move cursor to tool area.	Average precision motor action	330
l_6	Decide to click object (element O_Q) and simultaneously perform O_{22}^{ε}.	Decision-making action at sensory–perceptual level	
O_{22}^{ε}	Click object element O_Q.	Simple motor action	
-----	-----	-----	-----
O_{75}^{α}	Look at the object area and simultaneously	Simultaneous perceptual action with motor action	510
O_{76}^{ε}	Click OK to complete trail.	Simple motor action	

used by the users. For example, let us take a look at members of algorithm O_{18} and O_{22}. Both of them are related to the selection of element O_Q. However, these members of the algorithm are performed at two different stages in the task execution. As a result, the user performs the same members of the algorithm repeatedly to accomplish the same goal. However, the same goal can be accomplished without redundant performance of these members of the algorithm. Examining the algorithm on this basis helps us to remove some members of the algorithm and complete the task more efficiently.

Let us consider the other possible steps of analysis. It has been discovered that users spent significant time executing thinking actions. These kinds of actions are more complicated for users than perceptual actions. Therefore, the thinking actions should be approached first to understand any difficulty the users are facing. There is a high number of thinking actions at the initial stage of the task performance because of the comparison of the object and the goal area elements. However, at the later stage of the task performance, the duration of the same thinking actions is quite low. Hence, it can be suggested that improving users' instructions at the initial stage of the task performance can reduce the quantity and the duration of the thinking actions. Table 9.3 also presents time performance of each member of the algorithm. This makes it possible to assess complexity of the considered strategy of the task performance. The algorithm in Table 9.3 is developed based on experimental data and the experts' analysis. In order to discuss the most efficient strategies of task performance we utilize the term "perfect algorithm."

During independent learning, the user can shift from less efficient to more efficient strategies. At this period of learning, he or she develops his own understanding of "good strategy" and permissible deviation from this strategy. At the same time, there is a best possible strategy of task performance. Such strategy in most cases is unknown to the users. Therefore, users might never achieve such strategy. Due to multiple performances and independent training process, the user shifts from less efficient to more efficient strategies. He or she develops his or her own understanding of "good strategy" and permissible deviation from this strategy. Formation of such strategies can be explained from the activity self-regulation perspective. The final stage of activity self-regulation is the evaluative stage, which includes such function blocks or mechanisms as "subjective standard of successful result," "subjective standard of admissible deviation," "negative evaluation of result," and "positive evaluation of result." Based on repetitive task performance, the user gradually develops his or her own understanding of a "good" standard of task performance. Emotionally motivational factors, such as level of aspiration, are important at this stage. Users experience emotional satisfaction when they select a particular strategy of task performance. As a result, the user stops improving his or her performance based on his or her subjective criteria. If this strategy becomes habitual, then users may resist any changes in strategies of task performance. Therefore, the subjective criteria of success are important mechanisms in developing adequate strategies of performance. The designer should take into account how often this task is performed by a user and decide how close real strategies of task performance should be to the ideal strategy. The perfect algorithm can be efficient according to some criteria but it can be more complex for a user. It is critically important to consider the fact that a user may be accustomed to different modes of task performance. This is also an important factor in the optimization of task performance. Algorithmic description of task with temporal data of performing various members of an algorithm is an efficient tool in finding the design solution. However, quantitative assessment of task complexity is an additional important tool for this purpose. This stage of analysis will be discussed in the following chapter.

In conclusion to this chapter, we present the fragment of the perfect algorithm of task performance (Table 9.4).

As described in Table 9.4, the algorithm is developed based on the expert's analysis. It describes the perfect strategy of task performance that in this case is not achieved by any of the users. We can see that the perfect algorithm is significantly shorter than the real algorithm of task performance. The real algorithm has 76 members and the perfect algorithm has only 56 members and has shorter performance time. In a real situation, quantity

TABLE 9.4
Perfect Algorithm

Algorithm	Description	Actions Obtained from Action Classification Table	Time (ms)
O_1^{α}	Look at goal area and initial state of object area.	Simultaneous perceptual actions (3 actions)	1030
$O_2^{\alpha th}$	Find out differences between goal area and object area.	Thinking actions based on visual information (4 actions)	1650
$O_3^{\alpha th}$	Find out differences between goal area and object area and simultaneously perform O_4^{ε}.	Thinking actions based on visual information (2 actions)	740
O_4^{ε}	Move cursor closely to object area.	Simple motor action	
l_1	Decide to click object (element O_Q) and simultaneously perform O_5^{ε}.	Decision-making action at sensory perceptual level	330
O_5^{ε}	Click object element O_Q.	Simple motor action	
O_6^{α}	Look at goal area to evaluate color of elements.	Simultaneous perceptual action	370
l_2	Decide to click blue icon tool.	Decision-making action at sensory–perceptual level	420
O_7^{α}	Look at tool area and simultaneously perform O_8^{ε}.	Simultaneous perceptual action with motor action	400
O_8^{ε}	Move cursor to tool area	Precise motor action	
l_3	Decide to click object (element O_Q) and simultaneously perform O_{22}^{ε}.	Simultaneous perceptual action with motor action Decision making action at sensory perceptual level	330
O_9^{ε}	Click object element O_Q.	Simple motor action	
O_{10}^{α}	Look at tool area and simultaneously perform O_{11}^{ε}	Simultaneous perceptual action with motor action	420
O_{11}^{ε}	Move cursor to blue color tool.	Precise motor action	
O_{12}^{α}	Look at object area and simultaneously perform O_{13}^{ε}	Simultaneous perceptual action with motor action	400
O_{13}^{ε}	Click blue color tool.	Simple motor action	
l_4	Decide to click object (element O_D) and simultaneously perform O_{14}^{ε}.	Decision-making action based on information from memory	840
-----	-----	-----	-----
O_{55}^{α}	Continue looking at object area and simultaneously perform O_{56}^{ε}.	Simultaneous perceptual action with motor action	370
O_{56}^{ε}	Click finish for feedback.	Simple motor action	

of members of an algorithm and performance time cannot be the criteria for optimization of task performance. For example, an algorithm with shorter performance time can be more complex for a user than the one with the longer performance time. Hence, quantitative analysis of task performance based on comparison of complexity measures is a useful tool for the development of optimal strategies of task performance. The comparison of real and perfect algorithms demonstrates that a number of members of an algorithm that correspond to thinking components are reduced. Relative quantities of members of an algorithm that are involved in the thinking process are reduced. However, perceptual components of work are increased. As a result, duration of performance of separate members of algorithm is also reduced. The comparison of these two algorithms will not be discussed in detail. In Section 9.5, we consider evaluation of task complexity based on the data presented in Table 9.3.

9.5 Evaluation of Task Complexity of Computer-Based Task

The purpose of quantitative task complexity evaluation is to estimate cognitive effort during task performance. Based on measures of complexity, it is possible to optimize human performance and design solution. Complexity is a multidimensional concept that requires multiple measures for its evaluation. Having multiple measures allows enhancing and redesigning tasks based on each measure and their comparison. This is done by getting a clear understanding of the task's design shortcomings and various aspects of difficulty in task performance. In some tasks, certain measures might be more important than others. Some of these measures can be of zero value if they are not important for a particular task. The measures that have a value of zero in some cases can be useful because they give an idea about the specific characteristics of the task.

As we discussed before, for evaluation of complexity, computer-based tasks in cognitive psychology were recommended to use such measures as task solving time, number of different transitions, total number of system states, and number of production rules. However, the suggested units of measure are inadequate from a mathematical point of view because they are noncommensurable units. Such units of measure cannot be compared relative to each other.

In SSAT, basic rules were developed for complexity evaluation of time intervals for various elements of activity. These rules can be divided into three groups: (1) rules that describe possibility to perform elements of activity sequentially or simultaneously; (2) rules for evaluation of complexity activity elements; (3) rules for the evaluation of complexity of activity elements that can be performed simultaneously. These rules were described before and we do not consider them here. The general list of complexity measures and their psychological meaning (for HCI tasks) is presented in Table 8.2. In our example, we will utilize only some measures of complexity from Table 8.2.

In this study, the factor associated with complexity evaluation of simultaneously performed actions is ignored because the main elements of activity are cognitive and should be performed sequentially. Most mouse movements are associated with low or average concentration of attention. Only the last mouse movement stage (slowing phase of movement), when the pointer approaches the tool, requires the third category of complexity (high level of concentration of attention) because icons on the screen are small. In this task, the icons in the goal, tool, and object areas are fairly large (50 × 50 pixels on a 1024 × 768 resolution screen of 17 in.). Therefore, the slowing stage (adjustment phase) of movement is simpler and considered to be of the second category of complexity and ballistic stage (acceleration stage) as the first category of complexity. This means that the ballistic stage of movement requires a low level of concentration of attention and the slowing stage of movements requires an average level of concentration of attention.

Therefore, they are the first and second levels of complexity of motor components and the third level of complexity of cognitive components of activity in the task. According to the described rules, the combination of the first and second categories of complexity of motor elements of activity, with the third category of complexity of cognitive components of activity, does not increase complexity of the time interval. Such time intervals would still be the third category of complexity. Therefore, simultaneous performance of cognitive and motor actions can be ignored. As a result, it is not necessary to develop the time structure of activity in graphical form in order to gain better understanding of the combination of elements of activity in time. In Table 9.3, algorithmic description is combined with time structure description.

In this example, complexity evaluation is performed only for one most preferable and most frequently used strategy of task performance. Hence, there is no need to calculate probabilistic characteristics of task, because all elements of task for this strategy have a probability of 1. The calculation of complexity is performed based on the aforementioned procedures. Table 9.5 presents measures of complexity and performance time of different components of activity for a computer-based task, which were described in the previous sections.

Complexity is a multidimensional phenomenon. If a designer changes some features of the task, measures that are associated with these features can also change. The time of task performance is (T) 26.5 s: 14.7 s is devoted to afferent operators (sensory-perceptual actions, T_α); 7.5 s is related to thinking or analysis of the problem (T^{th}); and 4.3 s is associated with logical conditions L_g (decision-making). Analysis of temporal data of task performance demonstrates that motor actions in most cases are performed simultaneously with perceptual actions. Performance time of efferent operators (motor actions) T_{ex} requires 10 s; 9.3 s of these 10 s motor actions are performed simultaneously with perceptual actions.

Motor actions, however, are seldom combined with thinking and decision-making actions because of the logical structure of the considered task. While the analysis of the task and decision-making take place, the mouse is stationary or performs jittery (tremor) movements. Goal-directed motor actions are registered only when the subject moves the mouse more than the width of the icon (in this case, 50 pixels). When a motor activity dominates, various strategies of performance are possible. For example, if the operator performs

TABLE 9.5

Measures of Task Complexity for Computer-Based Task (the First Strategy)

Name of Measures	Value of Measures
Time for algorithm execution T (total time of task performance)	26.5
Time for performance of logical conditions L_g (decision-making)	4.3
Time for performance of afferent operators T_α (sensory-perceptual actions)	14.7
Time for performance of efferent operators T_{ex} (motor activity)	10
Time for discrimination and recognition of distinctive features of task approaching threshold characteristics of sense receptors $'T_\alpha$	0
Total time for performance in goal area T_{gol} (cognitive component)	4.68
Total time for performance in object area T_{obj} (cognitive component)	6.8
Total time for performance in tool area T_{tool} (cognitive component)	5.01
Proportion of time for cognitive activity in the goal area to total time of task performance N_{gol}	0.18
Proportion of time for cognitive activity in the object area to total time of task performance N_{obj}	0.26
Proportion of time for cognitive activity in the goal and object areas to total time of task performance N_{golobj}	0.43
Proportion of time for cognitive activity in the tool area to total time of task performance N_{tool}	0.19
Proportion of time for logical conditions to total time for task performance N_l	0.16
Time for performance of operators associated with thinking process based on external features presented through interface elements $T^{\alpha th}$	7.5
Time for performance of operators associated with thinking T^{th}	0.28
Time for performance of operators associated with thinking process based on data extracted from memory $T^{\mu th}$	0
Proportion of time for performance of operators associated with thinking process based on external features presented through interface elements $N^{\alpha th}$	0.28
Proportion of time for performance of operators associated with thinking process based on data extracted from memory $N^{\mu th}$	0
Proportion of time for thinking components of activity to total time of task performance ΔT_{th}	0.28
Proportion of time for logical components of work activity depending largely on information selected from long-term memory rather than external features presented through interface elements L_{ltm}	0
Proportion of time for retaining current information in working memory N_{wm}	0
Proportion of time for discrimination and recognition of distinct features of task approaching threshold characteristics of sense receptors Q	0
Proportion of time for efferent operators N_{mot} (motor activity)	0.4
Proportion of time for afferent operators N_α (sensory-perceptual activity)	0.55
Order Scale of Complexity (X_r)	
[a] *Algorithm* → 3.	
[b] *Member of algorithm* → 20 efferent operators as O^ε_4, O^ε_5, O^ε_7, etc., belong to the first category of complexity; 9 operators as O^ε_8, O^ε_{14}, O^ε_{21}, etc., belong to the second category of complexity. All members of algorithm associated with cognitive components of activity have the third category of complexity.	

motor actions of simple or average complexity, decision-making actions could take place at the same time for future motor activity. In other words, the subject combines ongoing motor actions with cognitive actions needed for future motor components of activity.

The width of the icon in our task is wider than the width of the regular icons. This difference means that movements of the pointer to the particular icon require less precision. As a result, 20 of the efferent operators belong to the first category of complexity and 9 efferent operators are of the second category of complexity. Therefore, the third category of complexity of motor operators does not exist in this task. From the 10.01 s associated with motor activity, 9.3 s motor actions are performed simultaneously with perceptual actions with the third category of complexity. When motor activity elements with one to two categories of complexity are combined with perceptual actions of the third category of complexity, the developed formalized rules that dictate this interval of time is related to the more difficult category. Hence, only a 0.71 s (10.01 – 9.3) interval of time associated with the motor activity can be related to the second category of complexity and 9.3 s are related to the third category. Therefore, based on the evaluation of cognitive components, almost all periods for task performance are of the third category of complexity. Combining motor activity with cognitive activity does not change the complexity of activity elements. Hence, this whole task is of the third category of complexity (see measure—X_r, Table 9.5). Other complexity measures are now considered. The fraction of time for logical conditions N_l (decision-making) is 0.16, for operators associated with thinking components of activity N^{th} is 0.28, and for afferent operators N_α (sensory-perceptual actions) is 0.55. It is interesting to evaluate measures that involve memory. For example, the fraction of time for performing operators associated with thinking process based on data extracted from memory equals 0. This means that when thinking, the user utilizes externally presented onscreen data first versus operating with data extracted from memory ($N^{th} = N^{\alpha th}$), which makes the task execution easier. The fraction of time for retaining current information in working memory N_{wm} is also 0, which means that the user does not keep intermediate data in memory during task performance. The same can be observed during the performance of logical conditions (decision-making). $N_l = 0.16$ ($L_{ltm} = 0$) when decision-making is also based on external information. There is no need for operating with visual information, which requires functioning sensory-perceptual processes in the threshold area. Therefore, the fraction of time for discrimination and recognition of distinct features of task approaching threshold of sense receptors (Q) equals 0. All of these make the task much easier and increases its usability. The fraction of time for cognitive activity in goal area N_{gol} is 0.18, in tool area N_{tool} is 0.19, and in object area N_{obj} is 0.26. In spite of the fact that the general goal was given in advance and the specific goal of task was presented externally in a ready form, users spent approximately the same fraction of time in the goal area as in the tool area. This is because goal interpretation, acceptance, and its transformation are important for task performance.

The goal cannot be considered simply as externally given in a ready form standard to which human performance is directed. Interpretation and acceptance of the goal, transformation of the goal or goal formation, modification of the goal, and utilization of its elements as subjective standards of success are significant elements of activity. The fraction of time for cognitive activity in the goal and object areas combined (N_{golobj}) is 0.43, which is devoted to orienting activity and is associated with the creation of a mental model of the situation. Even a short discussion of these measures demonstrates that they can accurately describe the internal structure of cognitive activity during performance of HCI tasks. If one changes the structure of the task, cognitive measures of complexity also change.

In the considered example, only one strategy of complexity is evaluated. When complexity of task in general needs to be determined, one would evaluate probability of utilization of each strategy and, based on this data, evaluate complexity of the whole task. Through evaluation of complexity of each strategy, specialists can determine which strategy is the best one. The algorithmic method suggested by SSAT allows describing standardized and individual strategies of activity performance. This means that we can evaluate complexity of individualized strategies considering that users can utilize various strategies.

Knowing the probability of using each strategy, we can calculate the average task complexity. Activity strategies change during the skills acquisition process. Therefore, the task complexity can be assessed at various stages of task acquisition.

The study demonstrates that complexity of task is a key characteristic of any system. This allows psychologists and ergonomists to capture the basic characteristics of task without ambiguous verbal description of its content at the final stage of task analysis. Human performance of computer-based tasks can be optimized utilizing quantitative measures of complexity. This chapter describes a general method and the principles of measuring task complexity of computer-based tasks. Task complexity evaluation is based on the assumption that the more complex a task is, the higher the probability that it will be difficult to perform and that mental effort and possibility of errors will also increase. The proposed measures and procedures are based on SSAT, which was developed by Bedny (1987, 1997). Activity is a multi-dimensional system. In this regard, the complexity of such systems cannot be reduced to a single numeric value. The presence of multiple complexity measures reflects the specifics of various aspects of activity performance and makes optimization of activity possible based on these complexity measures. Not all measures have the same value for different activity dimensions. Some of these measures might have a zero value for a particular task. Hence, professionals can select an adequate set of measures for assessment of task complexity depending on the data presented for the task. One important aspect of complexity assessment is that it can use not just experimental but analytical procedures as well. Hence, this method can be used at a design stage before a system has been constructed, which would bring down the cost of the project by preventing possible design errors at the early stages of the project. We can evaluate complexity of individualized strategies of task performance and optimize them based on quantitative procedures.

In this work, one suggestion was to use typical elements of activity as units of measures, which are classified according to their duration and developed criteria. Suggested measures take into account simultaneous and sequential performance of activity elements and probability of their occurrence. These measures allow evaluation and prediction of mental efforts during computer task performance, conducting comparative analysis of design solutions, and prediction of dynamics of skill acquisition, possibility of errors, and so on. This work shows that SSAT suggests a brand-new approach to usability evaluation and optimization of HCI.

Finally, the method proposed in this study is compared with GOMS (goals, operators, methods, and selection rules) method because both methods attempt to describe task performance algorithmically and determine complexity.

Whereas GOMS uses a computer-like algorithm, SSAT utilizes the concept of *human algorithm* that is based on totally different principles. GOMS's production system is a collection of *if... then* rules, but people carry out decision-making using more than two alternatives, and the probability of alternatives can vary between 0 and 1. Terms like goal, action, and operation have different meanings in SSAT. The method of classification of basic units of analysis and principles of their extraction from activity flow is also different. GOMS does not suggest any analytical tool for calculating task complexity. There is no discussion of elements of activity being executed sequentially and also in parallel and these elements appearing with various probabilities in activity structure. GOMS also does not consider that activity elements can appear with various probabilities. GOMS does not discuss the issue of commensurable units of complexity measurement. Units of roughly *equal size* are examples of noncommensurable units of measurement. It should be noted that describing cognition and human activity in general as a simple sequence of primitive operators or production system units contradict not only the theory of activity but also cognitive psychology. Measurement of task complexity will be discussed in more details in the second book of this serious (Bedny, 2014).

10

Introduction to Human Reliability Assessment

10.1 Method of Human Reliability Assessment of Computer-Based Tasks

The systemic-structural activity approach has developed principles of activity description as a multidimensional system and offered diverse models of task performance that capture various aspects of the activity structure (Bedny and Karwowski, 2003). This section describes a new method of reliability assessment of human performance that derives from the systemic-structural activity (SSAT) approach. Accuracy and reliability are two important characteristics of the system that are not the same. Accuracy refers to the precision with which a goal of the system is achieved, whereas reliability refers to failures of the system. An operator is a system component, and the accuracy and reliability of his or her performance influence the efficiency of the entire functioning system. There are a number of publications that cover a range of methods utilized for reliability analysis. However, there has been no attempt made to assess human performance reliability when a user interacts with a computer. The material presented here is the first attempt on using the task modeling method for this purpose. Moreover, a suggested method of reliability assessment can be used for reliability assessment of human performance in any system. In this chapter, we describe some basic concepts human reliability assessment of computer-based tasks from the SSAT perspective. Some methods that have been described in our previous chapters have been significantly modified and adapted to create a reliability assessment procedure that utilizes three stages including determining what kind of errors can occur and which errors can be considered as failures, what their probabilities are, and how these errors and/or failures can be reduced. This method not only permits to quantify human errors but also gives a qualitative description of such errors and suggests ways to reduce the number of errors and system failures. Existing cognitive psychology methods of assessing the precision of operator's performance is not clearly separated from the methods of operator's reliability assessment. These two methods of assessment are similar but not identical. The basis for each approach is a method of analyzing human error but their purpose is different. Accuracy characterizes the precision with which the goal of task is achieved. Reliability refers to failures of performance and how the probability of failure can change over time or in stressful situations. Human performance can be precise but not reliable. Not all errors can be considered as failures. Some errors can be recoverable or have a relatively small effect on functioning of personal or technical components of the system. Other errors are associated with hazardous accidents, nonadmissible losses of time, and so on. Only the last category of errors can be categorized as failure. We will use the term *errors* to evaluate the precision of human performance and *failures* to evaluate human reliability. When accuracy declines and falls below acceptable level, it becomes an operator's error. If as a result of operator's errors the system cannot function and achieve its goal or goal achievement is conveyed by unacceptable losses, it is considered a failure. Therefore, the main criterion for distinguishing between errors and failures is their consequences for the system. There are errors or failures caused by technical components of the system and errors caused by operator's erroneous actions. Of course, such division is relative. For example, an operator can perform wrong actions because the system design is not adequate or he or she does not possess the required skills.

Evaluation of reliability has its own specifics. It is important to trace errors to see if they are transformed into failures in various conditions of the functioning system. In some cases, a number of errors may increase to the point when their combined effect leads to a failure. As a result of fatigue, an operator can make serious mistakes that can be observed as a failure. In this case, an operator becomes an unreliable component of the system according to the time parameters because failures have appeared as a result of inadequate stamina. Another aspect of reliability is the operator's ability to avoid failure in emotionally stressful conditions. If an operator makes unacceptable mistakes under stress, he/she is not reliable in terms of emotional stability. Another factor that affects the reliability of human performance is the properties of cognitive processes. For example, from time to time, system requirements have increased demands on memory. In these circumstances, a particular operator may commit grave errors or failures. Another factor may be due to the fact that the operator works in distractive environment, and so on. This means that the operator does not have the required physical background or the resistance of cognitive processes to external destructions. In the absence of the extreme factors, an operator is functioning properly, and when they are presented with such factors, he or she functions inadequately, meaning that an operator can meet all the requirements under normal conditions but not in the presence of overload or extreme conditions.

Commonly, we discuss the assessment of reliability but do not consider accuracy. Evaluation of reliability always involves evaluation of operator's performance in cases of overload and extreme conditions. In assessing the reliability, we always consider unfavorable factors that may confront an operator or a technical system when they can still keep functioning. When assessing performance, the described accuracy factors often are not considered.

One of the authors of this book was a gymnast, so we will discuss an episode from his experience. He knew two excellent gymnasts, who were often competing with each other for the first place. After going through a few apparatuses, if one of them heard an announcement that he was holding the first place and could win a competition, he would perform poorly at the next event and loose the first position. In contrast, the other gymnast, who was at the second place, performed well and moved to the first place. The second gymnast, who was a friend of the first one, would smile and say, "Watch him at the next events, he'll make mistakes and I'll win." The first gymnast had a high level of precision, but he was not reliable. In the heat of the competition and under stress, he would not be able to sustain the required level of performance. Outstanding gymnasts always demonstrate a high level of reliability.

Similarly, sometimes an increase in equipment precision can be achieved only by increasing the system complexity by introducing new components of the system, and it might result in lower system reliability. This means that technical components of the system can provide a high precision but be unreliable. Reliability can also be reduced when one begins to approach system limits. For example, at low and average speeds of the flight, firing an aircraft's guns can produce very precise result, but if the pilot approaches the maximum permitted speed, firing accuracy may be sharply reduced. Hence, the technical components of the system also can be determined by demands to its reliability and precision.

Reliability is often overlooked in software design. When the software process is tested after introduction of required changes, it is usually tested for precision first to make sure the outcome of the improved and/or changed process is as should be expected. It is usually compared with the outcome of the old version of the same software to make sure that the precision level stays the same or enhanced. It is often the case that the improved process performs much faster than the old version when a lot of data has to be processed, but it might unexpectedly fail when there are no data coming in or the input is rather small. The improved version of the process might be more complicated and perform faster most of the times, but it is not consistently reliable. All such extreme conditions should be tested to make sure that the improved process reliability is adequate.

As has been discussed earlier, we can talk about human errors only after discovering his or her erroneous cognitive and behavioral actions. The concept of cognitive and behavioral actions is the central one for the analysis of errors and failures and therefore for the precision and reliability assessment. Hence, the reliability of task performance requires analysis of actions performed by an operator. An individual can transform or modify the object of activity (mental or physical) according to his or her goal by using various actions. To discover causes of failures or errors, it is necessary to determine what cognitive or behavioral actions are erroneous. For this purpose, a professional has to determine the goal of actions, feedback influences, the type of actions, their interdependence, goal of task and possible strategies of task performance, and so on (Bedny, 2004, 2006; Bedny and Sengupta, 2005; Bedny et al., 2010; Sengupta et al., 2008). In SSAT where activity is considered as a structurally organized system, reliability assessment is performed utilizing the parametric study of activity and systemic methods of activity analysis. Among them is a morphological approach that describes logical and space-temporal organization of human actions and quantitative evaluation of reliability at its final stage.

In our farther example (Section 10.3) at the first stage of analysis (qualitative stage), we utilized the traditional, objectively logical method of study; at the second stage, we used the algorithmic description of activity (morphological analysis); and at the final stage, we conducted the reliability assessment (quantitative stage). Feedback from the latter stages of analysis has been used to reconsider earlier levels of analysis estimating the reliability of computer-based tasks. Task-specific models of human activity were utilized during this analysis, which allowed eliminating experimental methods of study and substitute it with analytical procedures. An enhanced method of task performance was suggested based on the analytical description of the new method of work.

The computer-based task *receiving an order* has been selected as an object of study. Such operations are considered to be an indirect labor in the manufacturing process. Examples of this kind of work can be found in transportation, warehousing, distribution of instruments and raw materials, etc. There are also office functions that can be considered as indirect labor. With the arrival of the Internet, a lot of new companies emerged whose main function is gathering orders and delivering products to the customers.

In the considered process, the task *receiving an order* was the first one for the morning shift. Out of four stages of task analysis developed in SSAT, we have selected the following three: qualitative, algorithmic, and quantitative stages. The quantitative stage includes two basic methods: evaluation of task complexity and reliability assessment. In this chapter, we used reliability assessment as a quantitative stage of task analysis. The third stage of SSAT task analysis, called time structure analysis, has been omitted in this study, and the objectively logical method has been selected as a qualitative method of study.

10.2 Error Analysis in Computer-Based Tasks

Presently there is a significant trend toward computerization of the human–machine system, and therefore, analysis of human errors when performing computer-based tasks becomes very important. We can outline several major factors that affect human errors in the HCI system: quality of developed software, user qualifications (novice versus highly trained user), individual features of a user, reliability of technical components of computerized system, protection of computer system against viruses, etc.

In assessing errors, we need to consider not only the factors associated with lost productivity but also their impact on workers. As an example, consider a real situation when an error is associated not only with lower productivity but also with emotional stress.

We have to pay attention to the fact that analysis of user errors as a result of stress should be performed directly in the production environment. Therefore, the main research methods would be observation and interview. Let us consider the task *receiving orders.*

This task is performed at the beginning of the shift. Upon successful completion of this task, 50 operators that use computers in their work are given the task to be performed during the shift. Failure in performing this task leads to other employees losing about an hour of work. Observations and interviews of users have shown that stress is caused by two factors. One factor was the loss of time, and the other factor was associated with an unpleasant emotional state caused by the disruption of the usual mode of work. It was found that the users were experiencing mental states such as irritability, anxiety, anger, and frustration. Most users are experiencing negative emotional feelings in relation to a colleague who, in their opinion, is guilty of the situation.

It was found that stress and errors are caused by a combination of two factors—time limit and a negative emotional state. Individual differences in reaction to stress and failure to perform the job timely also were discovered in a considered situation. Some workers responded more aggressively to failure than others. Moreover, this aggression was clearly personified. For example, one worker accused his colleague even when he was responsible for the existing failures.

The operator, who failed in performing the task *receiving orders,* also demonstrated aggressiveness and blamed the programmers who created the software for this task, but in reality, he was the direct perpetrator of failure. It becomes obvious that the significance of an error that was made by only one user can be explained by the fact that his error not only affects the performance of a large number of employees but also resulted in their negative emotional state.

Thus, utilizing relatively simple methods of study, practitioners can assess the consequences of errors that are accompanied not only by loss of productivity but also by the negative emotional state of users.

Computer specialists, as well as computer system users, work in groups. Therefore, there are errors related to individual users and errors that depend on the group factor. In the latter case, we will talk about errors related to social factors.

Errors caused by social factors became particularly relevant in recent years. About 30 years ago, computer departments, even in very large companies, accounted approximately for up to 50 people. All functions were performed by the same group of people. Today, computer departments of large companies rose to up to 3000–4000 people.

Various functions such as protection of systems, technical support, and database (DB) software development are divided between the groups. Often it is difficult to even figure out which group should be addressing a particular issue because functions are not always precisely distributed between groups. As a result, many functions are overlapping in various groups. In such circumstances, some groups are trying to transfer responsibilities to each other. This often leads to intergroup conflicts. Errors related to social factors may also lead to intragroup conflicts.

The other factor is that each of these groups has its own terminology and understanding of the computing environment and the tasks they perform.

Terminology, which is used by these groups, overlaps very little for the groups that communicate with each other directly and is completely different for the groups that typically do not interact directly. Existence of different groups with their own specific goals, terminology, and specific interaction gives rise to the problem of intragroup and intergroup communication and social interaction, which requires error analysis generated by social factors. Social factors can be divided into three components: culture, preferred group strategy of activity, and history. The cultural component examines the structure of the group, cultural factors, and prevailing social norms. The behavioral factor evaluates activity of the group in the current situation as a result of the group's performance. In considering the social factor, specialists also analyze the way of group formation and individual goals of the group, strategies to achieving them, the existing rules and regulations produced by the group, etc. Social norms and group rules can vary between the groups. This can cause intergroup conflicts. History includes an overview of the changes of cultural norms and rules of behavior over time.

The factor called social interaction plays a leading role in group performance. As we discussed before, social interaction can be presented as *subject ↔ tools ↔ subject*.

Social interaction should be distinguished from *subject ↔ tool ↔ object*. Social interaction requires understanding a partner's goals, their verbal expressions, possible responses, motivation, etc. The main purpose of social interaction is exchanging information.

It must be remembered that mutual understanding is formed within the group much easier than between groups. This factor is especially important in interacting with a group of computer users. Most conflicts among these groups are due to the lack of intergroup understanding. One can identify interaction through a computer and direct interaction with each other. This situation is illustrated in Figure 10.1.

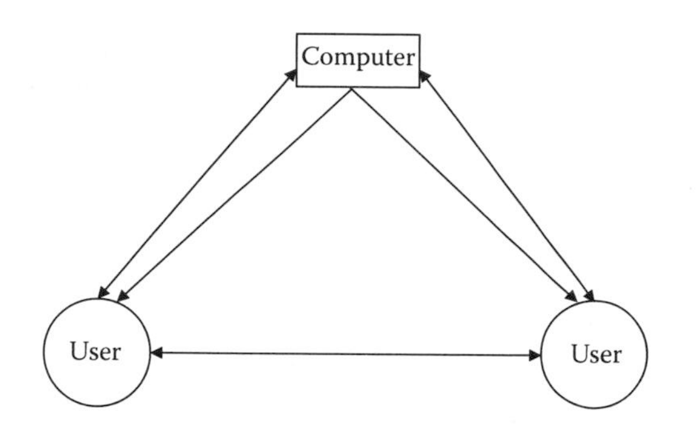

FIGURE 10.1
Social interaction in the *human* → *computer* system.

The earlier figure shows that in *human* → *computer* groups, there are direct and indirect communication and mutual understanding. Indirect communication is performed via a computer. Communication through a computer and related understanding can be very complex and can be verbal and nonverbal. Not only communication with other people through a computer but also the interaction with a computer may have a dialogical character. There are intra- and intergroup communication in the *human* → *computer* group. In the intragroup communication, the role of direct interaction is increased. In intergroup communication, the role of communication via a computer increases. Analysis of errors during social interaction is important in all such situations.

Analysis of the cases listed in the literature and our own observations suggests that there is a special kind of error associated with the social factor. This group of errors is often associated with communication between users of computer systems and software developers for these systems. Particular attention should be paid to the balanced relationship between direct communication and computer communication.

Social interaction occurs at the level of objectively defined meanings. However, their interpretation is dependent on past experience and significance of information to the users. According to the model of self-regulation, users have to create not identical but adequate goals, activate the structure of related knowledge, create a conceptual or stable model, and formulate adequate dynamic models of situation that include not just verbally logical (SA) but also imaginative, conscious, and unconscious components. These functional mechanisms are constructed during interaction with other functional mechanisms of activity regulation according to a goal of activity. Cognitive processes are combined in various ways during different stages of activity regulation. This combination of cognitive processes is performed as a system of goal-directed cognitive actions and operations. Working or operative thinking, which is a basic mechanism of gnostic dynamic, plays a leading role in developing a mental model. Thus, in contrast to cognitive psychology, which concentrates attention on what is happening in the brain, at this stage of analysis, SSAT concentrates its effort on how users interact with objects or with each other. With some modification, SSAT also uses cognitive analysis only at one stage. Thus, task analysis in most cases involves several stages and levels of analysis.

Feedback during the process of social interaction is important for its analysis because it provides information about the result of interaction. Based on such feedback, users realize the efficiency of interaction and can correct and regulate their joint activity. The success of group performance depends on how well social feedback is interpreted. Emotionally evaluative (significance) and motivational factors (inducing motivational mechanism) play an important role in the process of self-regulation of joint activity. Effectiveness of social interaction often largely depends on mechanisms of regulation of activity. This factor is virtually ignored in cognitive psychology.

Social interaction is a very complex process of mutual influences of subjects on each other. We cannot discuss here in detail this complex type of human activity. However, even this short analysis gives some general ideas about error reduction. Let us consider some examples.

Later, we shortly present errors caused by inadequate intergroup relationship.

If computer users are waiting for information that is required for further processing or development of report that summarizes the balance, and so on, and information is not received, then users, according to instructions, need to call the service department to determine the reasons for the delay of information. Users and service departments are two groups that have quite different terminologies and operate with nonoverlapping information. Their background and work experience are so different that it is incredibly difficult for them to understand each other. Based on exchanging information, they create a completely different mental model of the situation and as a result cannot correctly understand each other. Moreover, these two groups do not realize it. For example, users tend to know the name of the file or report that they expect to receive and the service department only knows which systems and which programs should provide them. When users start to ask questions, specialists in the service department are completely stumped for an answer. Paradoxically, users have to ask for help from programmers that could help to formulate the required questions understandable for both groups' terminology.

Let us consider the category of errors caused by the intergroup relationship. A computer program can be written or initiated by one programmer but must be completed or changed by another programmer, which can be the source of errors. Typically, such category of errors is associated with lack of correct information. A new specialist does not receive the necessary task-specific instructions or does not have the necessary knowledge and skills to perform it. Sometimes a previously engaged programmer cannot correctly explain his/her work to a new programmer.

Let us consider in an abbreviated manner the error analysis at the business organization level. One of the important issues of error analysis on the business organizational level is relation between the formal structure of business organization and users' opportunity to perform their individual duties.

Later, we present an example of possible causes of errors at the business organizational level.

1. Errors in software design and coding:
   a. Misinterpretation of requirements
   b. Poor testing
   c. Poor or no documentation
   d. Poor quality of user interface
2. Errors caused by the users or by the organization that utilize the computer system:
   a. Organization provided wrong specifications for the computer system.
   b. Key-entry errors during the data population process.
3. Errors that emerge during the system implementation or upgrade. This kind of error is usually a result of either miscalculation in technical requirements or lack of adequately trained personnel:
   a. Hardware/operating system failure during the implementation
   b. Poor system architectural design
   c. Incompatibility of hardware/operating system/DB with the business needs or inadequate training of staff to use the newly implemented system

A separate group of errors is related to the integration of various components of a computer system and coordination of functioning of such components that we will call *system consolidation errors*. Mergers and acquisitions are very popular these days. These processes lead to consolidation and integration of computer systems and DBs of the companies involved in such process. If the information has to flow from one of the consolidated company DB to the other company DB, the tables in both DBs should have an identical structure. The common error is to update one of such DBs to a new version or to change its structure without making the same changes in the DB that shares real-time information with the first one.

Let us consider an example of system consolidation errors. One company uses the system for the human resources department. It bought a company that uses the same system for both human resources and payroll departments. The payroll system has to be updated every 2 months because of the tax updates. The human resources system does not have to be updated that frequently. After the acquisition, both companies shared their information. The payroll system had to be updated. Every such update includes changes in the software and in the structure of the DB. It has been decided not to update the human resources system of the first company to the same version.

Figure 10.2 depicts the structure of the payroll deduction table with the names of its columns, the size of the columns, and their description. This figure shows the table structure and the characteristics of its columns.

Figure 10.3 depicts the fragment of the payroll deduction table with the data stored in this table. Each column in this table holds data for various types of deductions. Columns in this figure correspond to the rows in Figure 10.3. We choose this table as an example because update of the payroll system included structural changes of this table, and as a result, the human resources system was unable to access this table.

During the update of the payroll system, a new column with tax-related information has been added to the payroll deduction table. This change in the table structure made it different from the identical table in the human resources system, which led to the inability of these two systems to share information. This is an example of system consolidation errors. Taxation changes do not affect the human resources system. So it has been decided to leave it out of the update. However, these two systems share real-time information and their table should be in sync. Otherwise the information flow is disrupted.

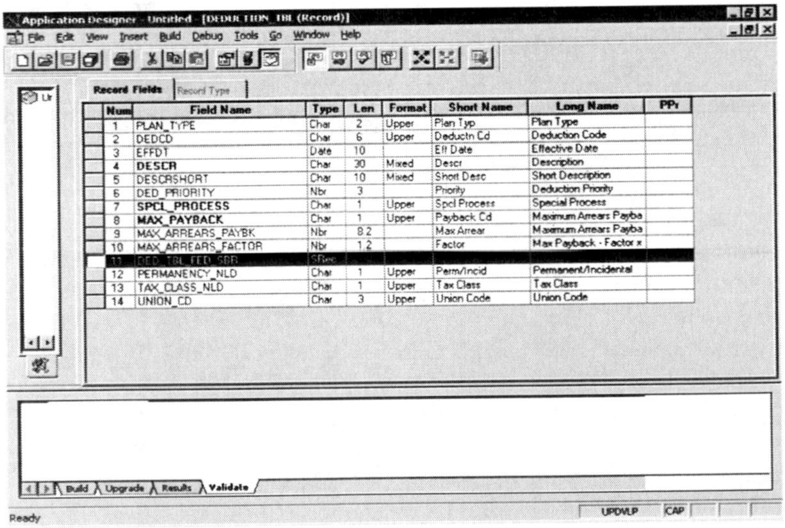

Application Designer - Untitled - [DEDUCTION_TBL (Record)]

File Edit View Insert Build Debug Tools Go Window Help

Record Fields | Record Type

Num	Field Name	Type	Len	Format	Short Name	Long Name	PPr
1	PLAN_TYPE	Char	2	Upper	Plan Typ	Plan Type	
2	DEDCD	Char	6	Upper	Deductn Cd	Deduction Code	
3	EFFDT	Date	10		Eff Date	Effective Date	
4	**DESCR**	Char	30	Mixed	Descr	Description	
5	DESCRSHORT	Char	10	Mixed	Short Desc	Short Description	
6	DED_PRIORITY	Nbr	3		Priority	Deduction Priority	
7	**SPCL_PROCESS**	Char	1	Upper	Spcl Process	Special Process	
8	**MAX_PAYBACK**	Char	1	Upper	Payback Cd	Maximum Arrears Payba	
9	MAX_ARREARS_PAYBK	Nbr	8.2		Max Arrear	Maximum Arrears Payba	
10	MAX_ARREARS_FACTOR	Nbr	1.2		Factor	Max Payback - Factor x	
11	DED_TBL_FED_SBR	SRec					
12	PERMANENCY_NLD	Char	1	Upper	Perm/Incid	Permanent/Incidental	
13	TAX_CLASS_NLD	Char	1	Upper	Tax Class	Tax Class	
14	UNION_CD	Char	3	Upper	Union Code	Union Code	

Build | Upgrade | Results | Validate

Ready UPDVLP CAP

FIGURE 10.2
Structure of the payroll deduction table.

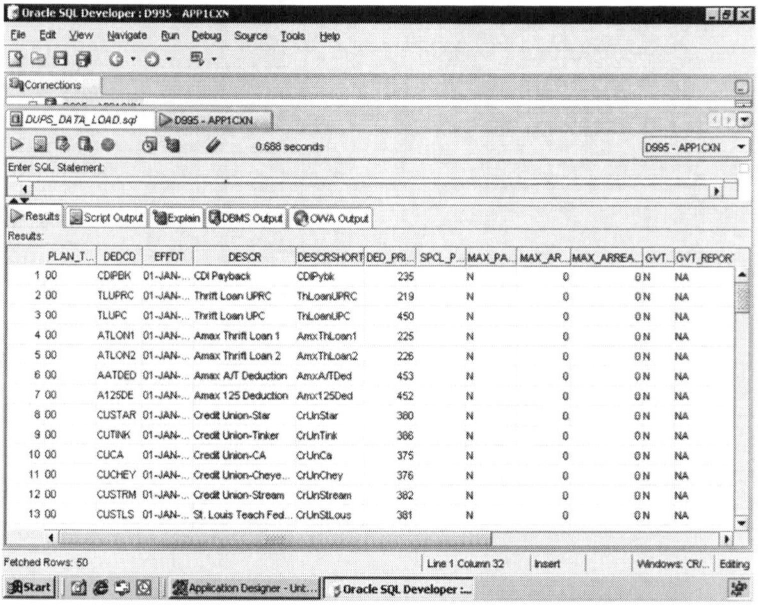
Oracle SQL Developer : D995 - APP1CXN
File Edit View Navigate Run Debug Source Tools Help
Connections
DUPS_DATA_LOAD.sql
D995 - APP1CXN
0.688 seconds
D995 - APP1CXN
Enter SQL Statement:
Results Script Output Explain DBMS Output OWA Output
Results:
PLAN_T... DEDCD EFFDT DESCR DESCRSHORT DED_PRI... SPCL_P... MAX_PA... MAX_AR... MAX_ARREA... GVT... GVT_REPOR
1 00 CDIPBK 01-JAN-... CDI Payback CDIPybk 235 N 0 0 N NA
2 00 TLUPRC 01-JAN-... Thrift Loan UPRC ThLoanUPRC 219 N 0 0 N NA
3 00 TLUPC 01-JAN-... Thrift Loan UPC ThLoanUPC 450 N 0 0 N NA
4 00 ATLON1 01-JAN-... Amax Thrift Loan 1 AmxThLoan1 225 N 0 0 N NA
5 00 ATLON2 01-JAN-... Amax Thrift Loan 2 AmxThLoan2 226 N 0 0 N NA
6 00 AATDED 01-JAN-... Amax A/T Deduction AmxA/TDed 453 N 0 0 N NA
7 00 A125DE 01-JAN-... Amax 125 Deduction Amx125Ded 452 N 0 0 N NA
8 00 CUSTAR 01-JAN-... Credit Union-Star CrUnStar 380 N 0 0 N NA
9 00 CUTINK 01-JAN-... Credit Union-Tinker CrUnTink 386 N 0 0 N NA
10 00 CUCA 01-JAN-... Credit Union-CA CrUnCa 375 N 0 0 N NA
11 00 CUCHEY 01-JAN-... Credit Union-Cheye... CrUnChey 376 N 0 0 N NA
12 00 CUSTRM 01-JAN-... Credit Union-Stream CrUnStream 382 N 0 0 N NA
13 00 CUSTLS 01-JAN-... St. Louis Teach Fed... CrUnStLous 381 N 0 0 N NA
Fetched Rows: 50
Line 1 Column 32
Insert
Windows: CR/...
Editing
Start
Application Designer - Unt...
Oracle SQL Developer :...

FIGURE 10.3
Fragment of the payroll deduction table.

The previously described error is related to system integration and/or consolidation errors and needs a system-level analysis.

Another kind of error that arises during the system integration/consolidation is related to the errors of data load and data population. When computer systems are consolidated, it is necessary to merge massive DBs together. Especially dedicated for this purpose, such process is usually facilitated by writing and testing software with DBs that have the same structure as the real ones. The typical error in such testing is to check only newly created DB. It is very important to compare the new DB with the sources of its population to make sure that the information has not been lost or corrupted. Consolidation of the systems always introduces additional risk of errors and failures. The businesses usually concentrate their attention only on reducing the cost of such consolidation.

The genetic method of study in AT has been first introduced by Vygotsky (1978). The essence of the genetic method is that psychological functions are studied dynamically during their development. This method in ergonomics can be used for task analysis. For example, scientists conduct task analysis in the process of task acquisition. Analysis of the acquisition process can predict the reliability and precision of task performance. Hence, error analysis is also very useful at the stage of task acquisition. Below, we demonstrate

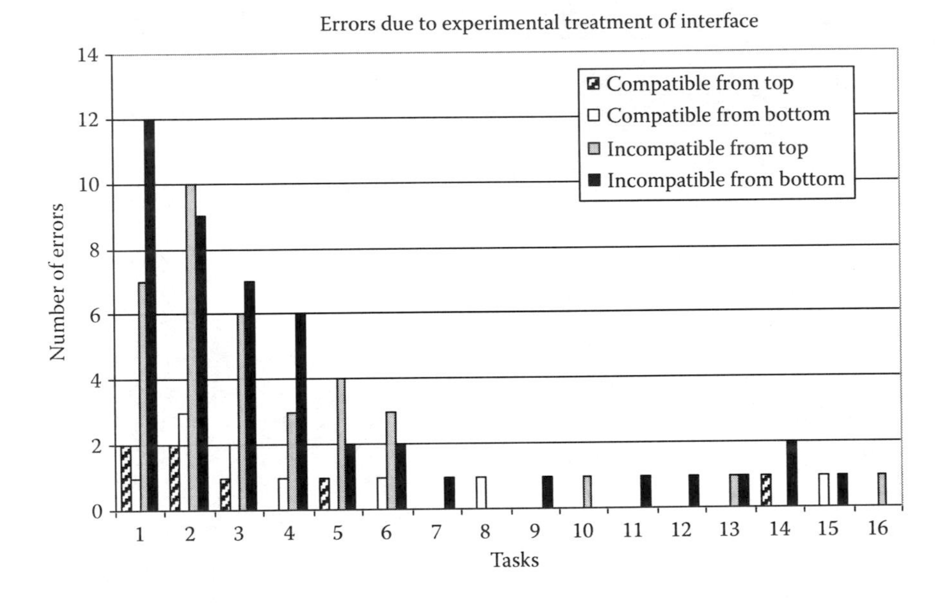

FIGURE 10.4
Errors due to the treatment of the interface for various groups.

how this method can be adapted for this purpose. The error analysis has been conducted based on the comparison of the number of errors at various stages of the skill acquisition process. Figure 10.4 depicts how the error rate has been changing across trials for groups who performed the tasks with different levels of complexity. Compatibility and tool arrangement and their combination were the main factors of task complexity in this experiment. The simplest task was compatible from the top and the most complex task was incompatible from the bottom. This figure demonstrates how the error rate changes across trials for groups who performed the tasks with different levels of complexity. There was a significant difference in the number of errors between the groups. This difference was the most significant at the first five trials.

It can be observed that after six trials, the number of errors due to the incompatibility of the interface dropped considerably across groups. However, there was a significant difference in the number of errors between groups.

Figure 10.5 shows different types of errors overall, in the first five sets, and in the last five sets.

One can observe a significant effect of the learning stage for more complicated tasks in particular. An analysis of variance showed a significant effect of compatibility ($F(1, 28) = 72.7$, $p < 0.001$), whereas no effect on the tool arrangement ($F(1, 28) = 2.1$, $p > 0.05$).

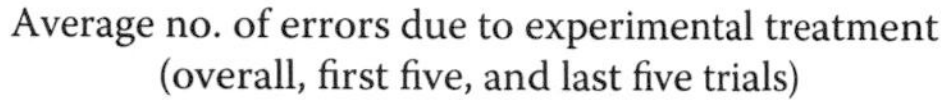

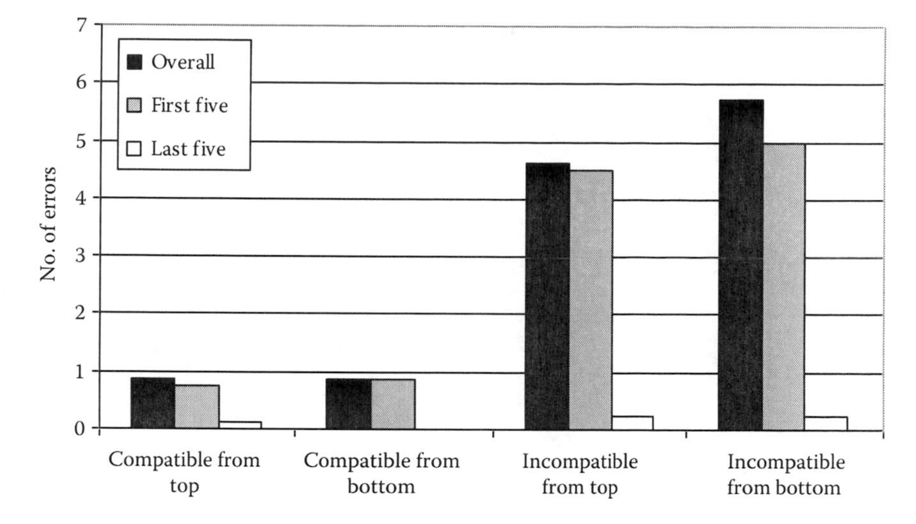

FIGURE 10.5
Average number of errors for different sets of tasks and errors in different groups across tasks.

The current section briefly presents error analysis principles that derive from SSAT. The presented method can be used for error analysis of man–machine and computer-based system and can be especially beneficial during consolidation of computer systems during mergers and acquisitions. The relation between the structure of human activity, activity strategy, and human errors should be analyzed.

11

Systemic-Structural Activity Approach to Reliability Assessment of Computer-Based Tasks

11.1 Objectively Logical Analysis of the Existing Method of Task Performance

Objectively logical analysis is the most common and simple qualitative method utilized for task analysis (Bedny and Karwowski, 2003). It may consist of a short verbal description of job and task performance and a brief description of technological processes, including description of major equipment, tools, raw material, and sequence of basic technological processes. Work conditions, relationship between computerized and noncomputerized components of work, potential for extreme situations, and so forth should be included as well. It is the first method that is utilized in this study.

The task *receiving the orders* is a computer-based task. The purpose of this task is to receive the file containing orders into the local computer system from another distant computer. This is the key task that is performed in the very beginning of the shift. Completion of this task allows creating work for 50–60 employees for the current shift. Without successful completion of this task, 50–60 employees with hourly pay cannot start their work.

The task *receiving orders* has been chosen as an object of study because of the frequent failures of this task performance (Bedny et al., 2010). The failure was more frequent on Mondays after 2 days off. In order to resolve the problem, the manager had to contact a computer specialist who had to re-create and resend the file from one computer system to another because the initially sent file has been lost or fragmented. Computer specialists who can re-create and resend the required file are located in another state and in a different time zone, which made this problem even harder to manage. The re-creation and resending of this file can take about an hour, which leads to idle time for 50–60 workers. This in turn causes significant financial loses. Therefore, this delay was classified as a failure.

The following are the steps involved in the performance of *receiving orders* task:

1. Turn the computer on.
2. Log into the system.
3. Follow the steps to receive the file.

The computer system in consideration uses UNIX operating system. The operator utilizes the command that allows him to see the list of files in the current directory. He or she then checks if the file that contains the orders is on the list. If the file has been received, the operator checks the date stamp on the file to make sure it is a new order file, not an old one. If the date stamp on the file is from today, then the task *receive the orders* is completed and the file can be processed using the software specifically designed for this purpose. As a result, the orders for the current shift are distributed among all workers. If the file name is not on the list or does not have a current date on its date stamp, then the operator should restart the communication software that facilitates the file transfer process. The communication software is designed to first send the file to a temporary holding area (separate region in the computer memory) to allow the receiving process to complete before the file is processed further. That has been done to make sure that the operator does not process a partially received file. Only after the receiving is completed the file is copied to the production directory where an operator can see it. In most cases, the file receiving process is completed long before the start of the morning shift, but in some cases, it was still running when the operator was restarting the communication software. It was leading to failures of the *receiving orders* task. The qualitative analysis of the task showed that the operator had no information if the file was coming or not. The operator could not understand why the same actions that in most cases gave the desired result lead to the failure in some. During phone conversations with computer specialists, the computer operator would state that he was just following his instructions as usual and did not understand why it lead to failure at this time. An operator would blame computer specialists and insisted that the computer-based informational system functioned incorrectly. It became necessary to determine what exactly is causing the system failure. The cause of the human–computer system malfunctioning can be explained only after uncovering breakdown in computer-informational system or uncovering the operator's erroneous actions. Therefore, the purpose of this study was to determine the cause of the *human–computer* system failure and to suggest system improvement that would allow reducing system failures and raising the reliability of task performance. At first, it was necessary to discover the cause of failures, and then assess the probability of system failures using the existing work method. As the second step, it was necessary to design a new, more efficient method with the corresponding evaluation of the reliability of its performance by utilizing the analytical methods of the study. If during calculation it would be discovered that the new method of performance is more reliable, then this method could be implemented. At the final stage, it was required to evaluate the reliability of the new method of performance after its implementation.

11.2 Algorithmic Description of Existing Method of Task Performance

According to the systematic-structural activity theory (SSAT), the next step of study is morphological analysis that includes algorithmic description of activity followed by breaking activity down into cognitive and motor actions and operations. Morphological analysis also involves the development of activity time structure. In our study, time structural analysis has been omitted. Therefore, after the qualitative stage of task analysis, the algorithmic description of the task has been performed.

Algorithmic analysis included table-symbolic and graphic-symbolic description of activity during task performance. The purpose of this second stage of analysis was to develop a human algorithm of task performance and to determine the logical organization of workers' cognitive and behavioral actions. This kind of algorithms differs from mathematical or computer algorithms. The main units of analysis used by such algorithms are actions, and members of the algorithm are classified according to psychological principles. To make it easier for the reader to comprehend the following material, we will briefly repeat some key points of the description of the human algorithm. A human algorithm models human performance and includes different kinds of operators and logical conditions. Operators include actions that transform objects, energy, or information. Subjectively, a member of the algorithm is perceived by a subject as a component of activity, which has logical completeness. Each member of the algorithm is described by a special symbol described before.

For example, afferent operators (involve receiving information) are designated by O^{α}, efferent operators (motor) are designated by O^{ε} and associated with the executive components of activity, and logical conditions (decision-making) are designated by the symbol l. This method is described in detail in previous chapters.

Algorithmic description can be utilized in table and/or graphic form. The first method presents human algorithm as a table. It utilizes a combination of verbal and symbolic description. The second method utilizes only geometric symbols. Each geometric symbol is designated to a particular member of the algorithm. The symbolical description that has been discussed in this section is depicted by these geometric symbols. These two methods of algorithmic description will be considered in more detail later.

As the first step, the table-algorithmic description has been conducted (see Table 11.1).

TABLE 11.1

Algorithmic Description of the Existing Method of Task Performance

Members of the Algorithm	Description of the Members of the Algorithm
O_1^{ε}	Type user name and password. Press "Enter."
O_2^{ε}	Type command "*ls -l*" and press "Enter."
O_3^{α}	Check to see if the order file is on the list.
$l_1 \overset{1}{\uparrow}$	If the file is on the list, go to O_4^{α}. If the name of the file is absent from the list, go to O_5^{ε}.
O_4^{α}	Check to see if the date stamp of the order file has a correct date.
$l_2 \overset{2}{\uparrow}$	If the file has today's date, then the received file is the expected file, go to O_8^{ε}. If the file has the old date, then go to O_5^{ε}.
$\overset{1}{\downarrow} O_5^{\varepsilon}$	Type "restore communication" and press "Enter." Repeat if necessary.
$O_6^{\alpha w}$	Wait for the completion of the restore of communication (until initial screen comes up).
$O_7^{\alpha w}$	Make a note of the time of completion of the restore. Wait for 5–7 min.
${}_1O_2^{\varepsilon}$	Type command "*ls -l*" and press "Enter"(the same as O_2^{ε}).
${}_1O_3^{\alpha}$	Check to see if the order file is on the list (the same as O_3^{α}).
${}_1l_1 \overset{1}{\uparrow}_1$	If the file is on the list go to ${}_1O_4^{\alpha}$. If the name of the file is absent from the list, go to ${}_1O_5^{\varepsilon}$ (the same as l_1).
${}_1O_4^{\alpha}$	Check to see if the date stamp of the order file has a correct date (the same as O_4^{α}).
${}_1l_2 \overset{2}{\uparrow}_2$	If the file has today's date, then the received file is the expected file, go to O_8^{ε}. If the file has the old date, then go to ${}_1O_5^{\varepsilon}$ (the same as l_2).
${}_1\overset{1}{\downarrow}_1 O_5^{\varepsilon}$	Type "restore communication" and press "Enter." Repeat if necessary (the same as O_5^{α}).
${}_1O_6^{\alpha w}$	Wait for the completion of the restore of communication (until initial screen comes up).
${}_1O_7^{\alpha w}$	Make a note of the time of completion of the restore. Wait for 5–7 min (the same as $O_7^{\alpha w}$).
${}_2O_2^{\varepsilon}$	Type command "*ls -l*" and press "Enter" (the same as O_2^{ε}).
${}_2O_3^{\alpha}$	Check to see if the order file is on the list (the same as O_3^{α}).
${}_2l_1 \overset{1}{\uparrow}_2$	If the file is on the list, go to O_4^{ε}. If the name of the file is absent from the list, go to O_9^{ε} (the same as l_1).
${}_2O_4^{\alpha}$	Check to see if the date stamp of the order file has a correct date (the same as O_4^{α}).
${}_2l_2 \overset{2}{\uparrow}_2$	If the file has today's date, then the received file is the expected file, go to O_8^{ε}. If the file has the old date, then go to O_9^{ε} (the same as l_2).

(Continued)

TABLE 11.1 (*Continued*)

Algorithmic Description of the Existing Method of Task Performance

Members of the Algorithm	Description of the Members of the Algorithm
$\overset{2}{\downarrow}_2 \overset{1}{\downarrow}_2 \overset{2}{\downarrow} O_8^{\varepsilon}$	Type command "interface-orders." The end of the considered task (**the goal achieved**) and the beginning of the following task.
${}_2\overset{1}{\downarrow}_2 \overset{2}{\downarrow} O_9^{\varepsilon}$	Call computer specialist (**Failure**).

In the next step, the table that described all actions utilized by human operator during task performance has been developed according to the presented algorithm of human activity (Table 11.2). Classification of actions is performed utilizing the SSAT classification principles described earlier.

This table helps the specialist to understand what kinds of actions are used during the performance of a particular member of the algorithm. The table

TABLE 11.2

Classification of Actions for the Existing Method of Task Performance

Members of Algorithm	Description of Members of Algorithm
O_1^{ε}	Two motor actions consisting of sequence of motor motions (pressing keys).
O_2^{ε}	The same as O_1^{ε}.
O_3^{α}	One perceptual action that represents searching for and identification of the name of the order file. The action is accompanied by visual screening operations.
l_1	Simple decision-making action with two outcomes.
O_4^{α}	Two perceptual actions.
l_2	The same as l_1.
O_5^{ε}	Two motor actions consisting of sequence of motor motions (pressing key).
$O_6^{\alpha w}$	Active waiting period that ends when initial screen comes up.
$O_7^{\alpha w}$	Active waiting period that ends depending on interval of time required for evaluation.
${}_1O_2^{\varepsilon}$	The same as O_2^{ε}.
${}_1O_3^{\alpha}$	The same as O_3^{α}.
${}_1l_1$	The same as l_1.
${}_1O_4^{\alpha}$	The same as O_4^{ε}.
${}_1l_2$	The same as l_2.
O_8^{ε}	Motor action that consists of sequence of motor motions (pressing keys).
O_9^{ε}	Consist of the sequence of verbal and motor actions. The sequence and number of actions are not precisely defined.

form of algorithmic description is required but not sufficient for reliability assessment. For further reliability assessment, it is important to transfer the table form of algorithmic description into the graphic form. This kind of model is an excellent visual aid for reliability assessment. The graphic form of algorithm of task performance is presented in Figure 11.1.

This model demonstrates in visual form probabilities of transition from one member of the algorithm to the next and possible strategies of task performance. Such model helps to calculate the probability of different elements of the algorithm when the human operator utilizes different strategies of performance and ultimately the probability of success and failure of task performance. Hence, during task analysis and calculation of the probability of successful and unsuccessful performance, the specialist utilizes the models of activity described earlier. During such analysis, switching from one model to another simplifies reliability analysis and helps to obtain the required information. Such table descriptions are also very helpful in working with experts. All questions can be presented to experts in the order related to the logic of the algorithmic description.

As it can be seen from the table and graphic algorithmic models of the task performance, it has two outcomes. One outcome is O_8^ε when the task has been successfully completed (goal-related member of the algorithm). The other outcome is O_9^ε (failure-related member of the algorithm). The algorithmic models show the activity strategies when after unsuccessful performance the human operator can repeat the part of the task that starts with the operator ${}_1O_5^\varepsilon$ (type "restore communication" and hit "Enter"). This leads to the second performance of O_5^ε and all following members of the algorithm. When O_5^ε is performed the second time on the left side (see Figure 11.1), it has subscription 1. If the same member of the algorithm is repeated the second time, the number 2 should be installed at the left of O_5^ε instead of 1 (${}_2O_5^\varepsilon$).

The existing method of task performance allows us to gather information about the probability of success and failure of the task performance. The empirical statistical analysis showed that the probability of success was 0.9 (probability of O_8^ε) and the probability of failure O_9^ε was 0.1. However, at this stage of analysis, the purpose of our study was to find out what was the impact of the outcome of each member of the algorithm on the final result of the task. The collected data allowed to determine the bottlenecks of the considered method of task performance and to use the obtained data on the probability of successful performances and failures of different members of the algorithm in calculation of the reliability of the new method of task performance. Expert opinion regarding probabilistic characteristics of each member of the algorithm was then corrected, taking into consideration the probability of O_8^ε and the probability of O_9^ε obtained through empirical statistical analysis. The initial probabilities required further calculations and the expert estimates corrected after the discussions are shown in Figure 11.1.

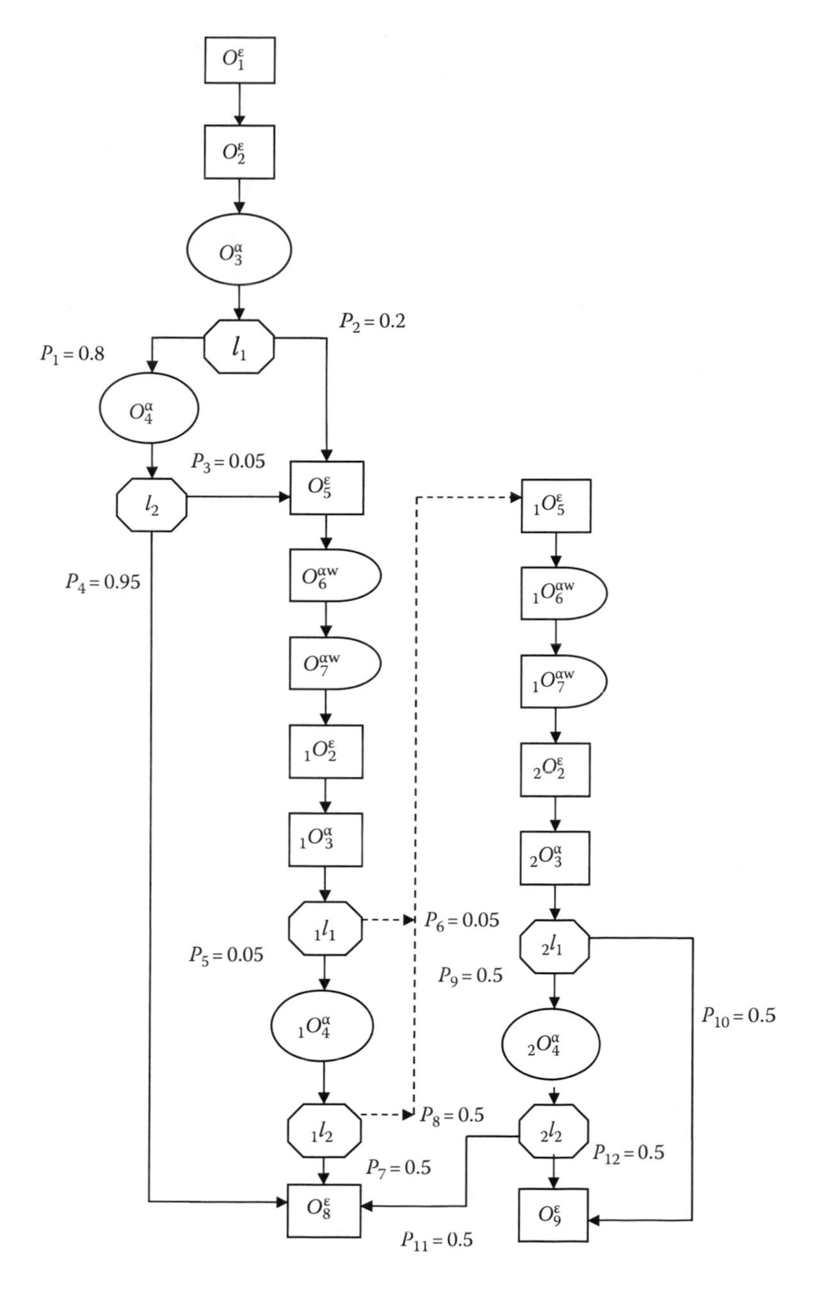

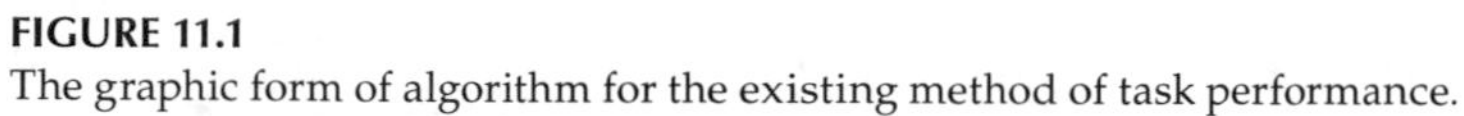

FIGURE 11.1
The graphic form of algorithm for the existing method of task performance.

It is important to emphasize that the analysis of the probabilistic structure of the task and the calculation of probabilities have been conducted using table and graphic algorithmic description of the existing method of task performance. Without these two algorithmic descriptions of task performance, it would be impossible to follow task flow. The questions that were presented to the experts also followed the logic of the sequence of the members of the algorithm. The probabilities were later adjusted, taking into consideration the interaction of the members of the algorithm, the specifics of the actions involved in their performance as described in Table 11.2, etc. Depending on the task specifics, it should be determined what branch of the algorithm should be analyzed first. In this step, the most informative branch of the algorithm that would allow obtaining the initial probabilistic data should be determined.

11.3 Analysis of Erroneous Actions and Failures for the Existing Method of Task Performance

The algorithmic activity description is critically important for the error analysis and for discovering their causes. It allows determining what members of the algorithm result in erroneous actions and what their objective causes are. Analysis of interdependence and interaction of actions are important in understanding the root causes of errors or failures. Thus, errors and failures are derived not only from separate erroneous actions but also from strategies of activity performance.

The analysis of the described algorithm of task performance demonstrates that errors leading to failures are associated with the following members of the algorithm:

1. Decision-making (l_1). If the name of the *order* file is on the list, go to O_4^α; if this name is absent, go to O_5^ε.
2. Executive operator (O_5^ε). Type "restore communication," hit "Enter." Repeat if necessary.
3. Decision-making (${}_1l_1$). Repeat performance of l_1.
4. Executive operator (${}_1O_5^\varepsilon$). Repeat performance of O_5^ε.

Let us analyze operators l_1 and O_5^ε. The decision l_1 is made based on O_3^α. Search for the name of the *order* file on the list of files on the screen, which includes a simple perceptual action O_3^α, as shown in Table 11.1. The error during the performance of this action has very low probability and we can ignore the possibility of its appearance in this task. The operator O_5^{ε} consists of sequential movements that include pressing the keys on the keyboard. These movements are integrated into two motor actions (see Table 11.2). These actions are very simple and if the errors occur, they can be corrected. Therefore, this kind of errors should not be considered as failures.

Now, we will analyze the decision-making action associated with l_1. There are two possible outcomes:

1. The file name *orders* is not on the list of file names on the screen. The computer operator makes the right conclusion that the communication is broken. Based on this conclusion, he or she makes the right decision to restore communication and performs O_5^{ε}.
2. The communication is not broken. The file has been transmitted later than usual and is accumulated in the temporary directory that the operator cannot see. There is no file name *orders* on the list of file names on the screen. The computer operator makes the wrong conclusion that the communication is broken and performs O_5^{ε}. When the computer operator performs O_5^{ε} while the file is coming to the temporary storage, he or she breaks the file transmission and this leads to system failure.

Let us consider the second round of performance of l_1 and O_5^{ε} (they are designated as ${}_1l_1$ and ${}_1O_5^{\varepsilon}$):

1. The initial performance of O_5^{ε} (type "restore communication" and hit "Enter") did not lead to the execution of this command. When the computer operator performed ${}_1O_5^{\varepsilon}$ (repeated O_5^{ε}), the communication line was restored and the file started accumulating in the temporary storage.
2. The initial performance of O_5^{ε} restored the communication and the file started accumulating in the temporary storage. The computer operator should wait till the file transmission is completed and the file appears into the production environment (performance of $O_6^{\alpha\omega}$ and $O_7^{\alpha\omega}$). However, there are cases when the file is bigger than usual or the transmission is slower than usual and 5–7 min waiting period is not sufficient or the computer operator just rushed and did not give it at least 5–7 min as required by instructions. Attempts to restore communication (to perform ${}_1O_5^{\varepsilon}$) again when the file was already coming to the temporary storage result in disruption of the transmission process and fragmentation of the file. This in turn leads to failure because ${}_1O_5^{\varepsilon}$ has been performed too soon.

Analysis of failures shows that erroneous actions that lead to fragmentation of the file are performed because the computer operator does not have accurate information about the state of communication and file transmission. The name of the *order* file might be missing from the list of the files on the screen but that does not always mean that the communication line is broken. The file might be in a process of coming to temporary storage and if O_5^{ε} or ${}_1O_5^{\varepsilon}$ is performed while the file is accumulating, it results in failure through fragmentation of the file and interruption of the transmission process.

As it has been demonstrated earlier, algorithmic models and analysis of the performed members of the algorithm allows understanding the causes of erroneous actions and system failures. For the task that has been analyzed, it can be concluded that the *erroneous actions* of the computer operator are caused by the flaw in the computer system design and by inaccurate instructions given to the operator. The operator makes an incorrect conclusion that from time to time the computer system just fails. However, the real problem has its root in the design of the human–computer interaction process.

11.4 Reliability Assessment of the Existing Method of Task Performance

After the analysis of the causes of system failures, it is possible to calculate the probability of reliable (successful) performance and failures. Hence, we move from error identification stage to error quantification stage. This stage is necessary for the assessment of the reliability of the existing task performance and for the later comparison of the results with the reliability of the improved method of task performance. In other words, this stage is necessary for switching to the error (failure) reduction stage. Reliability assessment stage will facilitate the qualitative comparison of two methods of task performance and will allow determining if the new method leads to a more reliable task performance.

There are three activity strategies of task performance: The main activity strategy is used when the order file is present with the right date stamp on the first attempt (see left branch of the algorithmic model graph in Figure 11.1); the second strategy is used when it is necessary to restore communication because the order file is absent or has the wrong date (see the middle branch in Figure 11.1); the third strategy is used when the right order file does not appear after the second attempt (the rightmost branch in Figure 11.1).

The probabilities for all three strategies are calculated in the following.

The probability ${}_1P_1$ of successful task completion or achievement of the goal-related member of the algorithm O_8^{ε} by utilizing the main activity strategy (see the left branch of the algorithmic model in Figure 11.1) can be calculated as follows:

$${}_1P_1 = P_1 \times P_4 = 0.8 \times 0.95 = 0.76,$$

where

P_1 is a probability of the successful outcome of the logical condition l_1

P_4 is the successful outcome of the logical condition l_2 (see left branch of the algorithm, Figure 11.1)

Let us now calculate the probability of the proper execution of the member of the algorithm O_5^{ε}. This member of the algorithm describes the scenario when the communication is broken and it becomes necessary to restore communication (perform O_5^{ε}, see Table 11.1 and Figure 11.1). The middle branch reflects the performance of the members of the algorithm necessary to restore communication and ultimately to achieve a goal-related member of the algorithm O_8^{ε}. This branch will be called the second strategy of activity.

The rightmost branch reflects the third strategy of task performance. The transition to these strategies of performance is facilitated by logical conditions ${}_1l_1$ and ${}_2l_1$. The transition to the third strategy is defined by a dash line. The third strategy describes the repetition of the second strategy one more time. It will be considered later. As can be seen in Figure 11.1, the second strategy starts with the performance of O_5^{ε}. Therefore, it was necessary to calculate the probability of O_5^{ε}. It can be determined based on the analysis of the initial probabilities of the logical conditions l_1 and l_2. As can be seen in Figure 11.1, the probability of l_1 is equal $P_2 = 0.2$ and the probability of l_2 is equal $P_3 = 0.05$. The probability of O_5^{ε} is a result of the following outcomes. The probability of the order file being present on the list of file names is $P_1 = 0.8$, then, $P_2 = 0.2$. Probability P_3 of the order file having the wrong date has been determined using expert analysis the same as the other two probabilities. Knowing P_1 and P_3, it is now possible to calculate the probability of the outcome of the logical condition l_2 on the right-hand side:

$${}_1P_2 = P_1 \times P_3 = 0.8 \times 0.05 = 0.04,$$

where

P_1 is a probability of the successful outcome of the logical condition l_1

P_3 is the unsuccessful outcome of the logical condition l_2 (on the right, Figure 11.1)

The probability of O_5^ε is the combination of the probabilities of two independent events ($P_2 = 0.2$ and ${}_1P_2 = 0.04$). Then, the probability of O_5^ε is equal to

$${}_1P_3 = P_2 + {}_1P_2 = 0.2 + 0.04 = 0.24,$$

where P_2 is a probability of the negative outcome of the logical condition l_1.

The calculation of the probabilities of other members of the algorithm is given later. These probabilities are also calculated using the initial probabilities determined based on the expert assessment. They are shown in Figure 11.1.

The probability of the existence of the order file with the current dates on its time stamp after the first attempt is $P_4 = 0.95$. The existence of the file name and the presence of the current dates on its time stamp are interdependent events. By definition, the probability of these two events occurring at the same time equals $P_1 \times P_4 = 0.8 \times 0.95$. Therefore, the probability of O_8^ε after the first attempt is equal to ${}_1P_1 = 0.76$.

Let us now calculate the probability of the second attempt to check the time stamp on the order file (${}_1O_4^\alpha$). The probability of ${}_1O_4^\alpha$ equals the product of the probability of O_5^ε and P_5:

$${}_1P_4 = {}_1P_3 \times P_5 = 0.175,$$

where P_5 is a probability of the successful outcome of the logical condition ${}_1l_1$.

The probability of the performance of the goal-related member of the algorithm O_8^ε after the second attempt can be calculated as a product of the probability of ${}_1O_4^\alpha$ and the probability P_7 that the order file has been discovered after the second attempt. Therefore, the probability of O_8^ε after the second attempt is

$${}_1P_5 = {}_1P_4 \times P_7 = 0.075,$$

where P_7 is a probability of the successful outcome of the logical condition ${}_1l_2$.

The following is the calculation of the probability of ${}_1O_5^\varepsilon$ at the top of the right branch in Figure 11.1:

$${}_1P_6 = {}_1P_3 \times P_6 = 0.24 \times 0.3 = 0.072,$$

where P_6 is a probability of the right-hand outcome of the logical condition ${}_1l_1$.

The combined probability of the outcome of the logical condition ${}_1l_2$ on the right-hand side is equal to

$${}_1P_7 = {}_1P_4 \times P_8 = 0.175 \times 0.5 = 0.075,$$

where P_8 is an initial probability of the negative outcome of the logical condition ${}_1l_2$.

The combined probability of ${}_1O_5^{\varepsilon}$ after the second attempt is calculated as

$$ {}_1P_8 = {}_1P_6 + {}_1P_7 = 0.15. $$

The combined probability of the right-hand outcome of the logical condition ${}_2l_1$ is equal to

$$ {}_1P_9 = {}_1P_8 \times P_{10} = 0.15 \times 0.5 = 0.075, $$

where P_{10} is an initial probability of the negative outcome of the logical condition ${}_2l_1$. It is the partial probability of O_9^{ε}.

The probability of ${}_2O_4^{\alpha}$ has been calculated as

$$ {}_1P_{10} = {}_1P_8 \times P_9 = 0.15 \times 0.5 = 0.075, $$

where P_9 is a probability of the successful outcome of the logical condition ${}_2l_1$.

Let us now complete the calculation of the probability of the goal-related member of the algorithm O_8^{ε}. The probability of the combined outcome of the logical condition ${}_2l_2$ that leads to O_8^{ε} is equal to

$$ {}_1P_{11} = {}_1P_{10} \times P_{11} = 0.075 \times 0.5 = 0.04, $$

where P_{11} is a probability of the successful outcome of the logical condition ${}_2l_2$.

The resulting probability of the goal-related member of the algorithm after all three strategies are applied can be expressed as follows:

$$ {}_1P_{13} = {}_1P_1 + {}_1P_5 + {}_1P_{11} = 0.76 + 0.075 + 0.04 = 0.88. $$

We can now finish calculating the probability of the system failure (member of the algorithm O_9^{ε}). The probability of the outcome of the logical condition ${}_2l_2$ that leads to O_9^{ε} is equal to the product of the probability of ${}_2O_4^{\alpha}$ and initial probability P_{12}. Therefore, the probability of O_9^{ε} is

$$ {}_1P_{12} = {}_1P_{10} \times P_{12} = 0.075 \times 0.5 = 0.04, $$

where P_{12} is a probability of the successful outcome of the logical condition ${}_2l_2$.

Hence, the resulting probability of the system failure can be expressed as follows:

$$ {}_1P_{14} = {}_1P_9 + {}_1P_{12} = 0.075 + 0.04 = 0.115. $$

So, the final probability of the goal-related member of the algorithm O_8^{ε} is equal to 0.88 and the probability of the system failure O_9^{ε} is 0.115. The sum of these probabilities is close to 1. This means that about 12 out of a hundred task performances end up as system failures, which is a pretty high rate of system failures considering the economic consequences of such failures. Analysis of empirical statistical data of failures and successful performances for the existing method of task performance demonstrates that the probability of failures was 0.1 and the probability of successful performances was 0.9. Therefore, the empirical result and the result obtained analytically are very similar.

Let us consider the new method of task performance.

11.4.1 Objectively Logical Analysis of the New Method of Task Performance

The following is the plan for designing the new method of task performance:

Qualitative stage of analysis

This stage of analysis includes a broad number of methods. In this study, only the objectively logical analysis has been used. It included the following steps:

1. The block diagram of the existing method of task performance has been developed.
2. The shortcomings of the existing method of task performance have been analyzed.
3. The new method of task performance has been proposed to overcome the existing shortcomings.
4. To analyze the new method, the block diagram of the new method of task performance has been developed.
5. The comparative analysis of two block diagrams has been performed.

   Algorithmic stage of analysis
6. The algorithmic analysis of the new method has been performed. This stage of analysis included the development of table-symbolic description, graphic-symbolic description, and classification table of actions during task performance.

   Quantitative stage of analysis
7. Initial probabilities for the new method of task performance have been estimated based on expert analysis utilizing probabilities determined for the existing method of task performance.
8. The probability of success and failure of the new method of task performance has been calculated.
9. Reliability of the existing and new methods of task performance has been compared.
10. The conclusion about the efficiency of the proposed improvement has been made.

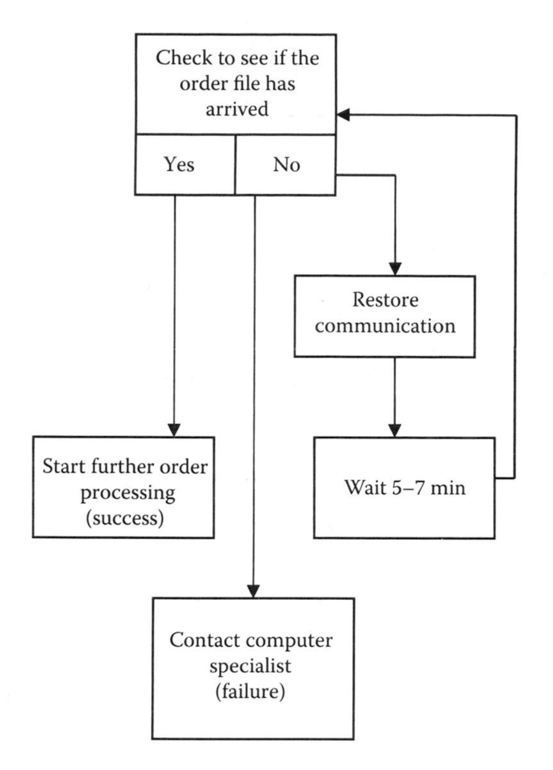

FIGURE 11.2
The block diagram of the existing method of task performance.

According to the described plan, the first step is to conduct comparative qualitative analysis of the existing and the new method of task performance based on the objectively logical analysis. Two block diagrams have been developed to demonstrate the flow of the existing and the proposed method of task performance. Figure 11.2 illustrates the existing method of task performance.

This diagram as well as the algorithmic description discussed earlier demonstrate that the main shortcoming of the existing method of task performance is that the computer operator does not know if the file is being accumulated or the communication is broken when the name of the file has not been found in the production directory. The new method of performance has been developed to overcome this shortcoming. The block diagram of the new method of task performance is presented in Figure 11.3. This block diagram shows that the new method of task performance allows the computer operator to determine if the file is still in a process of being transferred or the communication is broken. This new information allows the operator to avoid errors and successfully complete the task. The diagram demonstrates how the suggested improvement can increase the reliability of task performance.

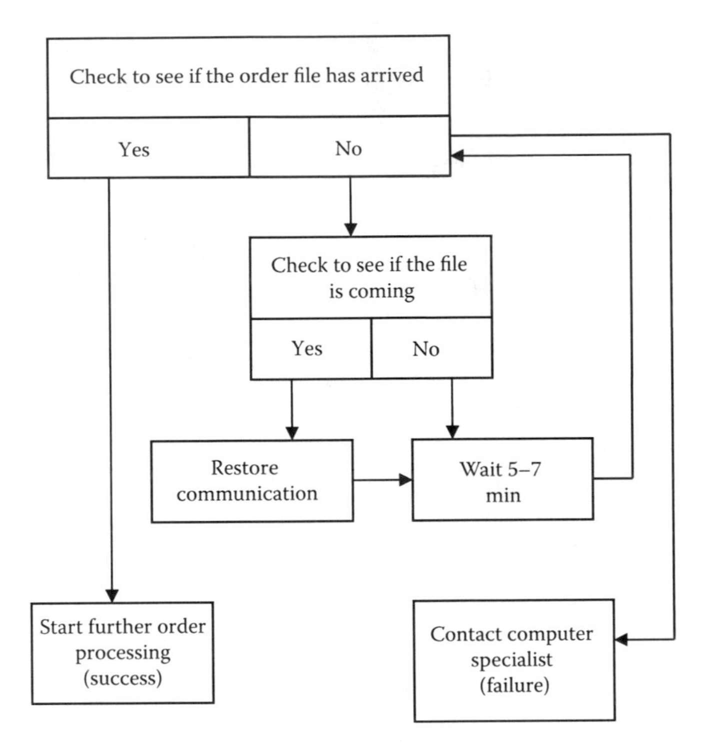

FIGURE 11.3
The block diagram of the new method of task performance.

11.5 Algorithmic Description of the New Method of Task Performance

According to the existing experimental procedures, in order to assess a new method of computer-based task performance, it is necessary to implement this new method first. It requires update of the existing software or development of a new one. Only after that the empirical evaluation of the suggested improvement becomes possible. SSAT allows to develop theoretical models of human performance and to conduct required qualitative and quantitative assessment of the suggested method by utilizing analytical methods. In this study, at the second stage of task analysis, the algorithmic description of activity has been used as a basic tool to study a new method of task performance. Based on the qualitative analysis, it has been suggested to introduce the new members to the algorithm. Performance of these new members of the algorithm allowed obtaining information about the file transfer to the temporary storage and determining when this transfer is completed. For instance, the computer operator had to type a new command "where_is_file" and check if the name of the order file is on the list and if the time stamp on the file has a current date. Table 11.3 illustrates the algorithmic description of activity for the new method of task performance.

The classification of the actions for the new method of task performance is presented in Table 11.4.

Further, the algorithmic-graphic model of task performance has been developed based on the data from Table 11.4 (algorithmic-table description). The algorithmic-graphic model for the new method of task performance is presented in Figure 11.4.

Researchers and experts collaborated to determine initial probabilities of the outcomes of logical conditions related to the success and failure of the task performance using developed algorithmic models of the task. Some probabilities were identical to the probabilities for the existing method of task performance. Other initial probabilities have been assigned based on the comparison of the actions performed by the computer operator in the existing and designed conditions. The probability of the outcomes for considered logical conditions can vary from 0 to 1. Only one logical condition in our algorithm has three logical outcomes. Therefore, this algorithm is not a deterministic but rather a probabilistic one (Bedny, 2000).

11.6 Evaluation of the Reliability of New Method of Task Performance

Calculation of the reliability of new method of task performance is the third stage of task analysis in this study. The initial probabilities have been assigned for the outcomes of the logical conditions of the new method of task performance based on the data gained by the comparative analysis of the models of the existing and new methods of task performance (see Figure 11.3). For example, the outcomes of the logical conditions l_1 and l_2 are the same for the new and the existing methods of task performance. All other probabilities have been calculated utilizing the initial probabilities. The following are the calculations of all probabilities required to assess the reliability of the new method of task performance.

The probability of the successful execution after the first attempt is calculated as

$$_1P_1 = P_1 \times P_4 = 0.8 \times 0.95 = 0.76,$$

where

P_1 is a probability of the successful outcome of the logical condition l_1
P_4 is a probability of the successful outcome of the logical condition l_2

TABLE 11.3

Algorithmic Description of the New Method of Task Performance

Members of Algorithm	Description of Members of Algorithm
O_1^{ε}	Type user name and password. Press "Enter."
O_2^{ε}	Type "*ls -l*" command and press "Enter."
O_3^{α}	Check to see if there is an order file on the list.
$l_1 \overset{1}{\uparrow}$	If the file is on the list, go to O_4^{α}. If the name of the file is absent from the list, go to O_5^{ε}.
O_4^{α}	Check to see if the date stamp of the order file has a correct date.
$l_2 \overset{2}{\uparrow}$	If the file has today's date, then the received file is the expected file, go to O_{13}^{ε}. If the file has the old date, then go to O_5^{ε}.
$\overset{1}{\downarrow} O_5^{\varepsilon}$	Type "where_orders" command and press "Enter."
O_6^{α}	Check to see if there is an "orders" file on the screen that reflects information stored in the computer temporary storage.
$l_3 \overset{3}{\uparrow}$	If there is an "orders" line on the list, go to O_7^{α}. If it is absent, go to O_{10}^{ε}.
O_7^{α}	Check the date stamp of the order file stored in computer temporary storage.
$l_4 \overset{4}{\uparrow}$	If the file in computer temporary storage has today's date, then go to $O_8^{\alpha w}$. If the file has the old date, then go to O_{10}^{ε}.
$O_8^{\alpha w}$	Wait until the order file disappears from the computer temporary storage and is transferred into the production environment.
O_9^{ε}	Type "return_to_production" and press "Enter" (automatic transition to production environment).
$_1O_2^{\varepsilon}$	Type command "*ls -l*" and press "Enter" (the same as O_2^{ε}).
$_1O_3^{\alpha}$	Check to see if an order file is on the list (the same as O_3^{α}).
$_1l_1 \overset{1(1-2)}{\uparrow_1}$	If the file is on the list, go to O_{13}^{ε}. If the name of the file is absent from the list, go to O_{14}^{ε}.
$\overset{2(2)}{\downarrow}\overset{3}{\downarrow}\overset{4}{\downarrow} O_{10}^{\varepsilon}$	Type "restore_communication" and press "Enter." Repeat if required.
$O_{11}^{\alpha w}$	Wait until the program that restores communication is completed (initial production screen reappears).
$O_{12}^{\alpha w}$	Make a note of the time of the restore completion. Wait for 5–7 min.
$_2O_2^{\varepsilon}$	Type command "*ls -l*" and press "Enter" (the same as O_2^{ε}).
$_2O_3^{\alpha}$	Check to see if an order file is on the list (the same as O_3^{α}).
$_2l_1 \overset{1}{\uparrow_2}$	If the ordered file is on the list, go to $_1O_4^{\alpha}$. If the name of the file is absent from the list, go to O_{14}^{ε} (the same as l_1).
$1O_4^{\alpha}$	Check to see if the date stamp of the order file has the current date (the same as O_4^{α}).

(Continued)

TABLE 11.3 (*Continued*)

Algorithmic Description of the New Method of Task Performance

Members of Algorithm	Description of Members of Algorithm
${}_1l_2 \overset{2(1-3)}{\uparrow}$	If the file has today's date, then the received file is the expected file, go to O^{ε}_{13}. If the file has the old date, then repeat O^{ε}_{10}–${}_1O^{\alpha}_4$. If the result does not change, go to O^{ε}_{14} (the same as l_2).
${}_1\overset{1(2)}{\downarrow}\overset{2}{\downarrow}O^{\varepsilon}_{13}$	Type "interface_orders" command. The end of this task (**the goal is achieved**) and the beginning of the following task.
${}_1\overset{1(2)}{\downarrow}{}_2\overset{1}{\downarrow}{}_1\overset{2(3)}{\downarrow}O^{\varepsilon}_{14}$	Contact computer specialist (**Failure**).

TABLE 11.4

Classification of Actions for the New Method of Task Performance

Members of Algorithm	Description of Members of Algorithm
O^{ε}_1	Two motor actions consisting of sequence of motor motions (pressing keys).
O^{ε}_2	The same as O^{ε}_1.
O^{α}_3	One perceptual action that represents searching and identifying the name of the order file. The action is accompanied by visual screening operation.
l_1	Simple decision-making action with two outcomes.
O^{α}_4	Two perceptual actions.
l_2	The same as l_1.
O^{ε}_5	Two motor actions consisting of sequence of motor motions (pressing keys).
O^{α}_6	The same as O^{ε}_2.
l_3	The same as l_1 and l_2.
O^{α}_7	Two perceptual actions.
l_4	The same as l_3.
$O^{\alpha w}_8$	Active waiting period based on visual information.
O^{ε}_9	Two motor actions consisting of sequence of motor motions (pressing keys).
O^{ε}_{10}	The same as O^{ε}_9.
$O^{\alpha w}_{11}$	The same as $O^{\alpha w}_8$.
$O^{\alpha w}_{12}$	The same as $O^{\alpha w}_{11}$.
${}_1O^{\varepsilon}_2$	The same as O^{ε}_2.
${}_1O^{\alpha}_3$	The same as O^{α}_3.
${}_1l_1$	The same as l_1.
${}_1O^{\alpha}_4$	The same as O^{ε}_4.
${}_1l_2$	The same as l_2.
O^{ε}_{13}	Two motor actions consisting of sequence of motor motions (pressing keys).
O^{ε}_{14}	Consists of sequence of verbal and motor actions. The sequence and number of actions are not precisely defined.

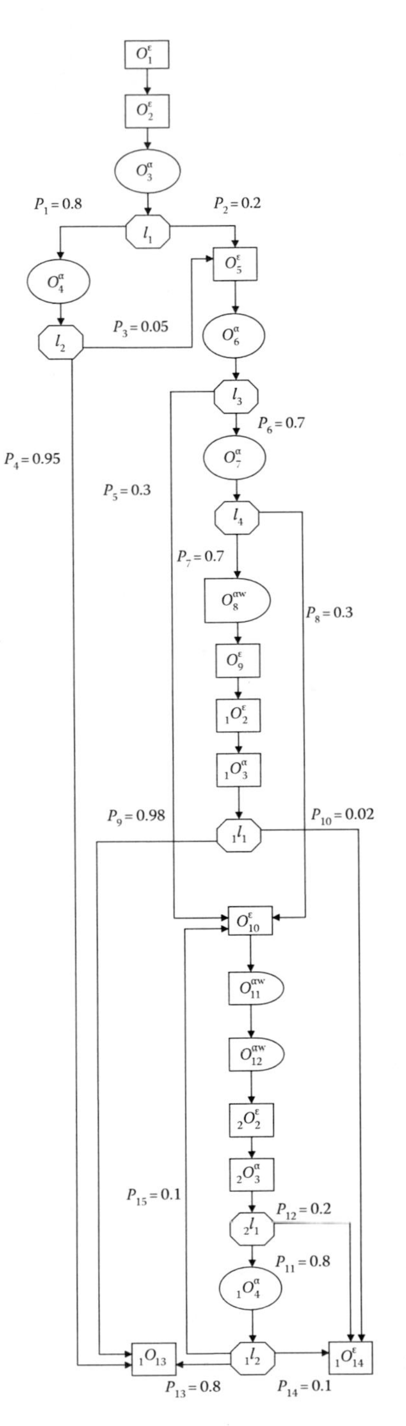

FIGURE 11.4
Algorithmic-graphic model for the new method of task performance.

Similarly, the probability of not receiving the order file after the first attempt is equal to

$$ {}_1P_2 = P_1 \times P_3 = 0.8 \times 0.05 = 0.04, $$

where

P_1 is a probability of the successful outcome of the logical condition l_1

P_3 is a probability of the negative outcome (on the right) of the logical condition l_2

Therefore, the probability of additional actions in order to receive the order file can be expressed as a sum of probabilities P_2 and ${}_1P_2$ of two independent events:

$$ {}_1P_3 = P_2 + {}_1P_2 = 0.2 + 0.04 = 0.24, $$

where P_2 is a probability of the negative outcome (on the right) of the logical condition l_1.

As has been expected, the sum of probabilities ${}_1P_1$ and ${}_1P_2$ is equal to 1.

In order to increase the reliability and decrease the rate of system failures, it has been suggested to introduce a new action O_5^ε. This action consists of executing the script "where_is_file" that would allow a human operator to check if the order file is in the transfer stage and is being accumulated in temporary storage.

The probability of the file being in the transfer state if it has not been found in the production directory is

$$ {}_1P_4 = {}_1P_3 \times P_6 = 0.24 \times 0.7 = 0.168, $$

where P_6 is a probability of the outcome of logical condition l_3 pointing downward in Figure 11.3.

The probability that the file located in temporary storage has the current date stamp can be calculated as

$$ {}_1P_5 = {}_1P_4 \times P_7 = 0.168 \times 0.7 = 0.118, $$

where P_7 is a probability of the outcome of logical condition l_4 that points downward in Figure 11.1.

If the order file has not been transmitted before the start of the shift and is not in the transfer state, the operator needs to restore communication by performing O_{10}^ε.

$$ {}_1P_6 = {}_1P_3 \times P_5 = 0.24 \times 0.3 = 0.072, $$

where P_5 is a probability of the left outcome of the logical condition l_3.

If the order file is present in the temporary storage but has an old time stamp, the operator needs to restore communication as well:

$$_1P_7 = {_1P_4} \times P_8 = 0.168 \times 0.3 = 0.05,$$

where P_8 is a probability of the right outcome of the logical condition l_4.

Therefore, the resulting probability of the operator O_{10}^{ε} can be calculated as a product of probabilities $_1P_6$ and $_1P_7$ as follows:

$$_1P_8 = {_1P_6} \times {_1P_7} = 0.072 \times 0.05 = 0.122.$$

The probability of checking the date stamp of the order file the second time $({_2O_4^{\alpha}})$ can be determined as

$$_1P_9 = {_1P_8} \times P_{11} = 0.122 \times 0.8 = 0.0976,$$

where P_{11} is a probability of the successful outcome of the logical condition $_2l_1$.

If the order file has a current date on its time stamp, then the partial probability of the goal-related operator O_{13}^{ε} can be calculated as

$$_1P_{10} = {_1P_9} \times P_{13} = 0.0976 \times 0.8 = 0.078,$$

where P_{13} is a probability of the successful outcome of the logical condition $_1l_2$.

If the operator determined that the file is being transmitted and waited until the file has been copied to the production directory, the probability of this event can be determined as follows:

$$_1P_{11} = {_1P_5} \times P_9 = 0.118 \times 0.98 = 0.116,$$

where P_9 is a probability of the successful outcome of the logical condition $_1l_1$.

Therefore, the resulting probability of the goal-related member of the algorithm O_{13}^{ε} can be expressed as the sum of

$$_1P_{13} = {_1P_1} + {_1P_{10}} + {_1P_{11}} = 0.76 + 0.078 + 0.116 = 0.956.$$

The following is the calculation of the probability of the system failure. The probability that the order file has been transmitted to the temporary storage but did not show up in production is very small and equal to

$$_1P_{12} = {_1P_5} \times P_{10} = 0.118 \times 0.02 = 0.002,$$

where P_{10} is a probability of the negative outcome of the logical condition $_1l_1$.

If the order file has not been transmitted even after the restore of the communication, the operator should call a computer specialist. The probability of this event is

$$_1P_{14} = {_1P_8} \times P_{12} = 0.112 \times 0.02 = 0.022,$$

where P_{12} is a probability of the negative outcome of the logical condition $_2l_1$.

The probability that the file is present after the second attempt but has an old date equals

$$_1P_{15} = {_1P_8} \times P_{11} \times P_{14} = 0.112 \times 0.8 \times 0.1 = 0.009,$$

where

P_{11} is a probability of the downward outcome of the logical condition $_2l_1$
P_{14} is a probability of the successful outcome of the logical condition $_1l_2$

Hence, the resulting probability of the system failure can be found as follows:

$$_1P_{16} = {_1P_{12}} + {_1P_4} + {_1P_{15}} = 0.002 + 0.022 + 0.009 = 0.033.$$

The described new method of task performance has been implemented. The gathered statistical data showed the probability of successful performance being equal to 0.96 and the probability of the system failure equal to 0.04. As it can be seen, the experimental data and the data obtained by using the analytical method give very close results. The comparison of the analytical data about the reliability of the new and existing method of task performance shows that the new method increases the reliability of task performance. The suggested analytical method of reliability assessment allowed predicting with high level of accuracy that the new method will lead to the increase in the reliability of task performance.

The material presented in this chapter shows that SSAT, with its precise units of analysis and systemic principle of study of human activity as a multidimensional system, is very useful in the reliability assessment of computer-based tasks. One of the most important aspects of the suggested approach is increased role of the analytical (theoretical) method of the reliability assessment. The experimental and expert evaluations are closely connected with analytical methods of activity description. The graphic-symbolic model is particularly important in the reliability assessment of computer-based task performance. This model allows to visualize the probabilities of the transfer from one member of the algorithm to the next. Such clear understanding of the task performance flow is extremely important in the reliability assessment. Cognitive and motor actions are evaluated against task-related errors and failures. This study demonstrates that the method of reliability assessment described in this chapter can be very effective at the early stages of the design process and might be used as a predictive tool to assess the reliability of task performance even when the real task does not exist yet.

The suggested method can also be applied to the study of any man–machine system where the operator does not have direct interaction with the computer. SSAT considers human activity as a complex system that can be described in terms of morphological, functional, and parametric characteristics. The morphological method of study with actions, operations, and members of the algorithm as units of analysis is central in this study.

12

Formalized and Quantitative Analysis of Exploratory Activity in HCI Tasks

12.1 General Characteristics of a Web-Survey Task

In Section 3.2, we considered orienting activity, which can be external and internal. In some cases, internal orienting activity is not sufficient to understand the situation. In such cases, the person also uses external orienting activity. Moreover, the subject must find ways to achieve the formulated goal of work activity in a not clearly defined or a little familiar situation. In cases where such activity becomes complex, with clearly identified external and internal components, such activity may also be called exploratory. Thus, in exploratory activity, the goal has not only cognitive functions. It is also connected with the achievement of a certain goal of work activity that requires the transformation of the situation. It should be noted that between the exploratory and orienting activity, there are no clear boundaries and they can be transformed into each other. If the subject is not sufficiently aware of the situation and does not know how to achieve the required goal of exploratory activity, such type of activity becomes important in the analysis of the situation, its reevaluation, and finding ways to achieve a required goal. The more complex and unfamiliar the situation is for the subject, the more important the external exploratory components of activity are. Exploratory activity has its roots in animals' behavior. All animals displayed an inborn exploratory reflex as orienting reactions to unfamiliar situations (Pavlov, 1927). Such reactions are often triggered almost automatically. For example, animals that entered the experimental room started walking around the room sniffing objects, etc. They perform various irregular movements before paying attention to the food. Human exploratory activity, which is much more complex and includes a conscious goal, also can be triggered almost automatically. In a stressful and difficult situation, such activity may acquire a chaotic character and be conducted in a wrong direction in accordance to the required goal of activity. It is therefore important that such activity be based on adequate hypotheses and move the subject to the established goal of activity. Orienting and exploratory activity includes cognitive and behavioral actions that can be redundant or erroneous.

Users have to perform multiple computer-based tasks. Some of them are performed only once. In most cases, computer-based tasks do not possess a rigorous standardized method of performance. In human computer interaction (HCI) tasks, a user often does not know in advance the sequence of actions he or she has to take, and even experienced users have to discover the details of the task performance through exploratory activity. Hence, continuous self-learning through explorative actions is an important component of users' professional activity. Even when there are standardized requirements for task performance, users still have some degree of freedom in task performance. HCI always involves self-learning, exploration, and individualized strategies of task performance (Sengupta and Bedny, 2008). The more complex the task is for the user, the more important self-learning and explorative strategies are during this task performance.

Explorative activity consists of correct and incorrect actions that provide users with information that helps to understand the system and to correct the performance strategies. The explorative stage allows to examine the situation and consequences of one's own actions when users often utilize *reversible errors*. Such errors can be eliminated without negative consequences for task performance. They perform an informational function. If the task is extremely difficult for the user, his or her goal-directed activity can be transformed into chaotic behavior. Explorative behavior is a self-regulative process that according to functional analysis of activity is the basis for learning, which in turn can be considered a transformation of strategies of performance (Bedny and Meister, 1997). The more complex the task is, the longer it takes the user to find a truly effective strategy. If the user performs similar tasks multiple times, he or she goes through intermediate strategies until he or she approaches the optimal one. Hence, learning and self-learning in particular are the transformation from a less efficient strategy to more efficient ones.

The user explores his or her options. It can be an external or internal mental exploration. A user can observe the result of externalized explorative actions on the screen and evaluate them as positive or negative. The explorative action is a cognitive function with the purpose of transforming the situation and evaluating the impact of this transformation. Internal explorative actions lead to increase in the duration of cognitive activity. If the task's complexity is increased, then the number of explorative actions also increases.

Explorative actions can be erroneous and lead to erroneous external changes and the user has to return to the previous or initial screen. We call cognitive and motor explorative actions that give undesirable result *abandoned* actions, and the goal of HCI design is to reduce them. The less abandoned actions are used, the better is the efficiency of task performance (Bedny and Bedny, 2012). Users correct their strategies of task performance based on the evaluation of the result of abandoned actions and select actions they evaluate as positive. HCI tasks include explorative activity that cannot be eliminated totally. In the production environment, some externalized explorative components of activity can lead to corruption of the database, deletion of important information, and other undesirable results. The goal of the design of HCI tasks is to eliminate or reduce the possibility of such actions.

An emotionally motivational factor is critically important in the learning and self-learning process. The user is particularly sensitive to the influence of feedback at the explorative stage of task performance when he or she promotes a hypothesis, formulates the goal, evaluates the difficulty and significance of the task, examines the consequences of his or her actions, etc. The feedback influences the user's emotionally motivational state. SSAT underlines the complex relationship of explorative, motivational, executive, and evaluative components of task performance. In this chapter, we demonstrate that analysis and reduction of abandoned actions are important methods of increasing the efficiency of computer-based task performance.

In this chapter, we have chosen web-survey task as the object of study. Electronic mail is an important communication media in the business environment. People are overwhelmed with the number of e-mails they have to read and/or reply to. The time someone has to read every e-mail is very limited. The time-restricted factor becomes especially important if the e-mail subjectively has a low priority and distracts from the main duties. There are group e-mails that are distributed to hundreds of employees to take surveys, etc., that are mandatory and are important for business organizations. Here we encounter a scenario when the e-mail is important for a sender and is not important for a receiver. The self-regulation concept of motivation (Bedny and Karwowski, 2007) considers personal importance or significance as an emotionally evaluative mechanism of the motivational process. In the model of activity self-regulation (see Chapter 3), the function block *assessment of the sense of input information* is a part of the emotionally evaluative stage of activity performance. Distributed e-mails that have low positive or, in some cases, even negative significance are perceived as sources that are a waste of time and money. A business e-mail designer should take into account the effectiveness of such e-mails because it saves companies a lot of money if they design properly.

In this work, we demonstrate that not only cognitive but also emotionally motivational aspects of user activity are important for task analysis. Cognitive aspects of activity should be studied in unity with emotionally motivational components. The more complex and lengthy supplemental task is, the greater the negative emotional effect it has on the user because he or she loses the connection with the main task of his or her work and it becomes more difficult to return to it. "Where was I?" is the first question of the user after completing a supplemental task. Distributed e-mails are the ones that are sent to multiple employees. Sometimes it is done at a regular frequency, once a quarter, once a year, etc. Every employee has to complete the distributed task that contains a questionnaire. These e-mails require careful reading and answering of a number of questions.

The questions can be wordy and numerous. Until the employee completes the questionnaire, the same e-mail keeps coming back. If such communication is composed without taking into consideration some psychological factors, it can cause confusion, loss of time, and a negative emotional state that affects productivity. We've chosen the e-mail-distributed task that has been associated with poor emotional and motivational state, because these conditions influence cognitive strategies.

In our example, the described task had to be performed by several thousand employees. Even a few minutes' reduction in the task performance time has a significant economic effect. An additional purpose of this study was to demonstrate the principles of SSAT that are instrumental in the enhancement of the HCI task design.

12.1.1 Task Description

As an example, we have chosen a real web-survey task. The task has been slightly modified for our research purposes. More than 5000 employees have received the e-mail with the following content (see Figures 12.1 and 12.2).

Each employee supposes to read this e-mail and fill up the questionnaire. As can be seen from Figures 12.1 and 12.2, this e-mail does not fit on one screen. As a result, an attached file cannot be observed without scrolling down. In this case, there is an interesting psychological factor associated with the motivation of employees. This e-mail-distributed task is not a high priority task.

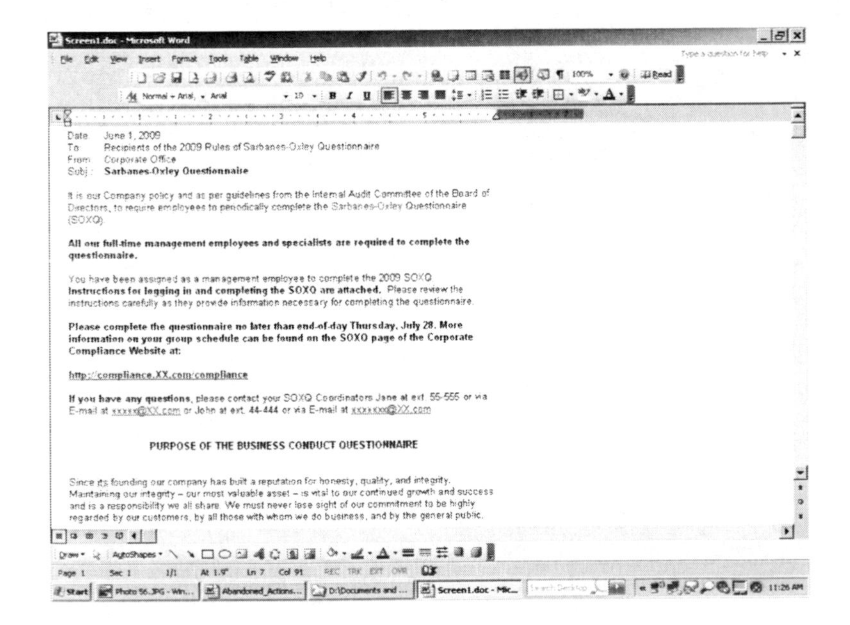
Screen1.doc - Microsoft Word

Date June 1, 2009
To: Recipients of the 2009 Rules of Sarbanes-Oxley Questionnaire
From: Corporate Office
Subj.: **Sarbanes-Oxley Questionnaire**

It is our Company policy and as per guidelines from the Internal Audit Committee of the Board of Directors, to require employees to periodically complete the Sarbanes-Oxley Questionnaire (SOXQ).

All our full-time management employees and specialists are required to complete the questionnaire.

You have been assigned as a management employee to complete the 2009 SOXQ. **Instructions for logging in and completing the SOXQ are attached.** Please review the instructions carefully as they provide information necessary for completing the questionnaire.

Please complete the questionnaire no later than end-of-day Thursday, July 28. More information on your group schedule can be found on the SOXQ page of the Corporate Compliance Website at:

http://compliance.XX.com/compliance

If you have any questions, please contact your SOXQ Coordinators Jane at ext. 55-555 or via E-mail at xxxxx@XX.com or John at ext. 44-444 or via E-mail at xxxxxxx@XX.com

PURPOSE OF THE BUSINESS CONDUCT QUESTIONNAIRE

Since its founding our company has built a reputation for honesty, quality, and integrity. Maintaining our integrity – our most valuable asset – is vital to our continued growth and success and is a responsibility we all share. We must never lose sight of our commitment to be highly regarded by our customers, by all those with whom we do business, and by the general public.

FIGURE 12.1
The first page of the e-mail.

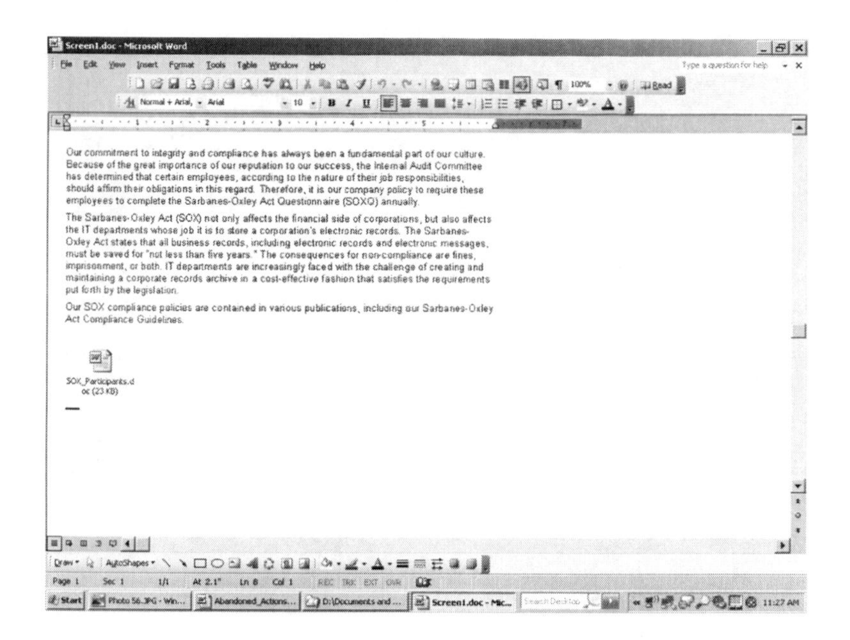

FIGURE 12.2
The second page of the e-mail.

It is considered by most employees as annoying, with low personal significance because it takes time from the main duties and as a result is accompanied by low motivation. However, it is a requirement to complete this survey. Moreover, the busier the employee is, the less he or she is motivated to take on this task. Often performance of such tasks causes irritation and negative emotional state. From the functional analysis perspective, this task has a negative significance for the employee (Bedny and Karwowski, 2007). This has been observed during discussion and analysis of this task performance. The self-regulation model of activity outlines the following motivational stages: (1) preconscious motivational stage, (2) goal-related motivational stage, (3) task evaluative motivational stage, (4) executive or process-related motivational stage, and (5) result-related motivational stage (Bedny and Karwowski, 2006). In the task under consideration, the goal-related motivational stage is in conflict with the executive stage because the task has to be completed but is boring and out of the scope of the main duties.

Observation demonstrates that employees attempt to complete this task as quickly as possible. They select a strategy to move through the content of the e-mail without carefully reading it and to get the questionnaire just as quickly as possible. For the employees, the most significant identification elements of the task are those that provide the direct link to the questionnaire, so they are looking for the links to the questionnaire. These most significant elements are the identification features or the task attributes. Our analysis of the e-mail and the user activity demonstrates that these attributes are not organized very well. The first page of the e-mail has only one link that immediately attracts attention

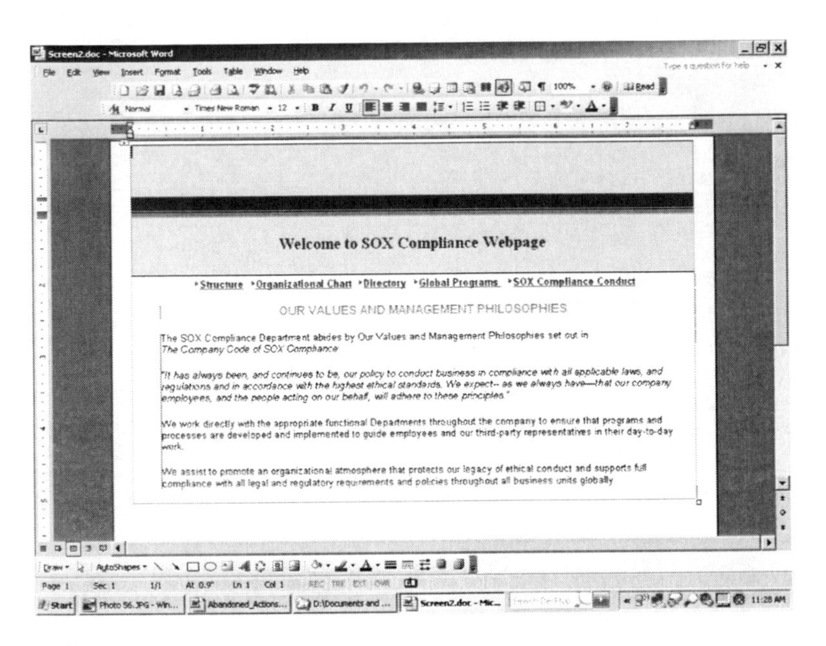

FIGURE 12.3
The screen after clicking the first link.

of the majority of employees. So they quickly click on the first link they see. When the webpage opens up, they do not see the expected login screen.

As an example, we present a webpage that opens up (see Figure 12.3).

The purpose of this page is to give some general information on the topic of the questionnaire. This webpage also includes links to some other pages with additional information. The employees again concentrate their attention only on available links. There are five of them in Figure 12.3 (see the top line of the screen). These links are the most significant identification features or attributes of the screen. The employees examine the links presented on the screen and click on the one that most probably leads to the login screen. Usually the fifth link is selected (SOX Sarbanes-Oxley Compliance Conduct) in accordance with the formulated goal. The next screen that opens up is demonstrated in Figure 12.4.

After looking at this screen, employees realize that they are on the wrong pass and start asking each other for help. It is usually just a waste of time and the employees decide to go back to the e-mail. They scroll down and discover an attachment at the bottom of the e-mail (see Figure 12.2).

Most of the employees click on the "Click here" link because it stands out by having two attention-attracting features: It is bold and underlined. There is also a motivational factor that plays a role here: a desire to find the link and to achieve the goal of activity (*cognitive, imaginative component of a desired future*). Opening the attachment and clicking on the tool "Click here" in the attachment (see Figure 12.5) eventually bring to the login screen the employees were looking for (see Figure 12.6).

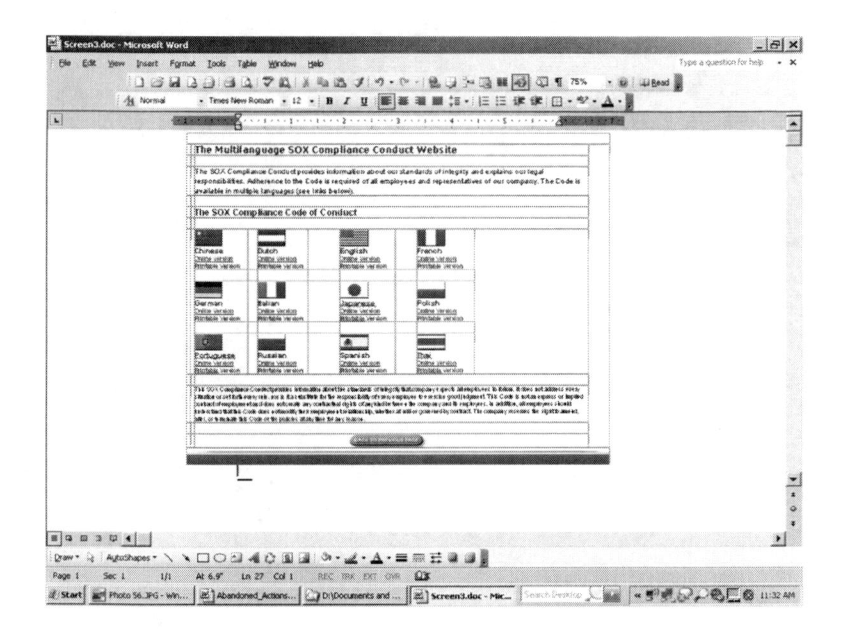

FIGURE 12.4
The screen after clicking the second link.

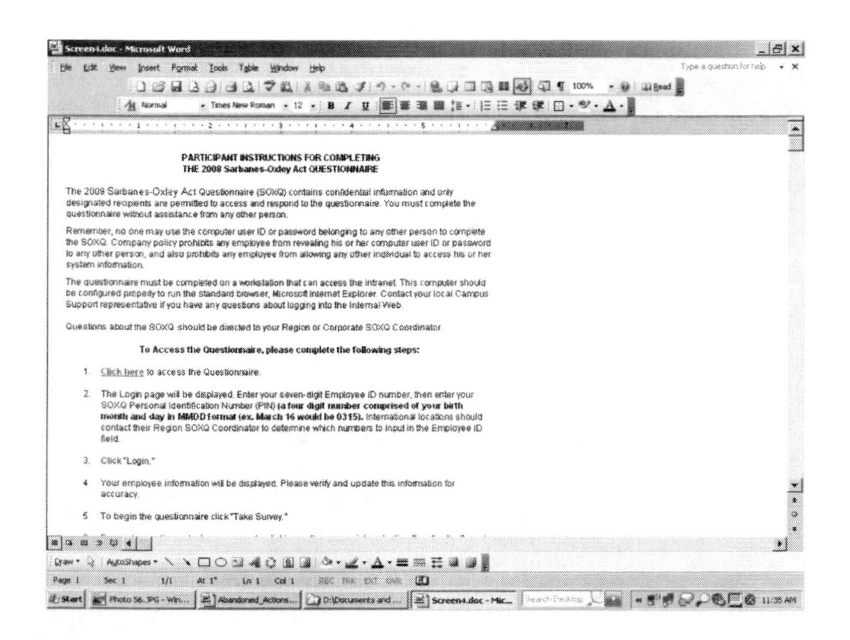

FIGURE 12.5
Attachment.

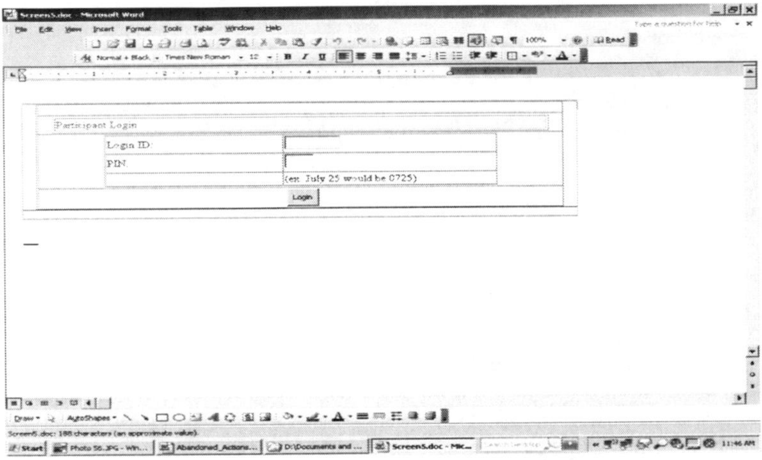

FIGURE 12.6
Login screen before improvement.

In order to login, they need to key in the login ID and PIN. Now they are confused again; they do not have the information they need to use in order to login. So the employees either go back to the attachment (Figure 12.5) or start asking each other for help again. Paragraph two of the attachment has the login information. Employees do not know in advance that this information will be required at the next step. Moreover, even if they would know, they still have to keep this information in working memory until they open the login screen and key it in. Keeping information in working memory is an undesirable factor for any task. The unnecessary mnemonic actions could be avoided if this information would be presented on the login screen itself.

So there are several factors that are leading to the inefficient strategies of task performance. One of them is low subjective significance of the task that results in the negative emotional-motivational association. The second factor is the inefficient task design where the identification features of the task are not adequate with the strategies of activity. The workers dropped a strategy of carefully reading a content on the screen. Developers of the e-mail had two broad goals (objective requirements) in mind: familiarization with the presented information and completion of the questionnaire. However, objectively defined goals (requirements) may not coincide with subjectively formulated or accepted goals (Bedny and Karwowski, 2007). Employees rejected the first goal and formulated their own general goal to complete the questionnaire as quickly as possible and return to their primary duty. In accordance with this goal, they had developed an adequate strategy of task performance.

Only after several failures to find a login screen did the employees gradually change their strategy and start paying more and more attention to its content. However, as will be shown further, that does not guarantee the success, because the structure of the task does not provide an understanding of its identification features necessary to carry out the actions. Qualitative analysis of this web questionnaire task demonstrates that the authors of the e-mail had a completely unsubstantiated mental picture of how it is going to be used by the employees. Moreover, there is no feedback information on how this e-mail has been utilized given to the sender of the e-mail for the future improvement of similar tasks.

There might be a deceptive impression that this task can be easily evaluated by experimental methods. In experimental conditions, it is very difficult to simulate the emotionally motivational state that arises during this task performance. It is clear that in experimental conditions, the subjects realize that they are under observation and the relevance of the task totally changes. In experimental conditions, an e-mail task would be seen as a test of a subject's ability to perform it. As a result, the strategies of the task performance might change completely. So the experimental methods would not be useful. Analytical methods such as analysis of self-regulation mechanisms and derived from them strategies of task performance and following algorithmic and quantitative analyses are a much better fit for studying this type of tasks. In this chapter we described analysis of possible strategies of task performance that derives from consideration of some functional mechanisms of activity self-regulation. In the following chapter we consider algorithmic analysis of a web-survey task.

12.2 Algorithmic Description of the Web-Survey Task

Algorithmic description of task in SSAT is a stage in morphological activity analysis which allows to describe the most preferable strategies of activity during task performance in the formalized manner (Bedny and Karwowski, 2003; Bedny and Meister, 1997). Every activity varies that is especially true for the performance of computer-based task performance. In order to analyze activity, there is no need to describe all possible strategies of task performance but rather consider the most typical ones and ignore minor variations in task performance. The most representative and important strategy of task performance should be selected. The combined probability of considered strategies should be equal to 1. This means that the selected strategies absorb all other strategies. Such approach is justified because the designed activity only approaches the real task performance and describes it with a certain level of approximation as does any model in the design process. This method is utilized in studying interchangeability of parts in the mass production.

Let us algorithmically describe the considered task. This task is multivariate and cannot be treated as deterministic. The objective of this study was not only to analyze the task as a whole but also analyze each member of the algorithm as a quasi-system. In order to do that, one would have to find the answers to the following questions:

What actions does the algorithm consist of? What operations are parts of the action? What is the duration of each action or operation? What is the level of concentration of attention during execution of a particular member of an algorithm (according to five-level scale developed in SSAT)? Can actions or operations be performed simultaneously or sequentially, depending on the level of concentration of attention? What is the probability of each member of the algorithm or its components? These data are used to develop the time structure of activity or quantitative measures of complexity of task performance, to evaluate the reliability of task performance, etc.

If the same screen is utilized differently, the next line of the table with the new member of the algorithm is used. Every time the same screen is utilized the same way, a compact description of an algorithm is possible. The next logical condition brings us to the place in the table where the first interaction with the screen was described, which means that the same member of the algorithm is used repeatedly. Despite the fact that the compact algorithm has fewer lines, the number of activity steps remains the same.

In the algorithmic analysis of task performance, the likelihood of the logical condition outcomes should be determined. It is possible to obtain such data through observation or experiment. In the study of operator's performance, it was shown that the experts can remember or estimate the likelihood of events (Kirwan, 1994). Hence, probability judgment can be utilized in our example where we assess the likelihood of only two possible outcomes within the precision of one or two digits after the decimal point. The experts' assessment of the probability of events is accurate enough for such approximation. We have used a modified table of transition from subjective judgments about the frequency of events to the quantitative data suggested by Zarakovsky and Pavlov (1987). Later, we present an algorithmic description of the considered task (Table 12.1).

In this table, the third column on the right presents a description of cognitive and motor actions as per SSAT methods (Bedny and Karwowski, 2007). Members of the algorithm and actions performed by the individuals are classified in a standardized manner. Each member of an algorithm usually includes from one to four motor or cognitive actions that are integrated by a high-ordered goal. The fourth column in the table has an action duration.

For the algorithmic description of the activity as shown in Table 12.1, we utilize various units of analysis. In the first and third columns of the table, there are psychological units of analysis, and in the second column, there are technological units of analysis. This combined utilization of units of analysis allows us to provide the most accurate description of the activity and its clear interpretation.

When performing computer-based tasks, users often need to read or print text. There are two strategies of reading: The first strategy involves careful reading of the entire text that is similar to reading a book (detail reading) and the second strategy involves browsing the text when certain parts of the

TABLE 12.1

Algorithmic Description of E-Mail-Distributed Task Performance

Members of the Algorithm (Symbolic Description)	Description of Members of the Algorithm	Classification of Actions	Time (in s)
O_1^α	Browse initial e-mail.	Successive perceptual actions (separate actions are not considered).	6
$\omega_1 \uparrow_{\omega_1}$	Always false logical condition ($p = 0.1$).	No actions.	0
O_2^ε	Move the mouse to the first link and click to open the next screen (Figure 12.2) ($p = 0.8$) (only this strategy would be considered further).	Motor action (mouse movement and click).	1.2
O_3^α	Search for the questionnaire on the opened screen (sequential examination of five choices; see subalgorithm description Table 12.2).	Five successive perceptual actions. Each action grasps a single meaningful output of verbal expression ($^*O_g^\alpha$; $^*O_1^\alpha$; $^*O_3^\alpha$; $^*O_5^\alpha$; $^*O_7^\alpha$; $^*O_9^\alpha$).	2.3
$l_1 \overset{(1-5)}{\uparrow}$	Choose one (see subalgorithm description Table 12.2).	Decision-making actions at a sensory-perceptual level (also performed five times) ($^*l_1$;$^*l_2$; $^*l_3$;$^*l_4$; $^*l_5$).	0.3
O_4^ε	Click on the selected choice (Code of Business Conduct).	Motor action (mouse movement and click).	1.2
O_5^α	Check opened *screen* (Figure 12.4). (There are two preferable strategies: select language or read the text on top of the screen.)	Successive perceptual actions during reading (separate actions are not considered) five or four successive perceptual actions.	5 Or 4 4
$l_2 \overset{2}{\uparrow}$	Select English language on the screen ($p = 0.8$), or go back to the e-mail (Figure 12.1; $p = 0.2$).	Decision-making actions at a sensory-perceptual level.	0.3
O_6^ε	Move mouse to a required position and click to select language (English).	Motor action (mouse movement and click).	1.2
O_7^α	Browse the text (no login is found; there is only a description of procedures).	Successive perceptual actions (separate actions are not considered).	3

(Continued)

TABLE 12.1 (*Continued*)

Algorithmic Description of E-Mail-Distributed Task Performance

Members of the Algorithm (Symbolic Description)	Description of Members of the Algorithm	Classification of Actions	Time (in s)
$O_{8}^{\alpha th}$	There is no questionnaire, and hence, I have to go back to the e-mail screen.	Deducing action.	0.35
$\overset{2}{\downarrow} O_{9}^{\varepsilon}$	Go back to the e-mail ($p = 0.2$) (see Figure 12.1).	Motor action (move mouse and click).	1.2
O_{10}^{α}	Look for another link (see Figure 12.1).	Thinking action based on visual information.	0.35
$\downarrow \omega 1 \underline{O_{11}^{\varepsilon}}$	Scroll down to the bottom of the e-mail and go to $\underline{O_{12}^{\alpha}}$.	Motor action (mouse movement click and hold) (this member of the algorithm partly overlaps with O_{12}^{α}).	2.5
$\underline{O_{12}^{a}}$	Notice the attachment at the bottom of the screen.	Simultaneous perceptual action (this member of the algorithm partly overlaps with O_{11}^{ε}).	0.35
O_{13}^{ε}	Double click on the attachment.	Motor action (mouse movement and double click).	0.25
O_{14}^{α}	Browse and detect the link (see Figure 12.4).	Successive perceptual actions (separate actions are not considered).	4
$O_{15}^{\alpha th}$	Here is the link to the questionnaire; hence, I need to click here.	Thinking action based on visual information.	0.35
O_{16}^{ε}	Click on the link.	Motor action (move mouse and click).	1.2
O_{17}^{α}	Examine the login *screen* (see Figure 12.6).	Three simultaneous perceptual actions.	0.3 0.25 <u>0.3</u> 0.85
$O_{18}^{\mu th}$	What should the login ID and PIN be?	Explorative thinking actions based on information extracted from long-term memory.	0.7
$O_{19}^{\mu th}$	I need instruction for the ID and PIN; hence, I go back to attachment (see Figure 12.5).	Logical thinking actions.	0.5
O_{20}^{ε}	Go back to the e-mail attachment.	Motor action (move mouse and click).	1.2
$O_{21}^{\alpha\mu}$	Read the instruction on how to login and keep information in working memory.	Successive perceptual actions combined with mnemonic action (separate actions are not considered).	11

(*Continued*)

TABLE 12.1 (*Continued*)

Algorithmic Description of E-Mail-Distributed Task Performance

Members of the Algorithm (Symbolic Description)	Description of Members of the Algorithm	Classification of Actions	Time (in s)
$O_{22}^{\varepsilon\mu}$	Click on the link to the questionnaire and keep information in working memory.	Motor action (mouse movement and double click) combined with mnemonic action.	1.2
O_{23}^{μ}	Recall instruction for login ID.	Mnemonic action.	0. 3
$O_{24}^{\varepsilon\mu}$	Recall and type the employee number and hit tab.	Combined action (executive action performed based on information extracted from memory; typing and recalling performed at the same time).	3.4
O_{25}^{μ}	Recall the instruction for the PIN.	Mnemonic action.	0.3
$O_{26}^{\varepsilon\mu}$	Recall and type employee birth date and hit login.	Combined action (executive action performed based on information extracted from memory; typing and recalling performed at the same time).	3
O_{27}^{α}	Read and answer the questions from the questionnaire.	—	

text are scanned (the user looks at a piece of text, captures the main idea of the fragment, and moves to the next piece of text). Browsing allows a user to get quickly acquainted with the main idea of a particular fragment of the text. After browsing the whole text, the user can return to some parts of the text for more details or move to another screen. Browsing the text usually requires simultaneous perceptual actions. Before describing the process of reading algorithmically, it is necessary to understand the relationship between these two strategies.

If necessary, segmentation of the reading text into separate verbal actions can be conducted. Each verbal action represents an elemental phrase, each of which represents a separate meaningful unit of information. Separate verbal–motor actions determine meaningful typing units such as typing a word or several interdependent words that convey one meaning. If the text is relatively homogeneous, there is no need to extract separate verbal actions. The duration of reading or printing of a text and the level of concentration when working with the text are the main criteria for the evaluation of this type of activity. In our study, the duration of reading or typing the text has been measured and its complexity has been evaluated based on the level of concentration of attention. Time performance for some simple cognitive and behavioral actions was taken from previous studies. For example, the duration of decision-making action at the sensory-perceptual level *if-then* has been estimated at 0.350 ms (Lomov, 1982; Myasnikov and Petrov, 1976). For simple motor actions such as a mouse click and for a mouse, movements (average) were assigned 0.1 and 1.1 s, respectively (see Card et al., 1983; Kieras, 1994). Other data were obtained from chronometrical studies. In more complicate cases, eye movement registration can be utilized (Bedny et al., 2008).

Let us consider some members of the algorithm (Table 12.1). The first step of task performance suggests that employees read all information on screen 1, which has two pages, and the attachment on the bottom of the second page. Our experimental data demonstrated that reading the whole first page takes 31 s in average (60 s for two pages). However, because of the low motivation to perform this task, employees are looking for the shortest way of task performance. Instead of reading the e-mail, they select a browsing strategy in order to find the link for the questionnaire quickly. It takes in average 6 s to find out the fist link available. After performance of O_1^α, there is always a false logical condition ω_1 that simply designates that there is a possibility of either O_2^ε or O_{11}^ε. This is the way to demonstrate transition from one member of the algorithm to another that does not involve any action. The probability of O_2^ε as per expert analysis is 0.9 (the first basic strategy) and of O_{11}^ε is 0.1 (the second basic strategy).

Let us consider the most probable strategy. After browsing the e-mail, employees quickly move their mouse to the first link they spot on the page and click. According to expert analysis, the probability of this strategy is 0.9. This step of task performance is described by two members of the algorithm (O_1^α and O_2^ε). This choice of action greatly affects the following strategy of task performance: O_3^α and l_1 are complex members of the algorithm and their detailed description can be found in Table 12.2. This table includes $^*O_{10}^\varepsilon$, which is the same as O_4^ε in Table 12.1. Stars on the left in Table 12.2 are used to distinguish symbols in Tables 12.1 and 12.2. This is an example of decomposition of activity during algorithmic description of task performance that helps in its understanding. In Table 12.2, we use two methods of symbolic description of the algorithm. The left-hand side of the table depicts the algorithm as a vertical formula. In parentheses, we continue the description of the second part of this algorithm as a horizontal formula. This formula demonstrates the sequence of members of the algorithm execution, going from left to right.

Before we describe activity algorithmically, we must find out the possible strategies of activity performance. They should be discovered at the preliminary stage of qualitative analysis of task performance. The simplest method is objectively logical task analysis. The more complex method is functional analysis when activity during task performance is considered as a self-regulative system. In our case, it is sufficient to use objectively logical analysis when specialists use such techniques as observation and discussion.

Table 12.2 describes the most preferable activity strategy when employees work with the second screen (Figure 12.4). Employees are expected to find the questionnaire or at least link to it on the second screen. Therefore,

TABLE 12.2

Subalgorithm of E-Mail-Distributed Task Performance

Member of the Algorithm (Symbolic Description)	Description of Members of the Algorithm
${}^{*}O_1^{\alpha}$	Read the first header.
${}^{*}l_1 \overset{1}{\uparrow}$	Can the first header lead to the questionnaire? If *yes*, click on it. If *no*, look at the second header.
O_2^{ε}	Click on the first header (this step is omitted).
$\overset{1}{\downarrow} {}^{*}O_3^{\alpha}$	Look at the second header and read it.
$*l_2 \overset{2}{\uparrow}$	Can this header lead to the questionnaire? If *yes*, click on it. If *no*, look at the second header.
O_4^{ε}	Click on the second header (this step is omitted).
$\overset{2}{\downarrow} {}^{*}O_5^{\alpha}$	Look at the third header and perform according to the formula (see the formula in the next row of the table).[a]
...	$\left({}^{*}O_5^{\alpha}\ {}^{*}l_3 \overset{3}{\uparrow}\ {}^{*}O_6^{\varepsilon} \overset{3}{\downarrow}\ {}^{*}O_7^{\alpha}\ {}^{*}l_4 \overset{4}{\uparrow}\ {}^{*}O_8^{\varepsilon} \overset{4}{\downarrow}\ {}^{*}O_9^{\alpha}\ {}^{*}l_5 \overset{5}{\uparrow}\ {}^{*}\boldsymbol{O}_{10}^{\varepsilon} \overset{5}{\downarrow} \omega \uparrow \right)^{a}$
${}^{*}\boldsymbol{O}_{10}^{\varepsilon}$	*Click on last selected choice—Code of Business Conduct* (the title of the last choice makes it possible to assume that one can find a link to the questionnaire).

[a] This is the second part of the subalgorithm described as a formula. The afferent operator $\boldsymbol{O}_5^{\alpha}$ in this table in the left column (vertical formula) and ${}^{*}\boldsymbol{O}_5^{\alpha}$ in parenthesis (horizontal formula) is the same member of the algorithm.

* The star symbol is used for demonstration differences between members of the algorithm in Tables 12.1 and 12.2. **Operators ${}^{*}\boldsymbol{O}_9^{\alpha}$ *and* ${}^{*}O_{10}^{\varepsilon}$ in Table 12.2 are the same as the operator in $\boldsymbol{O}_3^{\alpha}$ and $\boldsymbol{O}_4^{\varepsilon}$ in Table 12.1.

for employees who formulate such goal, the most significant and the most attractive are the five underlined headers (the identification features of this webpage; Figure 12.4). Employees assume that the five underlined headers can bring them to the next webpage and one of these links would open up the desired questionnaire. So after browsing the screen and discovering the headers, they start reading them from left to right. They guess that only the last choice implies the desired result and decide to click on it.

This strategy is described in Table 12.2. For the description of the further steps of task performance, we need to return to Table 12.1. The analysis of the performed algorithm shows that employees utilize five successive perceptual actions. Each perceptual action grasps a single meaningful output or verbal expression. The following members of the algorithm describe these perceptual functions (${}^{*}O_g^{\alpha}$; ${}^{*}O_1^{\alpha}$; ${}^{*}O_3^{\alpha}$; ${}^{*}O_5^{\alpha}$; ${}^{*}O_7^{\alpha}$; ${}^{*}O_9^{\alpha}$). The first member of the algorithm (${}^{*}O_g^{\alpha}$) is for receiving the information *there are a headers*. The other symbols are for possible perceptual steps associated with five possible perceptual analyses of the headers. The decision-making action at the sensory-perceptual level is selected from the following possible five (${}^{*}l_1$; ${}^{*}l_2$; ${}^{*}l_3$; ${}^{*}l_4$; ${}^{*}l_5$). Only the last efferent operator ${}^{*}O_{10}^{\varepsilon}$ has been selected because its title in contrast to four others gives the hope of finding the link to the questionnaire, and the employees clicked on the last header (simple motor action).

We do not go into a detailed description of this subalgorithm and just present it as a formula:

$$\left({}^{*}O^{\alpha^{*}}{}_{5} l_{3} \overset{3}{\uparrow} {}^{*}O^{\varepsilon}{}_{6} \omega_{1} \uparrow \overset{3}{\downarrow} {}^{*}O^{\alpha^{*}}{}_{7} l_{4} \overset{4}{\uparrow} {}^{*}O^{\varepsilon}{}_{8} \omega_{1} \uparrow \overset{4}{\downarrow} {}^{*}O^{\alpha^{*}}{}_{9} l_{5} \overset{5}{\uparrow} {}^{*}O^{\varepsilon}{}_{10} \overset{5}{\downarrow} \downarrow \omega_{2} \uparrow \right)$$

This formula demonstrates the sequence of the algorithm execution going from left to right. A logical condition with an arrow means that some steps might be omitted. The first member of this part of the algorithm is ${}^{*}O_{5}^{\alpha}$. It symbolizes the perceptual action *look at the third header.* If the value of logical condition ${}^{*}l_3$ is 0, then O_6^{ε} is performed. If ${}^{*}l_3$ is 1, then O_6^{ε} is bypassed and O_7^{α} is performed, etc. The first four logical conditions have a value of 1 and only the last one is valued at 0. Employees click on the last header. There are always false logical conditions ω_1 after each operator, which means that after making a selection, the switch is made back to Table 12.1. So employees perform six perceptual actions, one decision-making action, and the motor action *click*. The last member of the algorithm is an always false logical condition ω_2 that signals the switch back to the main algorithm. A formula description of the algorithm (as shown earlier) is suitable for simple tasks; otherwise it becomes difficult to read.

The next screen (see Figure 12.3) is the webpage that has rules of SOX Compliance in multiple languages because the company has offices in multiple countries. When employees open this screen, they once again hope to find the link to the questionnaire. Information presented on the screen does not meet their expectations and they explore the screen to find out its purpose. There are several preferable strategies of using this website. The first preferable strategy is to find the corresponding language and click on it, without reading information on the top of the screen; the second one is to read information at the top of the screen and decide to go back to the e-mail (screen 12.1); the third strategy is to carefully read information on the screen. This last strategy contradicts the dominating motivational state and its associated goal; the second one is most likely rejected because employees already made a number of steps and it is now subjectively quite risky to go back without checking the next screen. So they select the first one: "I made multiple steps before getting to this screen; it seems quite risky to go back without checking the next screen with the appropriate language. Maybe I have to return to this screen later on, so I better check what's on the next screen." The employees open up the screen, quickly look at it, find no link to the questionnaire, and go back to the e-mail. The probabilities of each of the two main strategies as per experts' estimate are presented in Table 12.1. The given probabilistic characteristics are based on a qualitative analysis of possible strategies of task performance. Analysis of the algorithm shows that there are two possible ways to go back to screen 1 (see Table 12.1, member of the algorithm O_9^{ε}). When calculating the time of task performance, we need to take into consideration these two possible strategies.

When employees get back to the e-mail (Figure 12.1), their goal is different because now they want to explore it in detail to find the link to the questionnaire; they eventually discover the attachment at the bottom of the e-mail (Figure 12.2) and open it (Figure 12.5). They browse the attachment, detect the link, and decide to click on it. The login screen before improvement opens up (Figure 12.6).

Employees examine this screen and find out that they do not have information necessary to login. This is an unexpected situation that activates users' explorative activity and they attempt to find related information in long-term memory. Employees realize that they do not possess such information and they have to return to screen 5 (see Table 12.1, $O_{18}^{\mu th}$ and $O_{19}^{\mu th}$). This analysis demonstrates that employees performed a number of unnecessary actions that do not advance them to the desired goal *to login,* and such actions are examples of abandoned actions. When employees return to screen 5, they carefully read the login instruction and keep information in working memory until completion of the login (see Table 12.1).

Once the algorithmic description is done, each member of the algorithm should be considered beginning with qualitative analysis. The possibility of errors and ways of their elimination should be analyzed. The performance time for each member of the algorithm, the level of concentration of attention (according to the five-point scale) during its performance, the fraction of time spent on perception of information, the fraction of time for decision making, etc., should be determined. Action that can be performed simultaneously or sequentially, work complexity, reliability of performance, etc., can be identified.

Analysis of Tables 12.1 and 12.2 demonstrates that the designer of this e-mail do not envision real execution strategies of this task and were never informed about the issues with this e-mail. The designer of this task assumed that the employees will open the e-mail and take the following steps: (a) read the text carefully, (b) open the attachment, (c) read the attachment carefully, (d) memorize information about the ID and pin, (e) click on the link, and (f) if any additional information is required, the link to the webpage that has SOX Compliance information can be utilized.

In fact, a completely different strategy has been observed (see algorithmic description in Tables 12.1 and 12.2). It is not a rational one but the users are looking for the shortest way to open the questionnaire. They click on the first link and they see and find themselves in the maize of the webpages they were not looking for. Such strategy contains a lot of unnecessary steps that we categorize as abandoned actions. Some of these actions require memorization and keeping information in working memory, deviate attention from the main elements on the screen, and produce irritation and a negative emotionally motivational state.

We have described the first main strategy that has some variations as has been shown in Table 12.1. There is also the second main strategy of this task performance, the probability of which is 0.1. From practical purposes, the second strategy can be neglected but its consideration might be useful for understanding the applied method of study. So we will consider it briefly. An always false logical condition ω_1 demonstrates a possibility of going directly to O_{11}^{ε} to perform the second part of the algorithm. Suppose we want to determine the time performance of a considered task and take into account two basic strategies. The first basic strategy has probability $p = 0.9$ and the second one $p = 0.1$. So task performance should be determined using the following formula:

$$T = \sum P_i t_i \tag{12.1}$$

where

P_i is the probability of the ith member of the algorithm
t_i is the performance time of ith member of the algorithm

Then the task performance time equals

$$T = P_1 \times Tst_1 + P_2 \times Tst_2, \tag{12.2}$$

where

Tst_1 and Tst_2 are performance times of the first and the second basic strategies
P_1 and P_2 are probabilities of performing these strategies

So, in our case, the task performance time equals

$$T = 0.9 \times Tst_1 + 0.1 \times Tst_2 \tag{12.3}$$

This example demonstrates that if necessary, we can determine all the required quantitative characteristics of the task under consideration based not only on the analysis of separate strategies but also on the analysis of all possible strategies of task performance. For simplification of our further discussions, we limit our discussion to the quantitative measures of abandoned actions for the first basic strategy.

At the last stage of task performance (see Table 12.1, $O_{23}^{\mu th}; -O_{26}^{\varepsilon\mu}$), the employee reads the instruction on screen 4; keeps them in working memory; goes to screen 5; recalls his or her employee number, birth date, and rule of transforming the birth date into the required format; and then keys the information in. This is a sequential mnemonic action that includes several mental operations.

12.3 Analysis of Abandoned Actions

In this section, we consider abandoned actions for the first main strategy. The most common abandoned actions for a particular task should be presented in algorithmic analysis. Depending on the purpose of the study, the performance time of abandoned actions can be considered or ignored.

An increase in the number of unnecessary explorative actions not only complicates and prolongs the performance time of the computer-based tasks but also has a negative effect from a technical point of view. The more switching from one screen to the other are performed, the more time delays are associated with such switching.

We will consider quantitative evaluation of the efficiency of the computer-based task performance using the measures we have developed for the assessment of abandoned actions. Efficiency measures derive from evaluation of time of task performance and duration of various types of abandoned actions. The following symbols will be utilized: A, general time for all abandoned actions; A^α, time required for afferent abandoned actions; A^ε, time required for efferent abandoned actions; A^l, time required for abandoned logical conditions; and A^μ, time required for abandoned actions associated with keeping information in working memory.

The first step in the assessment of task performance efficiency is the evaluation of task performance time (formula 1).

The time taken for all described abandoned actions can be determined as follows:

$$A = A^\alpha + A^\varepsilon + A^l + A^\mu + A^{th};\ A^\alpha = \sum P_i^\alpha \times t_i^\alpha;\ A^\varepsilon = \sum P_b^\varepsilon \times t_b^\varepsilon;\ A^l = \sum P_r^l \times t_r^l;$$
$$A^\mu = \sum P_j^\mu \times t_j^\mu;\ \hat{A}^{th} = \sum P_k^{th} \times t_k^{th},$$

where

$P_i^\alpha;\ P_b^\varepsilon;\ P_r^l;\ P_j^\mu;\ P_k^{th}$ are the probabilities of the *i*th abandoned action of the corresponding type

$t_i^\alpha;\ t_b^\varepsilon;\ t_r^l;\ t_j^\mu;\ t_k^{th}$ are the performance times of the *i*th abandoned action of the corresponding type

This time can be obtained based on existing studies or experimentally based on chronometrical analysis.

The next step in the evaluation of the task performance efficiency utilizes the following measures of efficiency:

$$\hat{A} = \frac{A}{T} \tag{12.4}$$

$$\hat{A}^\alpha = \frac{A^\alpha}{T} \tag{12.5}$$

$$\hat{A}^\varepsilon = \frac{A^\varepsilon}{T} \tag{12.6}$$

$$\hat{A}^l = \frac{A^l}{T} \tag{12.7}$$

$$\hat{A}^{\mu} = \frac{A^{\mu}}{T} \quad (12.8)$$

$$\hat{A}^{th} = \frac{A^{th}}{T} \quad (12.9)$$

These are measures of various types of abandoned actions. The less is the value of these measures, the more is the efficiency of performance. In any given study, all these measures or just the most suitable ones should be utilized. When mnemonic, thinking, or decision-making activities are performed simultaneously with motor components of activity, each type of activity is accounted for separately in the calculations.

Let us determine efficiency measures for the previously considered task.

The first step would be to define the time of task performance T and performance times for various types of abandoned actions. The performance time for each member of the algorithm is presented in the fourth column of Table 12.1. Using Table 12.2, we have evaluated performance time of the subalgorithm where the user performed in average six successive perceptual actions in sequence and one decision-making action at the sensory-perceptual level and a simplest motor action (click and release). Table 12.1 gives the performance time of O_3^{α} and l_1.

This example shows that if chronometrical measurements are utilized, the preferred strategies of activity performance should be identified first (Bedny and Karwowski, 2007; Bedny and Bedny, 2011). During chronometrical analysis, it is useful to collect subjective assessments of the performance pace by subjects involved in the experimental study. For measurements of duration of automotive mental operations, it is recommended to utilize methods developed in cognitive psychology (Sternberg, 1969b, 1975). In more complex tasks, eye movement registration might be necessary (Bedny et al., 2008).

When assessing the time performance of the first strategy, the outcomes of various options for logical conditions is taken into account, because they define the logic of the transfer to the individual members of the algorithm. Table 12.2 demonstrates the method of utilizing members of the algorithm O_3^{α} and l_1. The logical condition l_2 has two outputs. The probability of the first output is $P_1 = 0.8$, and for the second one, $P_2 = 0.2$. These two lead to the following two possible strategies of performance (see l_2).

The first strategy that has a probability of 0.8 can be described as

$$\text{STR}\,(1) = P_1 \times \left(tl_2 + tO_6^{\varepsilon} + tO_7^{\alpha} + tO_8^{\alpha th} + tO_9^{\varepsilon}\right)$$

The second strategy that has a probability of 0.2 can be described as

$$\text{STR}\,(2) = P_2 \times \left(tl_2 + tO_9^{\varepsilon}\right)$$

The performance time for the considered fragment of the algorithm is

$$MT = \text{STR (1)} + \text{STR (2)}$$

where MT is the mean time of these two strategies:

$$MT = P_1 \times \left(tl_2 + tO_6^{\varepsilon} + tO_7^{\alpha} + tO_8^{\alpha th} + tO_9^{\varepsilon}\right) + P_2 \times \left(tl_2 + tO_9^{\varepsilon}\right)$$

In brackets, there are performance times for the corresponding members of the algorithm.

Substituting symbols with corresponding values, we get the following result:

$$MT = 0.8 \times (0.3 + 1.2 + 3 + 0.35 + 1.2) + 0.2\ (0.3 + 1.2) = 5.14\ (\text{s})$$

The probability of each member of the algorithm except $l_2 - O_8^{\alpha th}$ equals 1. Summarizing the performance time of each member of the algorithm including MT (performance time of the fragment of the algorithm from l_2 up to O_9^{ε}) gives the task performance time (first main strategy):

$$T = 6 + 1.2 + 2.3 + 0.3 + 1.2 + 4.5 + 5.14 + 0.35 + 2.5 + 0.35 + 0.25 + 4 + 0.35 + 1.2 + 0.85 + 0.7 + 0.5 + 1.2 + 11 + 1.2 + 7 = 49.84\ (\text{s})$$

In this calculation, we ignored the second main strategy that has a low probability ($P = 0.1$).

At the next step, the performance time of all abandoned actions that are included in the following members of the algorithm is calculated:

$$A = O_2^{\varepsilon}; O_3^{\alpha}; l_1; O_4^{\varepsilon}; O_5^{\alpha}; l_2; O_6^{\varepsilon}; O_7^{\alpha}; \boldsymbol{O_8^{\alpha th}}; O_9^{\varepsilon}; O_{10}^{\alpha}; \boldsymbol{O_{15}^{\alpha th}}; O_{16}^{\varepsilon}; O_{17}^{\alpha}; O_{18}^{\mu th}; O_{19}^{\mu th}; O_{20}^{\varepsilon}$$

The combined time for abandoned actions should be determined taking into consideration that some abandoned actions or mental operations are performed simultaneously. All members of the algorithm that have a combination of several qualitatively different superscripts such as $O^{\mu th}$, l^{μ}, $O^{\varepsilon th}$, or $O^{\varepsilon\mu}$ are examples of combined actions where $O^{\mu th}$ is the combination of mnemonic and thinking actions or operations, l^{μ} is the decision-making action performed based on information extracted from memory and/or requires keeping information in working memory, $O^{\varepsilon th}$ is the combination of executive or motor actions with thinking actions or operations, and $O^{\varepsilon\mu}$ is the combination of executive or motor actions with mnemonic actions or operations.

So we are going to count the time when actions overlap separately for each type of these actions. The member of the algorithm $O_{19}^{\mu th}$ is an example of combining mnemonic and thinking actions. The total time for abandoned actions can be determined just summarizing their performance time. O_6^{ε}, O_7^{α}, and $O_8^{\alpha th}$ have a probability of 0.8. Hence, the time for all abandoned actions can be determined as follows:

$$A = 1.2 + 2.3 + 0.3 + 1.2 + 4.5 + 0.3 + 0.8 \times (1.2 + 3 + 0.35) + 1.2 + 0.35 + 0.35 + 1.2 + 0.85 + 0.7 + 0.5 + 1.2 = 19.79 \text{ (s)}$$

Therefore, the fraction of abandoned actions in the task performance time is

$$\hat{A} = \frac{A}{T} = \frac{19.79}{49.84} = 0.4$$

So abandoned actions take about 40% of this task performance time.

It is also possible to determine the fraction of perceptual, thinking, decision-making, and mnemonic abandoned actions in the task performance time by calculating the times for various types of abandoned actions: A^{α}, A^{ε}, A^{l}, A^{μ}, A^{th}.

Let us determine the performance time of the motor components of activity. All motor members of the algorithm have a probability of 1 excluding O_6^{ε} which has a probability of 0.8. The following efferent members of the algorithm are abandoned actions: $O_2^{\varepsilon}, O_4^{\varepsilon}, O_6^{\varepsilon}, O_9^{\varepsilon}, O_{16}^{\varepsilon}, O_{20}^{\varepsilon}$.

Therefore, the performance time for abandoned efferent actions (motor) is

$$A^{\varepsilon} = 1.2 + 1.2 + 0.8 \times 1.2 + 1.2 + 1.2 + 1.2 = 10.95 \text{ (s)}$$

The following are afferent abandoned actions (including reading): $O_3^{\alpha}, O_5^{\alpha}, O_7^{\alpha}, O_{10}^{\alpha}, O_{17}^{\alpha}$.

The performance time of the afferent abandoned actions is determined as

$$A^{\alpha} = 2.3 + 4.5 + 0.8 \times (3 + 0.35) + 0.85 = 10.33 \text{ (s)}$$

Sometimes the reading time should be considered separately.

Two existing logical conditions l_1 and l_2 are also abandoned actions.

Hence, the performance time for decision-making abandoned actions is

$$A^{l} = 0.3 + 0.3 = 0.6 \text{ (s)}$$

All members of the algorithm that include thinking actions are abandoned actions (O_8^{th}, O_{15}^{th}, $O_{18}^{\mu th}$, $O_{19}^{\mu th}$).

The performance time of these actions is

$$A^{th} = 0.8 \times 0.35 + 0.35 + 0.7 + 0.5 = 1.83 \text{ (s)}$$

Members of the algorithm that include mnemonic abandoned actions are $O_{18}^{\mu th}$ and $O_{19}^{\mu th}$ and their performance time is

$$A^{\mu} = 0.7 + 0.5 = 1.2 \text{ s}$$

The following are the coefficients for various types of abandoned actions:

$$\hat{A}^{\alpha} = \frac{A^{\alpha}}{T}; \quad \hat{A}^{\varepsilon} = \frac{A^{\alpha}}{T}; \quad \hat{A}^{l} = \frac{A^{l}}{T}; \quad \hat{A}^{\mu} = \frac{A^{\mu}}{T}; \quad \hat{A}^{th} = \frac{A^{th}}{T};$$

$$\hat{A}^{\alpha} = 0.207; \quad \hat{A}^{\varepsilon} = 0.22; \quad \hat{A}^{l} = 0.012; \quad \hat{A}^{\mu} = 0.024; \quad \hat{A}^{th} = 0.037$$

The sum of these five fractions is greater than the fraction of the combined abandoned actions because some of them are performed simultaneously. Because the second strategy has a low probability, we do not consider it at all. We calculate the previously described measures only for one basic strategy that has a probability of 0.9. However, if it is required, we can calculate these measures taking into consideration two basic strategies, one with a probability of 0.9 and the other with a probability of 0.1. If two main strategies are considered, the formula for the task performance time would be

$$T = 0.9 \times Tst_1 + 0.1 \times Tst_2$$

Using the same principle, it is possible to calculate measures that have more than two strategies. The analysis of the considered coefficients shows that some of them have insignificant value. The coefficients that describe the fraction of abandoned actions associated with logical conditions and thinking can be calculated together as an integral factor because the logical conditions that describe the decision-making process are one of the stages of the thinking process. The members of the algorithm, which include mnemonic, decision-making, and thinking actions, are critical points of any computer-based task, and therefore, specialists have to pay particular attention to these members of the algorithm. Analysis of such members of the algorithm as $O_{18}^{\mu th} - O_{23}^{\varepsilon\mu}$ demonstrates that there is unnecessary load on working memory. Some of these members include abandoned actions that should be eliminated. Moreover, even if the members of the algorithm do not include abandoned actions, the strategies of their performance still should be changed in order to reduce the temporal load on working memory as much as possible.

In our example, in order to reduce the load on working memory, information about the login and password should be transferred to the appropriate screen as shown in Figure 12.7.

Employees when on the login screen should be able to type the required login and password based on visually presented instruction. It is also desirable because of possible distraction, the need to answer phone calls while retaining information in working memory. Keeping information in working memory increases the likelihood that it can be forgotten and the user has to return to the original screen that contains information about the login and password (performance of $O_{18}^{\mu th} - O_{23}^{\varepsilon\mu}$ is repeated) and the number of unnecessary actions increases. The members of the algorithm, which includes

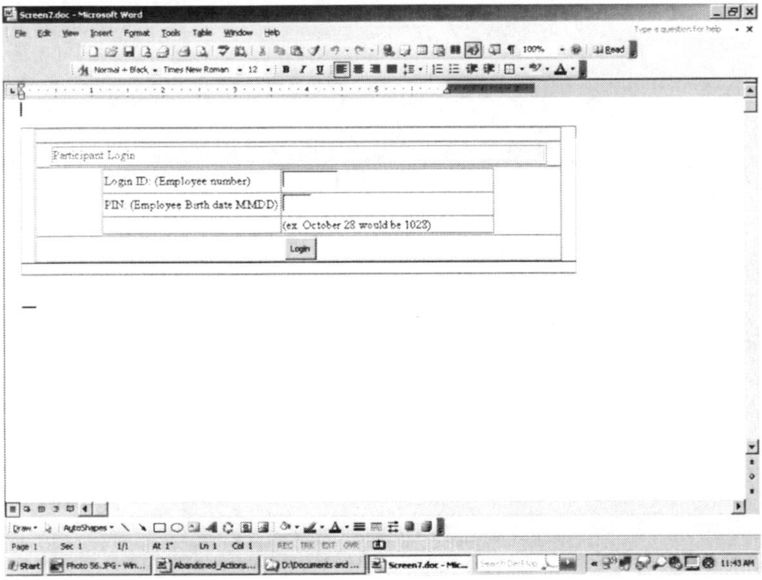

FIGURE 12.7
Login screen for the improved version.

thinking actions and/or logical conditions, should also be reduced as much as possible. It is best for the user to perform the required actions based on simple perceptual information.

A quantitative analysis of the abandoned action shows that they constitute a significant portion of the task, which means that the significant period of time is spent unproductively. Unproductive activity includes not only motor but also cognitive elements. Analysis of abandoned actions indicates that the objectively given goal *to take the web-survey questionnaire in accordance with the given e-mail* was subjectively reformulated into the goal *find where the login screen is*. When employees have a very low level of motivation to perform the task, such reformulation of the goal is understandable.

The strategies of activity performance of employees remind wondering in the maze. By trial and error, employees try to find a way out of the maze, but their actions are not blind trails and errors as described by Skinner. They act based on the formulated hypothesis that are evaluated and adjusted during task performance. Cognitive and behavioral actions are interconnected and perform cognitive, executive, and evaluative functions. Abandoned explorative actions test the formulated hypothesis and also perform these functions. The result of transformations on the screen is the source of information for further actions. This is an example of exploratory activity during interaction with a computer. This type of activity is critically important for computer-based tasks. If this task performance is inefficient, then the computer-based task has not been efficiently designed. The coefficients for various types of

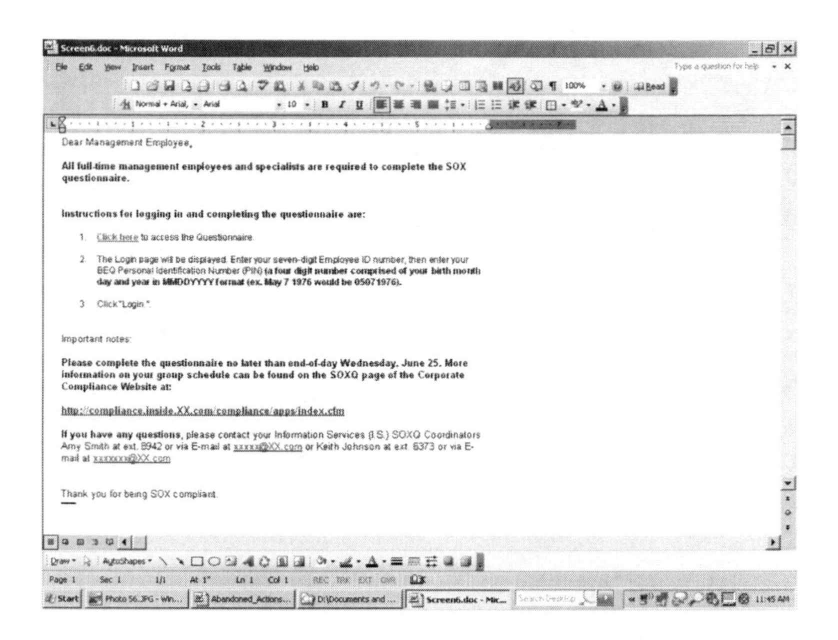

Screen6.doc - Microsoft Word

Dear Management Employee,

All full-time management employees and specialists are required to complete the SOX questionnaire.

Instructions for logging in and completing the questionnaire are:

1. Click here to access the Questionnaire.
2. The Login page will be displayed. Enter your seven-digit Employee ID number, then enter your BEQ Personal Identification Number (PIN) **(a four digit number comprised of your birth month day and year in MMDDYYYY format (ex. May 7 1976 would be 05071976).**
3. Click "Login."

Important notes:

Please complete the questionnaire no later than end-of-day Wednesday, June 25. More information on your group schedule can be found on the SOXQ page of the Corporate Compliance Website at:

http://compliance.inside.XX.com/compliance/apps/index.cfm

If you have any questions, please contact your Information Services (I.S.) SOXQ Coordinators Amy Smith at ext. 8942 or via E-mail at xxxxx@XX.com or Keith Johnson at ext. 6373 or via E-mail at xxxxxxx@XX.com

Thank you for being SOX compliant.

FIGURE 12.8
E-mail for the improved version of the web-survey task.

abandoned actions are useful in the analysis of the abandoned actions and exploratory activity in general. This information is useful for the design of the appropriate screen as shown in Figure 12.8.

Let us consider an algorithmic description of the optimized version of this task performance.

Qualitative and algorithmic analysis of the task at hand helps us to develop an improved version of the task under consideration that involves the use of more efficient strategy.

The most important information is now presented directly on the web-page containing a questionnaire in case employees need to get additional information.

Analysis of Table 12.3 shows that the algorithm of the task performance has changed. The quantity of the members of the algorithm is significantly reduced. Instead of 29 members of the algorithm, there are only 5. Most of them are easy to perform and their duration is reduced. Logical conditions are elimination, that is, the task does not require decision making. This implies that the initially considered version of the task was related to the algorithmic or rule-based tasks and the improved version belongs to skill-based tasks.

Table 12.3 presents an algorithmic description of the optimized version of task performance.

It is interesting to compare the performance strategy for the optimized algorithm when an employee performs $O_3^{\alpha} - O_6^{\varepsilon\mu}$ with $O_{23}^{\mu th} - O_{26}^{\varepsilon\mu}$ in a real algorithm. These members of the algorithm describe the same stage of task

TABLE 12.3
Algorithmic Description of E-Mail-Distributed Task Performance (Optimized Version)

Members of the Algorithm (Symbolic Description)	Description of Members of the Algorithm	Classification of Actions	Time (in ms)
O_1^{α}	Read the *e-mail* (see Figure 12.8).	Successive perceptual actions (separate actions are not considered).	6
O_2^{ε}	Click on the link to the questionnaire.	Motor action (mouse movement and double-click).	1.2
O_3^{α}	Read the login ID instruction on the *screen* (see Figure 12.7).	Perceptual action.	0.35
$O_4^{\varepsilon\mu}$	Recall and type the employee number and hit tab.	Combined action (executive action performed based on information extracted from memory; typing and recalling performed at the same time).	3
O_5^{α}	Read the PIN instruction on the screen (see Figure 12.7).	Perceptual action.	0.35
$O_6^{\varepsilon\mu}$	Recall and type the employee birth date and hit login.	Combined action (executive action performed based on information extracted from memory; typing and recalling performed at the same time).	3
O_7^{α}	Read and answer the questions from the questionnaire.	—	

performance. In the optimized algorithm, employees read the instruction for the login ID, recall the employee number, key it in, then read the instruction for the PIN, recall the birth date, transform the birth date into the required format, and key it in. In the optimized version of task performance, this stage includes two afferent operators: O_3^{α} and O_5^{α}. There are only two members of the algorithm $O_4^{\varepsilon\mu}$ and $O_6^{\varepsilon\mu}$ that are performed based on information from working memory. In a real algorithm, all four members of the algorithm $O_{23}^{\mu th} - O_{26}^{\varepsilon\mu}$ are performed based on information extracted from memory. If such type of action is repeated multiple times, it can lead to premature fatigue because regulation of actions based on information extracted from memory is more complicated than regulation of similar actions based on visual information.

The performance time of the optimized algorithm is $T = 13.9$ s. The time for executive (behavioral) actions $T\mu$ and $T\varepsilon$ is the same and equal to 6 s, and they are performed simultaneously. The time for afferent operators is $T\alpha = 6.7$ s. Abandoned actions are eliminated and other positive structural changes in activity performance can be observed.

Earlier, we considered the evaluation of explorative activity efficiency. In SSAT, there are also methods of quantitative evaluation of complexity and reliability of performing computer-based tasks. In the considered web-survey questionnaire, the workload on working memory and thinking mechanism could be assessed. Procedures of complexity evaluation are described in Bedny et al. (2008) where 20 main measures of complexity are suggested.

This study demonstrates that users are faced with a lot of computer tasks, which do not have strictly specified strategies of performance, and users often do not know in advance how the task should be performed. So they explore to find the way to perform the task. The more uncertain the task is, the more complicated is the explorative activity. In conditions when the uncertainty about the possible strategies is significantly increased, explorative activity can approach the chaotic mode. Emotional and motivational mechanisms are important in formation of such strategies. This situation has been encountered in the study of the web-survey questionnaire task. The concept of abandoned actions and a method of their analysis are important for the development of the methods of studying explorative activity in the performance of computer-based tasks.

There are various methods of reducing the number of abandoned actions. You can change the way you describe the task, change the location of its elements on the screen, develop special instructions, etc. In the task that was considered as an example, the access to the questionnaire should be facilitated. Employees first have to be able to open the questionnaire and, then if necessary, turn to the instructions or explanations. The most important information should be placed at the top of the screen. It is also necessary to eliminate if at all possible any situation when users have to keep intermittent information in working memory. Decision-making and thinking actions also have to be reduced as much as possible during analysis of abandoned actions.

The total elimination of explorative action usually is not achievable but can be reduced. It is necessary to reduce actions that include the processes of thinking, decision-making, and memory workload. It is desirable that the actions are performed based on the perceptual information. This is a recommendation for production tasks. In contrast to entertainment tasks, we often need to introduce explorative actions, which include decision making, thinking, etc. This leads to the conclusion that there is a need for quantitative analysis of explorative activity.

Not just cognitive and behavioral components are important in computer-based task performance. Emotionally motivational factors can totally change the strategies of task performance. In arising of unnecessary explorative activity emotionally-motivational mechanisms play a significant role. For example low motivation in task performance can cause development inefficient explorative activity. A low level of motivation and even negative motivation in the performance of the described task makes it difficult to study this task experimentally. If subjects are aware that they are involved in the experiment, their motivation could be different, and therefore, subjects would implement a different strategy of task performance. This suggests that in such cases, analytical methods of study are necessary. They includes qualitative, formalized and quantitative methods of study. At the qualitative stage of analysis the study of activity as self-regulative system is specifically useful.

Emotional and motivational components of activity are critical factors in the goal-formation process. The unity of cognitive and emotional aspects of activity is at the root of the goal-formation or goal acceptance process. The factor of significance is the mechanism that links the cognitive and emotionally motivational components of activity. There are no unmotivated goals. Goal and motive creates a vector that gives activity its directness and meaning.

In addition to the final goal of the task, there are intermediate goals. They are formed at different stages of the performance. In relation to the overall goal of the task, such goals have a subordinate role. The formation of subgoals depends on the specificity of the task, the methods of its presentation, and the user's individual features. The formation of intermediate goals is a critical component in the formation of activity strategies.

According to SSAT, there are different stages of the motivational process. In this study, the conflict between the process-related stage of motivation and the goal-related stage of motivation has been observed. On one hand, employees want to complete the task as quickly as possible because this is a mandatory task. This provides a positive goal-related stage of motivation. On the other hand, this task is perceived as not significant and distractive in relation to the main duties. As a result, the process-related stage of motivation is very low. This factor influences cognitive processing strategies. In most cases, employees are looking for the link to the questionnaire they can click on. Sending to the users an ineffectively designed e-mail survey accompanied by negative emotional and motivational factors, lead to cognitive strategies that sharply increase unwanted explorative actions.

Any computer-based task should be considered in terms of how important it is to the user. It is necessary to determine not only the significance of the task as a whole but also the significance of its individual elements in order to determine what possible cognitive strategies will be selected by users. The most significant elements of the task should be identified and optimized because these elements have a decisive importance for preferred strategies. Therefore, we cannot agree with the opinion of some experts in the field of HCI that the emotional-motivational factor is not important in the production environment. Emotionally motivational components of activity tightly connect with cognitive components of activity. Humans are not simple logical devices and always have predilection to events, situation, or information. In the activity approach, the design process always includes analysis of emotionally motivational components of activity in both the production and entertainment environment.

Conclusion

Two different approaches to task analysis are formulated in cognitive psychology. The first one integrates a number of techniques that strive to specify exactly what actions should be performed. The second approach describes a number of techniques that define what actions should not be performed during task performance. If the first approach emphasizes the importance of discovering and describing one best method of task performance, the second approach insists that a performer should independently decide how a task should be performed within existing constraints. In fact the first method is not adequate for analysis of flexible user activity, and the second approach rejects ergonomic principles of design in the human-computer interaction (HCI) field all together. Isolation and opposition of these two approaches are due to the fact that in ergonomics and psychology, there are no methods for analysis and description of human flexible activity. An effective approach to the analysis and description of flexible human activity is possible only when an efficient approach to the study of principles of activity self-regulation is developed. Such approach has been developed in the framework of the systemic-structural activity theory (SSAT). Studies of self-regulation in SSAT take their roots in the works of outstanding scientists Anokhin and Bernshtein. They introduced the notions of feedback and self-regulation into physiology and psychology almost a decade before this idea appeared in cybernetics. SSAT assimilates these theoretical ideas and developed basic principles of explanation of activity and cognition as a recursive system with multiple feedforward and feedback interconnections. It has been demonstrated that traditional concepts of self-regulation outside of SSAT are not useful when analyzing work activity. The control theory and the models of self-regulations derived from it are too mechanistic to be applied to human behavior and cognition. The process of self-regulation also cannot be reduced to analyzing separate psychological mechanisms such as motivation, volitional processes, and goal as an external standard, etc. Without developing adequate psychological models of self-regulation that include various mechanisms of activity regulation and their interaction, there is no theory of self-regulation. Usually all models of self-regulations outside of SSAT consist of one loop structured system, where feedback is created only after the performance of motor responses, implying that people can correct their behavior only after committing real errors. Feedback is usually associated with performance of selected response. Such feedback has been often labeled as knowledge of result. This means that errors were already made. Protection from errors is reduced to immediate feedback during task execution. However, such errors can have undesirable consequences. There is a need to prevent them. The concept of immediate feedback is often used in such cases. Immediate feedback is applicable only for motor responses. Such response can be selected inadequately. Hence, response can be correct while the entire activity would be incorrect. According to SSAT, a person can perform not only behavioral but also cognitive actions and use cognitive feedback. Only utilizing cognitive and mental feedback can prevent undesirable errors. Hence, the concept of self-regulation should be reexamined. Developed in systemic-structural activity theory (SSAT) approach to studying self-regulation allows predicting and describing preferable strategies of human performance. Qualitatively described strategies at the next stage of analysis can be analyzed utilizing a formalized method such as algorithmic description of activity. The algorithms can be deterministic and probabilistic. The last one can describe multivariate activity. Developed in SSAT quantitative principles of activity assessment allows to evaluate activity by using quantitative measurements.

In SSAT, self-regulation of activity is a conscious goal-directed process, with multiple loops and connections with diverse mechanisms of activity regulation. People can operate with internal images and meanings the same as with a real material object and therefore commit not real errors but errors in the mental plane. Feedback and feedforward connections are used not only in motor but also in cognitive activity. So real errors can be forecasted and prevented, because self-regulation activity is a flexible and adaptive system that can transfer a situation according to its goal that is developed during the process of self-regulation. Contemporary task analysis and specifically task analysis in HCI cannot be performed without understanding the principles of activity self-regulation and their derived strategies of task performance. Preferable strategies of task performance can be predicted based on analyzing the mechanisms of self-regulation. Contradictions between variability of human performance and requirements of a standardized design can be eliminated by utilizing such concepts as self-regulation, strategies, and a range of tolerance, human algorithm. In contemporary complex HCI systems, an operator utilizes flexible strategies of task performance, and without understanding the principles of activity self-regulation, task analysis of such tasks cannot be efficient. Multiple examples of application developed from SSAT models of self-regulation to study computer-based tasks are presented.

The book demonstrates new task analysis and design principles in the field of HCI. The purpose of design is the creation of documentation, including design models of activity during task analysis, according to which it is possible to produce new products, software, goods, method of performance, etc. The basic units of analysis in the design process are cognitive and behavioral actions and their smaller components are called psychological operations. As was demonstrated in the book, the concepts of cognitive and behavioral actions are critically important for task analysis and design. The SSAT concept of action is significantly different from the understanding of this concept outside of SSAT. SSAT considers cognitive and behavioral actions as basic elements of activity during task performance. A standardized classification of cognitive and behavioral actions and their operations is presented. The principles of cognitive action extraction in task analysis are described. The possibility of using eye movement analysis for this purpose is considered. Special attention was given to the dependency of eye movement strategies on activity goal and motivation. A nontraditional and more efficient method of utilizing the MTM-1 system for describing flexible motor components of a task was considered. It is shown that the MTM-1 system presents a standardized language for the description of motor components of activity and, therefore, can be used for creating models of work activity that are necessary for ergonomic design of computer-based tasks.

In the book, morphological analysis of activity during performance of HCI tasks is discussed. Morphological analysis is an important stage of studying any complex systems. This method of study describes an arrangement of various elements of a holistic object under investigation and considers an object as a structurally organized system. Morphological analysis is not quantitative but very useful for the description of an object's structure. In this book, we present a morphological method of analyzing flexible human activity developed in SSAT and adapted for the study of HCI tasks. The possibility of utilizing a deterministic and probabilistic algorithmic description of computer-based tasks was demonstrated. Adapted for this purpose, the time structure analysis is also described in this section. As presented in the book, morphological analysis of computer-based tasks demonstrates that there is an efficient method of formalized description of flexible human activity and therefore contradiction between instruction-based and constraint-based approaches in task analysis is eliminated. The presented material clearly shows the possibility of creating nonquantitative formalized models of human activity during interaction with a computer. The creation of such a method is a prerequisite to quantitative analysis of computer-based tasks.

The book presents basic quantitative methods of task analysis of computer-based tasks. It is well known that despite extensive research efforts over the years, there are no well-developed quantitative task assessment methods. It is particularly relevant for quantitative methods of evaluation of tasks in user–computer systems. The measures of complexity evaluation of computer-based tasks are described. Such quantitative methods as a human reliability assessment of computer-based task and quantitative analysis of explorative activity in HCI tasks are presented in the final part of the book. In this book, we prove that extremely flexible human activity can be described by utilizing formalized models and can be further evaluated quantitatively. Quantitative methods of task evaluation of computer-based tasks are brought to the level of practical application. They can be used in assessing the efficiency of users' performance and evaluation of computer interface design solutions. Material presented in this book is adapted for the analysis and design of HCI systems. Systemic-structural activity theory is an alternative psychological framework which can be successfully used in human factors, ergonomics and work psychology. At the same time this approach does not reject cognitive psychology. It utilizes cognitive psychology as possible stage of activity analysis.

Glossary

Basic Definitions of Self-Regulation in the Framework of Systemic-Structural Activity Theory

Currently psychology is widely used in the field of human–computer interaction (HCI). The success of any theoretical and applied research significantly depends on the proper use of basic terminology and basic concepts of the underlying theory. Nevertheless, the same terminology has completely different meaning in various fields of psychology. Moreover, terminology often does not have sufficient theoretical justification. A clear understanding of the basic terminology in psychology is particularly important for HCI. Much of the misunderstanding between researchers and practitioners in various fields of specialization derives from the misinterpretation of terminology. This can be explained by the fact that activity theory (AT) has its roots in Russian psychology. Many of the Russian–English translations fail to capture the original meaning of some AT terms.

There is general AT, applied AT, and systemic-structural AT (SSAT). The most rigorously developed terminology that can be utilized for studying human work can be found in SSAT. This terminology is not only justified theoretically, but is also standardized.

Terminology is essential in this book because it is devoted to the topic of activity self-regulation. So we give here only basic terminology related to this area of research.

Activity: A goal-directed system, where cognition, behavior, and motivation are integrated and organized by the mechanism of self-regulation toward achieving a conscious goal (activity has a recursive, loop-structured organization).

Conscious and unconscious levels of self-regulation: These levels are interdependent. Goal and verbally logical components of activity play the leading role at the conscious level of self-regulation whereas imagination, intuition, and nonverbalized meaning are important at the unconscious level of self-regulation.

Conscious and unconscious meaning: These two aspects of meaning are important mechanisms in activity regulation. Conscious meaning is associated with verbal aspects of goal directed thinking process that is accomplished by thinking actions. Non-verbalized meaning, on the other hand, is involved in unconscious level of thinking that is achieved by unconscious thinking operations. Thinking operations are organized by a goal-directed set and are not included in conscious activity. Such operations are important components of unconscious reflection of reality.

Evaluative stage of self-regulation: The self-regulation includes the following blocks:

1. Subjective standard of successful result is responsible for the development of subjective criteria for success, which might deviate from objective requirements.
2. Subjective standard of admissible deviation. Subjects define which errors are significant and which are not. If these deviations do not exceed subjective tolerance, subjects do not correct their actions.
3. Positive or negative evaluation of result is the final evaluation of the result based on subjective and objective criteria.
4. Information about interim and final result is the subjective interpretation of obtained data at various stages of self-regulation.

External and internal contours of self-regulation: An external contour includes feed-forward and external feedback from external receptors during activity performance. External feedback provides a meaningful interpretation of events.

An internal contour includes feed-forward and feedback in proprioceptive systems of motor activity that are typically unconscious. An internal contour of self-regulation can also be performed by using nonconscious internal mental operations. The interrelation between these two contours has a dynamic character. Some internal components of regulation can be transferred to external contours, enabling more exact conscious control of behavior.

Function block: This represents a coordinated system of subfunctions with specific purposes in activity structure. A function block is a functional mechanism that has specific relation with other functional mechanisms. Each function block mediates a particular function in the regulation of activity. Examining the relationship of function blocks is critical to the understanding of activity regulation. The content of a function block can change, but the purpose of each function block in the self-regulative model is constant.

Function block *afferent synthesis*: This block provides analysis, comparison, and synthesis of all data that an organism needs in order to perform adoptive response in given circumstances. The main stimulus that causes a reaction never exists in isolation. It interacts with supplementary environmental stimuli that influence what information is extracted from memory that is relevant to the response, current motivational state, and the response itself.

Function block *assessment of task difficulty*: This block is dedicated to the component of activity that involves awareness of the objective complexity of a task, as well as some intuitive assessment of its complexity. The more complex the task is, the greater is the probability that the task will be difficult for a subject. A subject can evaluate the same task as more or less difficult depending on his or her past experience or individual features. Therefore, the cognitive effort and inducing motivational components of activity depend on a task difficulty. An individual may under- or overestimate an objective complexity of a task, and this influences the strategies of task performance. An incorrect assessment of the complexity can result in inadequate personal sense or motivation to sustain the efforts for completing the task.

Function block *conceptual model*: The block responsible for developing a broad and relatively stable mental model, which serves as a general framework for understanding various situations relevant to particular professional duties. Although this model is general and is stored in long-term memory, it is more specific than past experience. Imaginative components are one of the distinguishing characteristics of this model and play an important role in its functioning.

Function block *formation of a program of task performance*: This block involves development of the program of execution of actions directed to achieving the accepted goal. This mechanism represents information regarding the method to be used in achieving the task goal and may or may not be conscious. This program is developed prior to the task or action performance and can be modified during that performance. The performance program can comprise hierarchically organized subprograms, some of which can be conscious and the other ones unconscious.

Function block *making decision about correction*: is involved in analyzing the corrections of self-regulative process and modification of the goal.

Function block *goal*: An integrative mechanism of self-regulative process that interacts with motivation and creates the vector *motive → goal*. This vector gives direction to the self-regulation process. Studies show that different individuals may have an entirely different understanding of a goal, even if objectively identical situation or instructions are given. Hence, we distinguish between *subjective* and *objective* interpretation of the goal. A goal cannot be considered as something externally given to the subject in a ready-made form. A goal is always associated with some stage of activity and includes stages such as goal recognition, goal interpretation, goal reformulation, goal formation, etc.

Function block *meaning*: This block is involved in the interpretation of input information (can be extracted not only from external data but from memory as well). It provides relationship not only of a sign and its referent but also of a sign and activity. Function block *meaning* in its study of the relationship between a sign and its meaning considers not only an individual but also the culture created by human activity. If the function block *meaning* is associated with a conscious goal it is a conscious level of self-regulation. When block *meaning* works together with a function block set, it is involved in unconscious level of activity self-regulation.

Function block *motivation*: The block responsible for the development of inducing components of motivation. While block *sense* refers to emotionally evaluative components of activity, motivational block determines activity goal directness and energetic components for attaining a specific goal. Function block *sense* and function block *motivation* are intimately connected, but sometimes, emotionally evaluative components of activity can be in conflict with inducing or motivational components.

Function block *orienting reflex*: The block that creates conditions for a heightened receptivity of the organism to sudden changes in the situation that is accomplished by the development of a complex, short-lived, and transitory physiological processes, the change of an activation level in the neural system with a general inhibition of conquering ongoing activity.

Function block *past experience*: This block analyzes the general background of a subject that also influences the strategies of performance and therefore can be considered as a functional mechanism. It includes general and professional knowledge of a subject, knowledge of culturally accepted norms of behavior, and customs that describe how a community functions. Past experience is acquired through activity that evolves over time within a culture. The interaction of past experience and new input information results in the assessment of the meaning of the immediate input information. Past experience includes not only cognitive but also emotionally motivational components and evaluation of task difficulty.

Function block *sense*: The block that covers emotional-evaluative aspects of activity and personal significance of its various components. Personal significance within the goal-directed activity leads a person to interpret the meaning of the presented information and transfer it into the subjective sense.

Function block *set*: The block characterized by the role it plays in the formation of the purposeful behavior. A set is responsible for the creation of the internal state of an organism that determines the purposefulness of human behavior, but this state is not conscious. A set creates a predisposition to processing incoming information in a particular way, or predisposition to performing particular actions. A set can be transferred into a conscious goal and vice versa. Therefore, a set performs similar functions at the unconscious level of self-regulation as a goal at the conscious level of self-regulation.

Function block *subjectively relative task conditions*: The block that analyzes the creation of a *dynamic model of the situation*. It is involved in the creation of the holistic mental model of reality and includes two subblocks: *operative image*, which to a large extent provides unconscious dynamic reflection of the situation, and *situation awareness*, which includes a logical and conceptual subsystem of the dynamic reflection of a situation in which an operator is aware of processing information. These two subsystems of dynamical reflection overlap. A subject is also conscious of processing information in the overlapping part of the imaginative subsystem. Conscious and unconscious components of dynamic reflection can, to some degree, be transformed into each other.

Functional analysis: An activity analysis performed based on various functional models of self-regulation. Its main units of analysis are functional mechanisms or function blocks. Its purpose is discovering basic strategies of task performance and evaluating and developing the most efficient strategies of task performance. It is a systemic, qualitative task analysis method that considers activity as a self-regulative system.

Functional macroblocks: This represents the decomposition of activity at the macro level of analysis and description of activity as a whole. The function block in this analysis has a complex architecture. At this stage of analysis, we do not apply chronometrical studies of very short duration cognitive processes because each function block requires considerable time for its realization. This method of analysis of activity is particularly important for the description of prescribed and real strategies of performance and their acquisition during the training process. The same components of activity can be analyzed utilizing different function blocks depending on their role in the process of self-regulation.

Functional mechanism or function block: The main units of analysis in self-regulative models of activity are functional mechanisms or function blocks. A mechanism that facilitates the integration of cognitive processes or actions for a particular purpose in a self-regulation process. Functional mechanisms can be considered as a subsystem with specific regulatory functions within activity structure. The term functional mechanism can be used when it is considered separately from other mechanisms. When functional mechanisms are described in relation to other functional mechanisms by using feed forward and feedback connections the term function block should be used. Thus, each function block represents integration of cognitive processes that are involved in a certain stage of activity regulation.

Functional microblock: A product of the decomposition of activity at a micro level. Chronometrical studies, which are applied for describing the functional structure of cognitive and motor actions, are important at this stage of analysis. Actions described as self-regulative systems are comprised of various functional microblocks. Each microblock describes psychological microprocesses or stages of actions' regulation.

General model of self-regulation: Model that describes the process of self-regulation of activity during task performance. It covers all stages of self-regulation, including executive stage of activity associated with the transformation of situation (object) according to the goal of activity. Behavioral actions and their relationship with cognitive actions are important at this stage of analysis.

Orienting activity: An activity that is explorative, or gnostic, in nature and, therefore, very flexible. The main characteristic of orienting activity is its dynamic reflection of the situation. Dynamic reflection of the situation, developing dynamic mental model and interpretation of the situation, is the main purpose of orienting activity. One should distinguish between orientation as a stage of activity and orienting activity when reflection of reality is the main purpose of the activity. The model of self-regulation of orienting activity describes a type of activity where executive components of activity are significantly reduced.

Physiological self-regulation: Self-regulation based on homeostasis. The purpose of this type of self-regulation is to reduce the discrepancy between the optimal state of the physiological system and the real state of the system in order to reduce disturbances on the system and restore balance. Many physiological imbalances are corrected automatically. The structure of physiological self-regulation processes is completely predetermined.

Plan and program of performance: The content and sequence of various components of activity or separate actions (mental or behavior) by means of which activity or separate actions should be performed. We use the term *plan* when the subject deliberately and consciously determines the sequence of the elements of activity in a particular situation. The term *program* is used in situations when planning is unconscious and has a very short duration.

Program of performance block: The block involved in the execution of required activity according to plan. Realization of the developed program of performance does not always match such program.

Psychological self-regulation: Self-regulation that is a goal-directed process. A system can change its own structure based on its experience. Such a system can form its own goals and subgoals and its own criteria for an activity evaluation. Psychological self-regulation integrates cognitive, executive, evaluative, and emotional aspects of activity.

The main purpose of psychological self-regulation is continuing reconsideration of activity strategies or even changing the goal of activity when internal and external conditions or situations change.

Self-regulation: An influence on a system that derives from this system in order to correct its behavior or activity. It includes cognitive (informational) and motivational (energetic) components and has a loop-structured organization. It also includes orienting, programming, executive, and evaluative components and can be performed at conscious or unconscious levels that interact with each other during the process of self-regulation.

Self-regulative model of formation and goal acceptance: Model that describes the process of goal formation and acceptance of the goal from the standpoint of self-regulation.

Situated system of activity: Activity is constructed or adapted to situations due to mechanisms of self-regulation. It includes flexible reconstructive strategies (situated components) and preplanned and preprogrammed (prespecified) components.

Strategy: A plan or a program of performance that is responsive to external contingencies, as well as to the internal state of a system. Strategy is dynamic and adaptive in nature, enabling changes in goal attainment as a function of external and internal conditions of a self-regulative system.

Bibliography

Aaltonen, A., Hyrskykari, A., and Räihä, K. (1998). 101 Spots, or how do users read menus? In *Proceedings of CHI 98 Human Factors in Computing Systems*, pp. 132–139. New York: ACM Press.

Ackoff, R. (1980). Towards a system of systems concepts. In H. R. Smith, A. B. Carroll, A. G. Kefalas, and H. J. Watson (Eds.), *Management: Making Organizations Perform*, pp. 43–87. New York: Macmillan.

Adams, J. A. (1968). Response feedback and learning. *Psychological Bulletin*, 70, 486–504.

Adams, J. A. (1971). A closed-loop theory of motor learning. *Journal of Motor Behavior*, 3, 11–150.

Adams, J. A. (1987). Historical review and appraisal of research on learning, retention and transfer of human motor skills. *Psychology Bulletin*, 10(1), 47–74.

Aladjanova, N. A., Slotintseva, T. V., and Khomskaya, E. D. (1979). Relationship between voluntary attention and evoked potentials of brain. In E. D. Khomskaya (Ed.), *Neuropsychological Mechanisms of Attention*, pp. 168–173. Moscow, Russia: Science Publishers.

Anderson, J. R. (1985). *Cognitive Psychology and Its Application*, 2nd edn. New York: Freemen.

Anderson, J. R. (1993a). Problem solving and learning. *American Psychologist*, 48(1), 35–44.

Anderson, J. R. (1993b). *Rules of the Mind*. Hillsdale, NJ: Lawrence Erlbaum Associates, Publishers.

Anokhin, P. K. (1935). *The Problem of Center and Periphery in the Physiology of Higher Nervous Activity*. Gorky, Russia: Gorky Publishers.

Anokhin, P. K. (1955). Features of the afferent apparatus of the conditioned reflex and their importance in psychology. *Problems of Psychology*, 6, 16–38.

Anokhin, P. K. (1962). *The Theory of Functional Systems as a Prerequisite for the Construction of Physiological Cybernetics*. Moscow, Russia: Academy Science of the USSR.

Anokhin, P. K. (1969). Cybernetic and the integrative activity of the brain. In M. Cole and I. Maltzman (Eds.), *A Handbook of Contemporary Soviet Psychology*, pp. 830–857. New York: Basic Books Inc., Publishers.

Austin, J. T. and Vancouver, J. B. (1996). Goal constructs in psychology: Structure, process, and content. *Psychological Bulletin*, 120(3), 338–375.

Ausubel, D. P. (1968). *Educational Psychology: A Cognitive View*. New York: Holt Rinehart and Winston.

Bainbridge, L. and Sanderson, P. (1991). Verbal protocol. In J. R. Wilson and E. N. Corlett (Eds.), *Evaluation of Human Performance: A Practical Ergonomics Methodology*, 2nd edn., pp. 169–201. Boca Raton, FL: Taylor & Francis Group.

Bakhtin, M. M. (1979). *Aesthetics of Verbal Creativity*. Moscow, Russia: Art Publisher.

Bandura, A. (1977). *Social Learning Theory*. Englewood Cliffs, NJ: Prentice-Hall.

Bandura, A. (1982). Self-efficacy mechanism in human agency. *American Psychology Journal*, 37, 122–147.

Bandura, A. (1989). Self-regulation of motivation and action through internal standards and goal system. In L. A. Pervin (Ed.), *Goal Concepts in Personality and Social Psychology*, pp. 19–86. Hillsdale, NJ: Lawrence Erlbaum Associates, Publishers.

Bandura, A. (1997). *Self-Efficacy: The Exercise of Control*. New York: W. H. Freeman.

Bandura, A. and Locke, E. A. (2003). Negative self-efficacy and goal effect revisited. *Journal of Applied Psychology*, 88, 87–99.

Bardin, K. V. (1982). The observer's performance in a threshold area. *Psychological Journal*, 1, 52–59.

Bardin, K. V. and. Voytenko, T. P. (1985). The phenomenon of simple discrimination. In Y. M. Zabrodin and A. P. Pakhomov (Eds.), *The Psychophysics of Discrete and Continual Tasks*, pp. 73–95. Moscow, Russia: Science Publishers.

Barness, P. M. (1980). *Motion and Time Study Design and Measurement of Work*. New York: John Wiley & Sons.

Baron, R. A. (1992). *Psychology*, 2nd edn. Boston, MA: Allyn and Bacon.

Bartlett, F. C. (1932). *Remembering: A Study in Experimental and Social Psychology*. Cambridge, England: Cambridge University Press.

Bedny, G. and Karwowski, W. (2006). General and systemic-structural activity theory. In W. Karwowski (Ed.), *International Encyclopedia of Ergonomics and Human Factors*, Vol. 3, pp. 3159–3167. London, U.K.: Taylor & Francis Group.

Bedny, G., Karwowski, W., and Bedny, M. (2001). The principle of unity of cognition and behavior: Implications of activity theory for the study of human work. *International Journal of Cognitive Ergonomics*, 5(4), 401–420.

Bedny, G., Karwowski, W., and Sengupta, T. (2006). Application of systemic-structural activity theory to design of human–computer interaction tasks. In W. Karwowski (Ed.), *International Encyclopedia of Ergonomics and Human Factor*, Vol. 1, pp. 1272–1286. Boca Raton, FL: CRC Press/Taylor & Francis Group.

Bedny, G. and Meister, D. (1997). *The Russian Theory of Activity: Current Application to Design and Learning*. Mahwah, NJ: Lawrence Erlbaum Associates, Publishers.

Bedny, G. and Meister, D. (1999). Theory of activity and situation awareness. *International Journal of Cognitive Ergonomics*, 3(1), 63–72.

Bedny, G. and Seglin, M. (1999a). Individual style of activity and adaptation to standard performance requirement. *Human Performance*, 12(1), 59–78.

Bedny, G. and Seglin, M. (1999b). Individual features of personality in former Soviet Union. *Journal of Research in Personality*, 33(4), 546–563.

Bedny, G., Seglin, M., and Meister, D. (2000). Activity theory: History, research and application. *Theoretical Issues in Ergonomics Science*, 1(2), 165–206.

Bedny, G., von Breven, H., and Synytsya, K. (2012). Learning and training: Activity approach. In N. M. Seel (Ed.), *Encyclopedia of the Science of Learning*, 1st edn., pp. 1800–1805. Boston, MA: Springer.

Bedny, G. Z. (1976). Adaption to job requirements based on individual style of performance. In V. G. Aseev (Ed.), *Personality and Individual Differences*, Vol. 3, pp. 34–45. Ircutsk, Russia: Irkutsk University Press.

Bedny, G. Z. (1979). *Psychophysiological Aspects of a Time Study*. Moscow, Russia: Economics Publishers.

Bedny, G. Z. (1981). *The Psychological Aspects of a Timed Study during Vocational Training*. Moscow, Russia: Higher Education Publisher.

Bedny, G. Z. (1987). *The Psychological Foundations of Analyzing and Designing Work Processes*. Kiev, Ukraine: Higher Education Publishers.

Bedny, G. Z. (2000). Activity theory. In Karwowski, W. (Ed.). *International Encyclopedia of Ergonomics and Human Factors,* Vol. 1, pp. 358–362. London, UK: Taylor and Francis Ltd.

Bedny, G. Z. (Ed.). (2004). Preface Special issue. *Theoretical Issues in Ergonomics Science,* 5(4), 249–253.

Bedny, G. Z. (2006). Activity theory. In W. Karwowski (Ed.), *International Encyclopedia of Ergonomics and Human Factor,* Vol. 1, pp. 571–576. Oxford, U.K.: CRC Press/ Taylor & Francis Group.

Bedny, G. Z. (2014). *Application of Systemic-Structural Activity Theory to Design and Training*. Boca Raton, FL: CRC Press/Taylor & Francis Group.

Bedny, G. Z. and Chebykin, O. Y. (2013). Application the basic terminology in activity theory. *IIE Transactions on Occupational Ergonomics and Human Factors,* 1(1), 82–92.

Bedny, G. Z. and Harris, S. (2005). The systemic-structural activity theory: Application to the study human work. *Mind, Culture, and Activity: An International Journal,* 12(2), 128–147.

Bedny, G. Z. and Harris, S. R. (2008). "Working sphere/engagement" and the concept of task in activity theory. *Interacting with Computers: The Interdisciplinary Journal of HCI,* 20(2), 251–255.

Bedny, G. Z. and Harris, S. R. (2013). Safety and reliability analysis methods based on systemic-structural activity theory. *Journal of Risk and Reliability* (Sage Publisher), 227(5), 549–556.

Bedny, G. Z. and Karwowski, W. (2003). A systemic-structural activity approach to the design of human–computer interaction tasks. *International Journal of Human-Computer Interaction,* 2, 235–260.

Bedny, G. Z. and Karwowski, W. (2004a). A functional model of human orienting activity. In G. Z. Bedny (invited editor). Special issue. *Theoretical Issues in Ergonomics Science,* 5(4), 255–274.

Bedny, G. Z. and Karwowski, W. (2004b). Activity theory as a basis for the study of work. *Ergonomics,* 47(2), 134–153.

Bedny, G. Z. and Karwowski, W. (2004c). Meaning and sense in activity theory and their role in study of human performance. *Ergonomia,* 26(2), 121–140.

Bedny, G. Z. and Karwowski, W. (2006). The self-regulation concept of motivation at work. *Theoretical Issues in Ergonomics Science,* 7(4), 413–436.

Bedny, G. Z. and Karwowski, W. (2007). *A Systemic-Structural Theory of Activity. Application to Human Performance and Work Design*. Boca Raton, FL: CRC Press/ Taylor & Francis Group.

Bedny, G. Z. and Karwowski, W. (2008a). Application of systemic-structural theory of activity to design and management of work systems. In W. W. Gasparski and T. Airaksinen (Eds.), *Praxiology and the Philosophy of Technology.* The International Annual of Practical Philosophy and Methodology #15, pp. 97–144. New Brunswick, NJ: Transaction Publishers.

Bedny, G. Z. and Karwowski, W. (2008b). Activity theory: Comparative analysis of eastern and western approaches. In O. Y. Chebykin, G. Z. Bedny, and W. Karwowski (Eds.), *Ergonomics and Psychology. Development in Theory and Practice*, pp. 221–246. London, U.K.: Taylor & Francis Group.

Bedny, G. Z. and Karwowski, W. (2008c). Time study during vocational training. In O. Y. Chebykin, G. Z. Bedny, and W. Karwowski (Eds.), *Ergonomics and Psychology. Development in Theory and Practice,* pp. 41–70. London, U.K.: Taylor & Francis Group.

Bedny, G. Z. and Karwowski, W. (2011a). Functional analysis of attention. In G. Z. Bedny and W. Karwowski (Eds.), *Human–Computer Interaction and Operators' Performance. Optimization of Work Design with Activity Theory*, pp. 307–330–185. Boca Raton, FL: CRC Press/Taylor & Francis Group.

Bedny, G. Z. and Karwowski, W. (2011b). Introduction to applied and systemic-structural activity theory. In G. Z. Bedny and W. Karwowski (Eds.), *Human–Computer Interaction and Operators' Performance. Optimization of Work Design with Activity Theory*, pp. 3–30. Boca Raton, FL: CRC Press/Taylor & Francis Group.

Bedny, G. Z. and Karwowski, W. (2013). Analysis of strategies employed during upper extremity positioning actions. *Theoretical Issues in Ergonomics Science*, 14(2), 175–194.

Bedny, G. Z., Karwowski, W., and Bedny, I. S. (2011a). The concept of task for non-production human–computer interaction environment. In D. B. Kaber and G. Boy (Eds.), *Advances in Cognitive Ergonomics*, pp. 663–672. Boca Raton, FL: CRC Press/Taylor & Francis Group.

Bedny, G. Z., Karwowski, W., and Jeng, O.-J. (2004). The situation reflection of reality in activity theory and the concept of situation awareness in cognitive psychology. In G. Z. Bedny (invited editor). Special issue. *Theoretical Issues in Ergonomics Science*, 5(4), 275–296.

Bedny, G. Z., Karwowski, W., and Sengupta, T. (2008). Application of systemic-structural theory of activity in the development of predictive models of user performance. *International Journal of Human–Computer Interaction*, 24(3), 239–274.

Bedny, G. Z., Karwowski, W., and Voskoboynikov, F. (2011b). The relationship between external and internal aspects in activity theory and its importance in the study of human work. In G. Z. Bedny and W. Karwowski (Eds.), *Human–Computer Interaction and Operators' Performance. Optimization of Work Design with Activity Theory*, pp. 31–62. Boca Raton, FL: CRC Press/Taylor & Francis Group.

Bedny, G.Z. and Voskoboynikov, F. (1975). Problems of how a person adapts to the objective requirements of an activity. In V.G. Asseev (Ed.), Psychological Problems of Personality (Vol. 2, pp.18–30), Irkutsk, Russia: Irkutsk University Press.

Bedny, I. S. (2006). General characteristics of human reliability in system of human and computer. *Science and Education* (Ukraine, Odessa), (N 1–2), 58–61.

Bedny, I. S. (2006). On systemic-structural analysis of reliability of computer based tasks. *Science and Education* (Ukraine, Odessa), 1–2(7–8), 58–60.

Bedny, I. S. and Bedny, G. (2011a). Analysis of abandoned actions in the email distributed tasks performance. In D. B. Kaber and G. Boy (Eds.), *Advances in Cognitive Ergonomics*, pp. 683–692. Boca Raton, FL: CRC Press/Taylor & Francis Group.

Bedny, I. S. and Bedny, G. Z. (2011b). Abandoned actions reveal design flaws: An illustration by a web-survey task. In G. Z. Bedny and W. Karwowski (Eds.), *Human–Computer Interaction and Operators' Performance. Optimization of Work Design with Activity Theory*, pp. 149–185. Boca Raton, FL: CRC Press/Taylor & Francis Group.

Bedny, I. S., Karwowski, W., and Bedny, G. (2012). Computer technology at the workplace and errors analysis. In K. M. Stanney and K. S. Hale (Eds.), *Advances in Cognitive Engineering and Neuroergonomics*, pp. 167–176. Boca Raton, FL: CRC Press/Taylor & Francis Group.

Bedny, I. S., Karwowski, W., and Bedny, G. Z. (2010). A method of human reliability assessment based on systemic-structural activity theory. *International Journal of Human–Computer Interaction*, 26(4), 377–402 (CRC Press/Taylor & Francis Group).

Bedny, I. S. and Sengupta, T. (2005). The study of computer based tasks. *Science and Education* (Ukraine, Odessa), 1–2(7–8), 82–84.

Beregovoy, G. T., Zavalova, N. D., Lomov, B. F., and Ponomarenko, V. A. (1978). *Experimental-Psychology in Aviation and Aeronautics*. Moscow, Russia: Science Publishers.

Bernshtein, N. A. (1935). The problem of the relationship between coordination and localization. *Archives of Biological Science*, 38(1), 1–34.

Bernshtein, N. A. (1947). *On the Structure of Movement*. Moscow, Russia: Medical Publishers.

Bernshtein, N. A. (1966). *The Physiology of Movement and Activity*. Moscow, Russia: Medical Publishers.

Bernshtein, N. A. (1969). Methods for developing physiology as related to the problems of cybernetics. In M. Cole, I. Maltzman (Eds.). *A Handbook of Contemporary Soviet Psychology*. Basic Books Publishers, pp. 441–451.

Bernshtein, N. A. (1967). The coordination and regulation of movements. Oxford, U.K.: Pergamon Press.

Bernshtein, N. A. (1996). On dexterity and its development. In M. L. Latash and M. T. Turvey (Eds.), *Dexterity and Its Development*, pp. 1–244. Mahwah, NJ: Lawrence Erlbaum Associates, Publishers.

Bloch, V. (1966). Level of wakefulness and attention. In P. Fraisse and J. Piaget (Eds.), *Experimental Psychology*, Vol. 3, pp. 97–146. Paris, France: University Press of France.

Boekaerts, M., Pintrich, P. R., and Zeidner, M. (Eds.). (2005). *Handbook of Self-Regulation*. San Diego, CA: Academic Press.

Broadbent, D. E. (1958). *Perception and Communication*. London, U.K.: Pergamon Press.

Bruner, J. S. (1957). On going beyond the information given. In J. S. Bruner (Ed.), *Contemporary Approaches to Cognition*, pp. 41–69. Cambridge, MA: Harvard University Press.

Brushlinsky, A. V. (1979). *Thinking and Forecasting*. Moscow, Russia: Thinking Press.

Bühler, K. (1934). *Theory of Language: The Representational Function of Language*. Jena, Germany: Fischer.

Bundura, A. (1989). Self-regulation of motivation and action through internal standards and goal systems. In A. Pervin (Ed.), *Goal Concept in Personality and Social Psychology*, pp. 19–86. Hillsdale, NJ: Lawrence Erlbaum Associates, Publishers.

Card, S. K., Moran, T. P., and Newell, A. (1983). *The Psychology of Human–Computer Interaction*. Hillsdale, NJ: Lawrence Erlbaum Associates, Publishers.

Carrol, J. M. (Ed.). (1987). *Interfacing Thought*. Cambridge, MA: MIT Press.

Carruth, D. W. and Duffy, V. G. (2008). Integrating cognitive and digital human models for virtual product design. In O. Y. Chebikin, G. Z. Bedny, and W. Karwowski (Eds.), *Ergonomics and Psychology: Developments in Theory and Practice*, pp. 29–40. Boca Raton, FL: CRC Press/Taylor & Francis Group.

Carver, C. S. and Scheier, M. F. (1998). *On the Self-Regulation of Behavior*. New York: Cambridge University Press.

Carver, C. S. and Scheier, M. F. (2005). On the structure of behavioral self-regulation. In M. Boekaerts, P. R. Pintrich, and M. Zeidner (Eds.), *Handbook of Self-Regulation*, pp. 42–84. San Diego, CA: Academic Press.

Chebykin, O. Y., Bedny, G. Z., and Karwowski, W. (Eds.). (2008). *Ergonomics and Psychology: Developments in Theory and Practice*. Boca Raton, FL: CRC Press/Taylor & Francis Group.

Chkhaidze, L. V. (1970). *Control of Movements*. Moscow, Russia: Physical Culture and Sport.

Cooper, C. S. and Shepard, R. N. (1973). Chronometric studies of the rotation of mental images. In W. G. Chase (Ed.), *Visual Information Processing*. New York: Academic Press.

Dainoff, M. J. (2008). Ecological ergonomics. In O. Y. Chebikin, G. Z. Bedny, and W. Karwowski (Eds.), *Ergonomics and Psychology: Developments in Theory and Practice*, pp. 3–28. Boca Raton, FL: CRC Press/Taylor & Francis Group.

Dekker, S., Hummerdal, D., and Smith, K. (2010). Situation awareness: Some remaining questions. *Theoretical Issues in Ergonomics Science*, 11(1–2), 131–135.

Diaper, D. (2004). Understanding task analysis for human–computer interaction. In D. Diaper and N. Stanton (Eds.), *The Handbook of Task Analysis for Human–Computer Interaction*, pp. 5–47. Mahwah, NJ: Lawrence Erlbaum Associates, Publishers.

Diaper, D. and Stanton, N. A. (2004). Wishing on a sTAr: The future of the task analysis. In D. Diaper and N. Stanton (Eds.), *The Handbook of Task Analysis for Human-Computer Interaction*, pp. 585–602. Mahwah, NJ: Lawrence Erlbaum Associates, Publishers.

Dmitrieva, M. A. (1964). Speed and accuracy of information processing and their dependence on signal discrimination. In B. F. Lomov (Ed.), *Problems of Engineering Psychology*, pp. 121–126. Leningrad, Russia: Association of Psychology Publishers.

Dobrinin, N. F. (1958). Voluntary and involuntary attention. In N. F. Dobrinin (Ed.), *Scientificworks of the Moscow State Pedagogical University*, Vol. 8, pp. 34–52. Moscow, Russia: Pedagogical Publishers.

Dobrolensky, U. P., Zavalova, N. D., Ponomarenko, V. A., and Tuvaev, V. A. (1975). In U. P. Dobrolensky (Ed.), *Methods of Engineering Psychological Study in Aviation*. Moscow, Russia: Manufacturing Publishers.

Drury, C. G. (1975). Applications of Fitts' law to foot pedal design. *Human Factors*, 17, 368–373.

Edwards, W. (1961). Behavioral decision theory. *Annual Review of Psychology*, 12, 473–489.

Eisenstadt, S. A. and Simon, H. A. (1997). Logic and thought. *Minds Mach*, 7, 365–385.

Endsley, M. R. (2000). Theoretical underpinnings of situation awareness: A critical review. In M. R. Endsley and D. J. M. Garland (Eds.), *Situation Awareness Analysis and Measurement*, Mahwah, NJ: Lawrence Erlbaum Associate Associates, Publishers. pp. 3–32.

Endsley, M. R. and Jones, D. (2012). *Design for Situation Awareness: An Approach to User-Centred Design*, 2nd edn. Boca Raton, FL: CRC Press/Taylor & Francis Group.

Farfel, V. S. (1969). *Physiology of Muscular Activity in Work and Sport*. Moscow: Medical Publisher.

Fafrowicz, M. and Marek, T. (2008). Attention, selection for action, error processing, and safety. In O. Y. Chebykin, G. Z. Bedny, and W. Karwowski (Eds.), *Ergonomics and Psychology: Development in Theory and Practice*, pp. 203–220. Boca Raton, FL: Taylor & Francis Group.

Fitts, P. M. (1954). The information capacity of the human motor system in controlling the amplitude of movement. *Journal of Experimental Psychology*, 47, 381–391.

Fitts, P. M. and Posner, M. I. (1967). *Human Performance*. Belmont, CA: Brooks/Cole.

Frege, G. 1948 (1892). Sense and reference. *The Philosophical Review*, 57, 207–230.

Frese, M. and Zapf, D. (1994). Action as a core of work psychology: A German approach. In H. C. Triadis, M. D. Dunnette, and L. M. Hough (Eds.), *Handbook of Industrial and Organizational Psychology*, pp. 271–340. Polo Alto, CA: Consulting Psychologists Press.

Galactionov, A. I. (1978). *The Fundamentals of Engineering-Psychological Design of Automatic Control Systems of Technological Processes*. Moscow, Russia: Energy Publishers.

Gal'perin, P. Y. (1969). Stages in the development of mental acts. In M. Cole and I. Maltzman (Eds.), *A Handbook of Contemporary Soviet Psychology*, pp. 249–273. New York: Basic Books Inc., Publishers.

Gal'perin, P. Y. (1973). Experience of development basic notions in psychology. *Questions of Psychology*, 2, 146–152.

Genisaretsky, O. I. (1975). Methodology of organization of activity system. In E. G. Yudin (Ed.), *Development of Automatic Systems in Design*. Moscow, Russia: Science Publishers.

Gibson, J. J. (1979). *The Ecological Approach to Visual Perception*. Boston, MA: Houghton-Mifflin.

Gilbreth, F. V. and Gilbreth, L. M. (1920). *Motion Study for Handicapped*. London, U.K.: George Routledge and Sons.

Goldberg, J. H. and Kotval, X. P. (1998). Eye movement-based evaluation of the computer interface. In S. K. Kumar (Ed.), *Advances in Occupational Ergonomic and Safety*, pp. 529–532. Amsterdam, the Netherlands: ISO Press.

Goldberg, J. H. and Kotval, X. P. (1999). Computer interface evaluation using eye movements: Methods and constructs. *International Journal of Industrial Ergonomics*, 24, 631–645.

Gollwitzer, P. M. (1996). The volitional benefits of planning. In P. Gallwitzer and J. P. Bargph (Eds.), *The Psychology of Action*, pp. 287–312. New York: The Gilford Press.

Gordeeva, N. D. and Zinchenko, V. P. (1982). *Functional Structure of Action*. Moscow, Russia: Moscow University Publishers.

Gurevich, K. M. (1970). *Professional Fitness and Basic Feature of Nervous System*. Moscow, Russia: Pedagogical Academy of Science.

Granovskaya, R. M. (1974). *Perception and Models of Memory*. Leningrad, Russia: Science Publishers.

Green, D. M. and Swets, J. A. (1966). *Signal Detection Theory and Psychophysics*. New York: Wiley.

Hacker, W. (1985). Activity: A fruitful concept in industrial psychology. In M. Frese and J. Sabini (Eds.), *Goal Directed Behavior: The Concept of Action in Psychology*, pp. 262–284. Hillsdale, NJ: Lawrance Erlbaum Associates, Publishers.

Hacker, W. (1986). *Work Psychology*. Bern, Switzerland: Huber.

Hacker, W. (1994). Moving from cognition to action? Control theory and beyond. In M. Frese (Ed.), *Applied Psychology: An International Review*, 43(3), 379–381. Hillsdale, NJ: Lawrence Erlbaum Associates, Publishers.

Hancock, P. A., Flach, J., Caird, J., and Vicente, K. J. (1995). *Local Application of the Ecological Approach to Human–Machine System*. Hillsdale, NJ: Lawrence Erlbaum Associates, Publishers.

Heckhausen, H. (1991). *Motivation and Action*. Berlin, Germany: Springer-Verlag.

Henderson, J. M. (1993). Visual attention and saccadic eye movements. In G. D'Ydewalle and J. Van Rensberggen (Eds.), *Perception and Cognition: Advances in Eye Movement Research*, pp. 37–50. Amsterdam, the Netherlands: Elsevier Science Publishers.

Hick, W. E. (1952). On the role of gain of information. *Quarter Journal Experimental Psychology*, 4, 11–26.

Hogan, J., Hogan, R., and Murtha, T. (1992). Validation of a personality measures of job performance. *Journal of Business and Psychology*, 7, 225–236.

Hoppe, F. (1930). Success and failure. *Psychological Studding*, 14, 1–62.

Hornof, A. J. (2001). Visual search and mouse pointing in labeled versus unlabeled two-dimensional visual hierarchies. *ACM Transactions on Computer–Human Interaction*, 8(3), 171–197.

Hull, C. L. (1951). *Essentials of Behavior*. New Haven, CT: Yale University Press.

Hyman, R. (1953). Stimulus information as a determinant of reaction time. *Journal of Experimental Psychology*, 45, 423–432.

Jack, G., Schmidt, J. G., and Smith, K. U. (March, 1971). Feedback analysis of eye tracking of auditory and tactical stimuli. *American Journal of Optometry and Archives of American Academy of Optometry*, 48(3), 204–209.

Jacko, J. A. (1997). An empirical assessment of task complexity for computerized menu systems. *International Journal of Cognitive Ergonomics*, 1(2), 137–147.

Jacko, J. A., Salvendy, G., and Koubek, R. J. (1971). Modeling of menu design in computerized work. *Interaction with Computer*, 7(3), 304–330.

Jacko, J. A. and Ward, K. G. (1996). Toward establishing a link between psychomotor task complexity and human information processing. *Computers and Industrial Engineering*, 31(1–2), 533–536.

Jacob, R. J. K. (1991). The use of eye movement in human–computer interaction techniques: What you look at is what you get. *ACM Transactions on Information System*, 9, 152–169.

Just, M. A. and Carpenter, P. A. (1976). Eye fixation and cognitive processes. *Cognitive Psychology*, 8, 441–480.

Just, M. A., Carpenter, P. A., and Miyake, A. (2003). Neuroindices of cognitive workload: Neuroimagining, pupillometric and event-related brain potential studies of brain work. *Theoretical Issues in Ergonomic Science*, 4, 56–58.

Kahneman, D. (1973). *Attention and Effort*. Englewood Cliffs, NJ: Prentice-Hall.

Kahneman, D. and Tversky, A. (1984). Choices, values and frames. *American Psychologist*, 39, 341–350.

Kamishov, I. A. (1968). The methodology of eye movement registration and determination of the operator's eye movement. *Questions of Psychology*, 4, 62–81.

Kanfer, R. (1996). Self-regulation and other non-ability determinants of skill acquisition. In P. M. Gollwitzer and J. P. Bargph (Eds.), *The Psychology of Action*, pp. 404–423. New York: The Gilford Press.

Karat, J. (1993). Software evaluation methodologies. In M. Helander (Ed.). *Handbook of Human-Computer Interaction*, pp. 892–904. North-Holland, Netherlands: Elsevier Science.

Karat, J., Karat, C.-M., and Vergo, J. (2004). Experiences people value: The new frontier for task analysis. In D. Diaper and N. Stanton (Eds.), *The Handbook of Task Analysis for Human–Computer Interaction*, pp. 585–602. Mahwah, NJ: Lawrence Erlbaum Associates, Publishers.

Karger, D. W. and Bayha, F. H. (1977). *Engineering Work Measurement*, 3rd edn. New York: Industrial Press.

Khilchenko, A. E. (1960). Study of flexibility of nervous system. *Physiological Journal*, 6(1), 21–28.

Kieras, D. E. (1993). Towards a practical GOMS model methodology for user interface design. In M. Helendar (Ed.), *Handbook of Human–Computer Interaction*, pp. 135–157. North-Holland, the Netherlands: Elsevier Science.

Kieras, D. E. (1994). Towards a practical GOMS model methodology for user interface design. In *Handbook of Human Computer Interaction.* (Ed.) M. Helender, pp. 135–156. Amsterdam: North Holland.

Kieras, D. E. (2004). GOMS models for task analysis. In D. Diaper and N. Stanton (Eds.), *The Handbook of Task Analysis for Human–Computer Interaction*, pp. 83–116. Mahwah, NJ: Lawrance Erlbaum Associates, Publishers.

Kieras, D. E. and Polson, P. G. (1985). An approach to the formal analysis of user complexity. *International Journal of Man-Machine Studies*, 22, 365–394 (in complexity HCI).

Kim, J. (2008). Perceived difficulty as a determinant of Websearch performance. *Information Research*, 13(4), Paper 379.

Kirwan B. (1994). *A Guide to Practical Human Reliability Assessment.* London, U.K.: Taylor & Francis Group.

Klatsky, R. L. (1975). *Human Memory. Structure and Processes.* San Francisco, CA: Freeman.

Kleinback, U. and Schmidt, K. H. (1990). The translation of work motivation into performance. In V. Kleinback, H.-H. Quast, H. Thierry, and H. Hacker (Eds.), *Work Motivation*, pp. 27–40. Hillsdale, NJ: Lawrence Erlbaum Associates, Publishers.

Klimov, E. A. (1969). *Individual Style of Activity.* Kazan, Russia: Kazahnsky State University Press.

Kochurova, E. I., Visyagina, A. I., Gordeeva, N. D., and Zinchenko, V. P. (1981). Criteria for evaluating executive activity. In V. Wertssch (Ed.), *The Concept of Activity in Soviet Psychology*, pp. 383–433. New York: M. E. Sharpe, Inc.

Konopkin, O. A. (1980). *Psychological Mechanisms of Regulation of Activity.* Moscow, Russia: Science Publishers.

Konopkin, O. A., Engels, I. L., and Stephansky, V. T. (1983). On shaping of subjective standards of success. *Problems of Psychology*, 6, 109–114.

Konopkin, O. A. and Zhujkov, Ju. S. (1973). About a person's ability to assess the probabilistic characteristics of alternative stimuli. In D. A. Oshanin and O. A. Konopkin (Eds.), *Psychological Aspects of Activity Regulation*, pp. 154–197. Moscow, Russia: Pedagogical Publishers.

Kosslyn, S. M. (1973). Scanning visual images: Some structural implications. *Perception and Psychophysics*, 14, 90–94.

Kotik, M. A. (1974). *Self-Regulation and Reliability of Operator.* Tallinn, Estonia: Valgus.

Kotik, M. A. (1978). *Textbook of Engineering Psychology.* Tallinn, Estonia: Valgus.

Kozulin, A. (1986). The concept of activity in soviet psychology. *American Psychologist*, 41(3), 264–274.

Krinchik, E. P. and Risakov, S. L. (1965). Influence of significance factor of signal on specificity of processing information. In V. P. Zinchenko, A. N. Leont'ev, and D. Y. Panov (Eds.), *Engineering Psychology*, pp. 155–159. Moscow, Russia: Moscow University Publishers.

Kuhl, J. (1992). A theory of self-regulation: Action versus state orientation, self-discrimination and some applications. *Applied Psychology: An International Review*, 41, 97–129.

Landa, L. M. (1976). *Instructional Regulation and Control: Cybernetics, Algorithmization and Heuristic in Education.* Englewood Cliffs, NJ: Educational Technology Publication (English translation).

Landa, L. M. (1984). Algo-heuristic theory of performance, learning and instruction: Subject, problems, principles. *Contemporary Educational Psychology*, 9, 235–245.

Landy, F. J. and Conte, J. M. (2007). *Work in the 21st century. An Introduction to Industrial and Organizational Psychology*. Second edition. Blackwell Publisher.

Langolf, C. D., Chaffin, D. B., and Foulke, S. A. (1976). An investigation of Fitts' law using a wide range of movement amplitudes. *Journal of Motor Behavior*, 8, 113–128.

Lazareva, V. V., Svederskaya, N. E., and Khomskaya, E. D. (1979). Electrical activity of brain during mental workload. In E. D. Khomskaya (Ed.), *Neuropsychological Mechanisms of Attention*, pp. 151–168. Moscow, Russia: Science Publishers.

Lee, T. W., Locke, E. A., and Latham, G. P. (1989). Goal setting, theory and job performance. In A. Pervin (Ed.), *Goal Concepts in Personality and Social Psychology*, pp. 291–326. Hillsdale, NJ: Lawrence Erlbaum Associates, Publishers.

Leont'ev, A. N. (1977). *Activity, Consciousness, Personality*. Moscow, Russia: Political Publishers.

Leont'ev, A. N. (1978). *Activity, Consciousness and Personality*. Englewood Cliffs, NJ: Prentice Hall.

Leont'ev, A. N. (1981). The problem of activity in psychology. In J. V. Wertsch (Ed.), *The Concept of Activity in Soviet Psychology*. New York: M. E. Sharpe, Inc.

Leplat, J. (1963). Sensorimotor connections. In J. Piaget and P. Fraisse (Eds.), *Experimental Psychology*, Vol. 1–2, pp. 375–427. Paris, France: University Press of France.

Lisina, M. I. (1957). Some methods for converting involuntary reactions into voluntarily reaction. *Reports of the Academy of Pedagogical Sciences*, 1, 24–35.

Locke, E. A. (1994). The emperor is naked. In R. Lord and P. E. Levy (Eds.), *Applied Psychology: An International Review*, 43, 367–370.

Locke, E. A. and Latham, G. P. (1990). Work motivation: The high performance cycle. In V. Kleinbeck et al. (Eds.), *Work Motivation*, pp. 3–26. Hillsdale, NJ: Lawrence Erlbaum Associates, Publishers.

Lomov, B. F. (1966). *Man and Machine*. Moscow, Russia: Soviet radio.

Lomov, B. F. (Ed.). (1982). *Handbook of Engineering Psychology*. Moscow, Russia: Manufacturing Publishers.

Lord, R. G. and Levy, P. E. (1994). Moving from cognition to action: A control theory perspectives. In M. Frese (Ed.), *Applied Psychology: An International Review*, 43, 335–398.

Maslov, O. P. and Pronina, E. E. (1998). Psychic and reality: Topology of virtual reality. *Applied Psychology*, 6, 41–49.

McLeond, R. W. and Sherwood-Jones, B. M. (1992). Simulation to predict operator workload in command system. In B. Kirwan and L. K. Ainsworth (Eds.), *A Guide to Task Analysis*, pp. 301–310. New York: Taylor & Francis Group.

Meas, S. and Gebhard, W. (2005). Self-regulation and health behavior: The health behavioral goal model. In M. Boekaerts, P. R. Pintrich, and M. Zeidner (Eds.), *Handbook of Self-Regulation*, pp. 343–368. San Diego, CA: Academic Press.

Meletinsky, E. M. (1976). *Poetry of Myth*. Moscow, Russia: Science Publishers.

Merlinkin, V. P. (1977). Some neuro psychological differences in development novice tumblers' skills. In E. A. Klimov (Ed.), *Neuropsychological Study of Personality*, pp. 72–87. Perm, Russia: Perm University Publisher.

Meulenbroek, R. G. J. and Rvan Galen, G. P. (1988). Foreperiod duration and the analysis of motor stages in a line-drawing tasks. *Acta Psychologica*, 69, 19–34.

Meyer, D. E. and Kieras, D. E. (1997). A computational theory of executive cognitive processes and multiple-task performance: Part 1. Basic mechanisms. *Psychological Review*, 104, 3–65.

Miller, G. A., Galanter, E., and Pribram, K. H. (1960). *Plans and the Structure of Behavior.* New York: Holt.

Miller, R. B. (1953). A method for man-machine task analysis (Report 53-137). Wright-Patterson AFB, OH: Wright Air Research and Development Command.

Morris, Ch. W. (1946). *Signs, Language and Behavior*. New York: Braziler.

Myasnikov, V. A. and Petrov, V. P. (Eds.). (1976). *Aircraft Digital Monitoring and Control Systems.* Leningrad, Russia: Manufacturing Publishers.

Nebilitsin, V. D. (1976). *Psychological Study of Individual Differences*. Moscow, Russia: Pedagogy.

Newell, A. and Simon, H. A. (1972). *Human Problem Solving*. Englewood Cliffs, NJ: Prentice-Hall.

Nojivin, U. S. (1974). On psychological self-regulation of sensory motor actions. In V. D. Shadrikov (Ed.), *Engineering and Psychology*, Vol. 1, pp. 206–210. Yaroslav, Russia: Yaroslav University.

Norman, D. and Bobrow, D. (1975). On data-limited and resource processing. *Journal of Cognitive Psychology*, 7, 44–60.

Norman, D. A. (1976). *Memory and Attention: An Introduction to Human Information Processing*, 2nd edn. New York: Wiley.

Norman, D. A. (1986). Cognitive engineering. In D. Norman and S. Draper (Eds.), *User Centered System Design: New Perspectives on Human–Computer Interaction*, pp. 31–61. Hillsdale, NJ: Lawrence Erlbaum Associates, Publishers.

Norman, D. A. (1988). *The Psychology of Everyday Things*. New York: Harper & Row.

Novan, D. and Gopher, D. (1979). On the economy of the human-processing system. *Psychological Review*, 86, 214–255.

Novikov, A. I., Sidorov, I. N., and Fedorov, I. V. (1980). Study of operator activity in conditions of destroying feedback connections. *Ergonomics*, 19, 32–39.

Novikov, A. M. (1986). *Process and Method of Formation of Vocational Skills*. Moscow, Russia: Higher Education.

Ormrod, J. E. (1990). *Human Learning: Theories, Principles and Educational Applications.* New York: Macmillan.

Oshanin, D. A. (1976). Dynamic operative system. *The Problems and Results in Psychology*, 59, 37–48.

Oshanin, D. A. (1977). Concept of operative image in engineering and general psychology. In B. F. Lomov, V. F. Rubakhin, and V. F. Venda (Eds.), *Engineering Psychology*. Moscow, Russia: Science Publishers.

Pashler, H. and Johnston, J. C. (1998). Attention limitations in dual-task performance. In H. Pashler (Ed.), *Attention*, pp. 155–190. East Sussex, U.K.: Psychology Press.

Patrick, J. (1992). *Training Research and Practices*. San Diego, CA: Academic Press.

Pavlov, I. P. (1927). *Conditioned Reflex*. London, U.K.: Oxford University Press.

Person, R. S. (1965). *Muscle-Antagonists in Human Movements*. Moscow, Russia: Medical Publishers.

Pervin, L. A. (1989a). Goal concepts, themes, issues and questions. In L. A. Pervin (Ed.), *Goal Concepts in Personality and Social Psychology*, pp. 173–180. Hillsdale, NJ: Lawrence Erlbaum Associates, Publishers.

Pervin, L. A. (Ed.) (1989b). *Goal Concepts in Personality and Social Psychology*. Mahwah, NJ: Lawrence Erlbaum Associates, Publishers.

Piaget, J. and Inhelder, B. (1966). *The Child's Conception of Space*. London, U.K.: Routledge & Kegan Paul.

Platonov, K. K. (1970). *Problems of Work Psychology.* Moscow, Russia: Medicine Publishers.
Ponomarenko, V. A. and Lapa, V. V. (1975). Impact of the intellectual assessment of the situation on nature of emotional reactions in pilots. *Space Biology and Medicine*, 1, 66–79.
Ponomarenko, V. A. and Zavalova, N. D. (1981). Study of psychic image as regulator of operator actions. In B. F. Lomov and V. F. Venda (Eds.), *Methodology of Engineering Psychology and Psychology of Work of Management*, pp. 30–41. Moscow, Russia: Science Publishers.
Powers, W. T. (1973). *Behavior: The Control of Perception.* Chicago, IL: Aldine Publishing Company.
Powers, W. T. (1978). Quantitative analysis of purposive systems: Some spadework at the foundation of scientific psychology. *Psychology Review*, 85(5), 417–435.
Pushkin, V. V. (1978). Construction of situational concepts in activity structure. In A. A. Smirnov (Ed.), *Problem of General and Educational Psychology*, pp. 106–120. Moscow, Russia: Pedagogy.
Pushkin, V. N. and Nersesyan, L. S. (1972). *Psychology of the Railroad.* Moscow, Russia: Transportation.
Rasmussen, J. and Goodstein, L. P. (1988). Informational technology and work. In M. G. Helendar (Ed.), *Handbook of Human–Computer Interaction*, pp. 175–201. Amsterdam, the Netherlands: Elsevier.
Rauterberg, M. (1996). How to measure of cognitive complexity in human–computer interaction. In R. Trappl (Ed.), *Cybernetic and Systems*, Vol. 2, pp. 815–820. Vienna, Austria: Austrian Society for Cybernetic Studies (compl HCI).
Rayner, K. (1992). Introduction. In K. Rayner (Ed.), *Eye Movement and Visual Cognition: Scene Perception and Reading*, pp. 1–7. New York: Springer-Verlag.
Rayner, K. (1998). Eye movement in reading and information processing: 20 Years of research. *Psychological Bulletin*, 124, 372–422.
Ritchey, T. (1991). Analysis and synthesis. On scientific method-based on a study by Bernhard Riemann. *Systems Research*, 8(4), 21–41.
Rubinshtein, S. L. (1940). *Problems of General Psychology.* Moscow, Russia: Academic Science.
Rubinshtein, S. L. (1957). *Existence and Consciousness.* Moscow, Russia: Academy of Science.
Rubinshtein, S. L. (1958). *About Thinking and Methods of Its Development.* Moscow, Russia: Academic Science.
Rubinshtein, S. L. (1959). *Principles and Directions of Developing Psychology.* Moscow, Russia: Academic Science.
Salmon, P. M., Stanton, N. A., Walker, G. H., Jenkins, D., Baber, C., and Mcmaster, R. (2008). Representing situation awareness in collaborative systems: A case study in the energy distribution domain. *Ergonomics*, 51(3), 367–384.
Salvendy, G. (2004). Classification of human motions. *Theoretical Issues in Ergonomics Science*, 5(2), 169–178.
Sanders, A. F. (1980). Stage analysis of reaction processes. In G. E. Stelmach and J. Requin (Eds.), *Tutorials in Motor Behavior*, pp. 331–354. Amsterdam, the Netherlands: North-Holland.
Sapir, E. L. (1956). Language and environment. *American Anthropologist*. 1912; 14(2), 226–242, April–June 1912.

Schacter, S. and Singer, J. E. (1962). Cognitive, social, and physiological determinants of emotional state. *Psychological Review*, 69, 379–399.

Schmidt, R. A. (1975). A schema theory of discrete motor skill learning. *Psychology Review*, 82(4), 225–260.

Schmidt, R. A. and Russell, D. G. (1972). Movement velocity and movement time as determinant of degree of preprogramming in simple movement. *Journal of Experimental Psychology*, 82(4), 225–260.

Schultz, D. and Schultz, S. (1986). *Psychology and Industry Today: An Introduction to Industrial and Organizational Psychology*. New York: Macmillan Publishing Company.

Schneider, W. and Shiffrin, R. M. (1977). Controlled and automatic human information processing: I. Detection, search, and attention. *Psychology Review*, 84, 1–66.

Seel, N. M. and Winn, W. D. (1997). Research on media and learning: Distributed cognition and semiotics. In R. D. Tennyson, F. Schott, N. M. Seel, and S. Dijkstra (Eds.), *Instructional Design. International Perspective*, Vol. 1, pp. 293–326. Mahwah, NJ: Lawrence Erlbaum Associates, Publishers.

Sengupta, T., Jeng, O. -J. (2003). Activity based analysis for drawing task. *The Proceeding of the XVth Trienial Congress of the Ergonomics Association and the 7th Joint Conference of Ergonomic Society of Korea/Japan Ergonomic Society*, pp. 455–458. Ergonomic Society of Korea.

Sengupta, T., Bedny, I. S., and Karwowski, W. (2008). Study of computer based tasks during skill acquisition process. *2nd International Conference on Applied Ergonomics Jointly with 11th International Conference on Human Aspects of Advanced Manufacturing*, Las Vegas, NV, July 14–17, 2008.

Sengupta, T., Bedny, I.S. (2008). Study of computer-based tasks during skill acquisition process. *Second International Conference on Applied Ergonomics jointly with Eleventh International Conference on Human Aspects of Advance Manufacturing* (July 14–17, 2008).

Shchedrovitsky, G. P. (1995). *Selective Works*. Moscow, Russia: Cultural Publisher.

Shepard, R. N. (1978). The mental image. *American Psychologist*, 33, 125–137.

Simon, H. A. (1999). *The Sciences of the Artificial*, 3rd revised edn. Cambridge, MA: MIT Press.

Smith, B. A., Ho, J., Ark, W., and Zhai, S. (2000). Hand eye coordination patterns in target selection. *Proceedings of the Symposium on Eye Tracking Research & Application*, Palm Beach Gardens, FL, pp. 117–122.

Sokolov, E. N. (1960). Neural models and the orienting reflex. In A. B. Braziere (Ed.), *The Central Nervous System and Behavior*, pp. 187–276. New York: Josiah Macy, Jr. Foundation.

Sokolov, E. N. (1963). *Perception and Conditioned Reflex*. New York: Macmillan.

Sokolov, E. N. (1969). The modeling properties of the nervous system. In M. Cole and I. Maltzman (Eds.), *Handbook of Contemporary Soviet Psychology*, pp. 671–704. New York: Basic Books, Publishers.

Sokolov, E. V. (1974). *Culture and Personality*. Moscow, Russia: Science Publishers.

Sternberg, S. (1969a). The discovery of processing stages: Extension of Donder's method. *Acta Psychological*, 30, 276–315.

Sternberg, S. (1969b). Memory-scanning, mental processes revealed by reaction-time experiments. *American Scientist*, 57, 421–457.

Sternberg, S. (1975). Memory scanning: New findings and current controversies. *Quarterly Journal of Experimental Psychology*, 27, 1–32.

Sternberg, S. (2008a). Identification of mental modules. In O. Y. Chebykin, G. Bedny, and W. Karwowski (Eds.), *Ergonomics and Psychology: Development in Theory and Practice*, pp. 111–134. Boca Raton, FL: CRC Press/Taylor & Francis Group.

Sternberg, S. (2008b). Identification of neural modules. In O. Y. Chebykin, G. Bedny, and W. Karwowski (Eds.), *Ergonomics and Psychology: Development in Theory and Practice*, pp. 135–166. Boca Raton, FL: CRC Press/Taylor & Francis Group.

Sternberg, S., Wright, C. E., Knoll, R. L., and Monsell, S. (1980). Motor programming and rapid speech: Additional evidence. In R. A. Cole (Ed.), *The Perception and Production of Fluent Speech*. Hillsdale, NJ: Lawrence Erlbaum Associates, Publishers.

Suchman, L. A. (1987). *Plans and Situated Actions: The Problem of Human–Machine Interaction*. Cambridge, U.K.: Cambridge University Press.

Swets, J. A. (1964). *Signal Detection and Recognition by Human Observers*. New York: Wiley.

Telegina, E. D. (1975). The interrelation of conscious and subconscious actions in the thinking process. *Questions of Psychology*, 2, 15–32.

Thomas, J. C. and Richards, T. (2012). Achieving psychological simplicity measures and methods to reduce cognitive complexity. In J. A. Jucko (Ed.), *The Human–Computer Interaction Handbook: Fundamentals, Evolving Technologies, and Emerging Application*, pp. 491–513. Boca Raton, FL: CRC Press/Taylor & Francis Group.

Tikhomirov, O. K. (1984). *Psychology of Thinking*. Moscow, Russia: Moscow University.

Tolman, E.C. (1932). *Purposive Behavior in Animals and Men.* New York: Century

Treisman, A. (1969). Strategies and models of selective attention. *Psychological Review*, 76, 282–299.

Turvey, M. T. (1996). Dynamic touch. *American Psychologist*, 51(11), 1134–1152.

UK MTMA. MTM-1. Analyst Manual. London: The UK MTM Association, 2000.

Uznadze, D. N. (1967). *The Psychology of Set*. New York: Consultants Bureau.

Vancouver, J. T. (2005). Self-regulation in organizational settings: A tale of two paradigms. In M. Boekaerts, P. R. Pintrich, and M. Zeidner (Eds.), *Handbook of Self-Regulation*, pp. 303–341. San Diego, CA: Academic Press.

Van Santen, J. H. and Philips, N. Y. (1970). Method and time study of mental work. *Work Study and Management Services*, 14(1), 21–27.

Vertegaal, R. (1999). The GAZE groupware system: Mediating joint attention in multiparty communication and collaboration. In *Proceedings of the ACM CHI' 99 Human Factors in Computing Systems Conference*, Pittsburgh, PA, pp. 294–301. New York: ACM Press.

Vicente, K. J. (1999). *Cognitive Work Analysis: Toward Safe, Productive, and Healthy Computer-Based Work*. Mahwah, NJ. Lawrence Erlbaum Associates, Publishers.

Viviani, P. (1990). Eye movement in visual search: Cognitive, perceptual and motor control aspects. In E. Kowler (Ed.), *Eye Movements and Their Role in Visual and Cognitive Processes*, pp. 353–375. Amsterdam, the Netherlands: Elsevier Science Publishers.

von Bertalanffy, L. (1962). General system theory: A critical review. *General Systems*, VII, 1–20 (Ann Arbor, MI).

Vygotsky, L. S. (1962). *Thought and Language*. Cambridge, MA: MIT Press.

Vygotsky, L. S. (1978). *Mind in Society. The Development of Higher Psychological Processes*. Cambridge, MA: Harvard University Press.

Welford, A. T. (1968). *Fundamentals of Skill*. London, U.K.: Methuen.

Wickens, C. D. and Hollands, J. G. (2000). *Engineering Psychology and Human Performance*, 3rd edn. New York: Harper-Collins.

Wickens, C. D. and McGarley, J. S. (2008). *Applied Attention Theory*. Boca Raton, FL: Taylor & Francis Group.

Wiener, N. (1948). *Cybernetics: Or Control and Communication in the Animal and the Machine*. Cambridge, MA: MIT Press.

Wiener, N. and Rosenblueth, A. (1950). Purposeful and non-purposeful behavior. *Philosophy of Science*, 17, 20–36.

Yarbus, A. L. (1965). *The Role of Eye Movements in the Visual Process*. Moscow, Russia: Science Publishers.

Yarbus, A. L. (1969). *Eye Movement and Vision*. New York: Plenum.

Yarovoj, I. N. and Maljuta, N.G. (1966). Regulation of motor actions during filing operation. In *Vocational Training*, (3), 23–30.

Young, M. S. and Stanton, N. A. (2002). Malleable attention resources theory: A new explanation for the effect of mental underload on performance. *Human Factors*, 44(3), 365–375.

Zabrodin, Y. M. (1985). Methodological and theoretical problems of psychophysics. In B. F. Lomov and Y. M. Zabrodin (Eds.), *Psychophysics of Discrete and Continual Tasks*, pp. 3–26. Moscow, Russia: Science Publishers.

Zabrodin, Y. M. and Chernishev, A. P. (1981). On the loss of information in describing the activity of human-operator by the transfer function. In B. F. Lomov and V. F. Venda (Eds.), *Theoretical and Methodological Analysis in Engineering Psychology, Psychology of Work and Control*, pp. 244–250. Moscow, Russia: Science Publishers.

Zaporozhets, A. V. (1969). Some of the psychological problems of sensory training in early childhood and the preschool period. In M. Cole and I. Maltzman (Eds.), *A Handbook of Contemporary Soviet Psychology*, pp. 86–120. New York: Basic Books, Inc., Publishers.

Zarakovsky, G. M. (1976). Evaluation of operator's workload based on operational-psychophysiological method. In V. A. Myasnikiv and V. P. Petrov (Eds.), *Aircraft Digital Monitoring and Control Systems*, pp. 523–540. Leningrad, Russia: Manufacturing Publishers.

Zarakovsky, G. M. (2004). The concept of theoretical evaluation of operator's performance derived from activity theory. Special Issue: Activity Theory. (G. Z. Bedny, Ed.). *Theoretical Issues in Ergonomics Science*, 5(4), 313–337.

Zarakovsky, G. M., Korolev, B. A., Medvedev, V. I., and Shlaen, P. Y. (1974). *Introduction to Ergonomics*. Moscow, Russia: Soviet Radio.

Zarakovsky, G. M. and Medvedev, V. I. (1971). Psychological evaluation of efficiency of man–machine system. In *Third Conference of Operator Reliability*, pp. 82–96. Leningrad, Russia: Psychology Society.

Zarakovsky, G. M. and Pavlov, V. V. (1987). *Laws of Functioning Man-Machine Systems*. Moscow, Russia: Soviet Radio.

Zavalova, N. D., Lomov, B. F., and Ponomarenko, V. A. (1971). Principle of active operator and function allocation. *Problems of Psychology*, 3, 3–12.

Zimmerman, B. J. (2005). Attaining of self-regulation: A social cognitive perspectives. In M. Boekaerts, P. R. Pintrich, and M. Zeidner (Eds.), *Handbook of Self-Regulation*, pp. 13–41. San Diego, CA: Academic Press.

Zinchenko, P. I. (1961). *Involuntary Memorization*. Moscow, Russia: Pedagogy.

Zinchenko, T. P. (1981). *Identification and Coding*. Leningrad, Russia: Leningrad University Publishers.

Zinchenko, V. P., Munipov, V. M., and Gordon, V. M. (1973). Study of visual thinking. *Questions of Psychology*, 2, 57–66.

Zinchenko, V. P. and Ruzkaya, A. G. (1962). Comparative analysis sense by touch and vision. Does hand teaches the eye? Presentation 11. *Selective work of the Academy of Pedagogical Sciences of USSR*, Vol. 3, pp. 17–29. Moscow, Russia: Academy of Pedagogical Sciences Publishers.

Zinchenko, V. P. and Vergiles, N. Y. (1969). *Creation of Visual Image*. Moscow, Russia: Moscow University.

Zinchenko, V. P., Vergiles, N. U., and Vuchetich, B. M. (1980). *Functional Structure of Visual Memory*. Moscow, Russia: Moscow State University Publishers.

Zwicky, F. (1969). *Discovering, Invention, Research-Through the Morphological Approach*. Toronto, Ontario, Canada: The Macmillan Company.

Index

D

M

Q

R

S

V

W